SYNTHESIS OF LUMPED ELEMENT, DISTRIBUTED AND PLANAR FILTERS

SYNTHESIS OF LUMPED ELEMENT, DISTRIBUTED AND PLANAR FILTERS

Joseph Helszajn DSc CEng FIEE

Heriot-Watt University

McGRAW-HILL BOOK COMPANY

London · New York · St Louis · San Francisco · Auckland · Bogotá · Guatemala
Hamburg · Lisbon · Madrid · Mexico · Montreal · New Delhi · Panama
Paris · San Juan · São Paulo · Singapore · Sydney · Tokyo · Toronto

Published by
McGRAW-HILL Book Company (UK) Limited
Shoppenhangers Road
Maidenhead, Berkshire, England SL6 2QL
Telephone Maidenhead (0628) 23432
Cables MCGRAWHILL MAIDENHEAD Telex 848484
Fax 0628 35895

British Library Cataloguing in Publication Data
Helszajn J. (Joseph)
  Synthesis of lumped element, distributed and planar
  filters
  1. Electronic equipment. Circuits
  I. Title
  621.3815′3

  ISBN 0-07-707166-2

Library of Congress Cataloging-in-Publication Data
Helszajn J. (Joseph)
  Synthesis of lumped element, distributed and planar filters /
  Joseph Helszajn
    p.   cm.
  Bibliography: p.
    Includes index.
    ISBN 0-07-707166-2
    1. Electric filters, Passive.   2. Electronic circuits.   I. Title.
  TK7872.F5H393   1990
  621.381′5324—dc20   89-34285

1234 CUP 9210

Typeset by Mid-County Press, London
and printed and bound at the University Press, Cambridge

To Deborah because she did not ask.

CONTENTS

PREFACE

Filter networks are essential building elements in many areas of electrical engineering. Such networks are used to separate or combine signals at different frequencies in a host of modern equipments. Filter-like networks are also met in the matching problem of transmission lines of different characteristic impedances or in that between resistive and complex loads. Although the physical realization of filters at audio, radio and microwave frequencies may vary, the network topology is common to all. The class of filters studied in this text is passive in that there is no energy source within the network and it is reactive in that dissipation is neglected. The signal is therefore reflected or transmitted but not dissipated. Filter circuits using lumped element reactances and distributed transmission lines are separately considered. In the design of distributed filter circuits an equivalence between lumped element capacitors and inductors in the s plane and open-circuited and short-circuited transmission lines in the t plane is established. The necessary and sufficient conditions for the realizability of one-port immittances is then the same provided the distributed circuit is commensurate. An additional element, the unit element (UE), not encountered in lumped element synthesis in the form of a distributed transmission line, is available in the construction of distributed lines. The description of some microwave planar parallel line filter circuits is included in the text to illustrate some applications. The text therefore primarily consists of a study of the amplitude squared transfer function approximation problem, the definition of the unitary condition, the development of Darlington's method, the study of positive real functions and immittances, the synthesis of an immittance function as a one- or two-port low-pass network and the construction of suitable frequency transformations between the low-pass prototype and the other filter circuits. The use of immittance inverters to realize a circuit using two kinds of elements by using only one kind is given special attention. The use of UEs and t-plane capacitors and inductors in the synthesis of filters using distributed lines is also given some attention.

The four basic types of frequency-selective filters are the low-pass, high-pass, band-pass and band-stop ones. The low-pass one passes signals from zero frequency to some determined cutoff frequency and reflects all signals above that frequency. The high-pass filter prevents transmission below some cutoff frequency and transmits all signals above it. The band-pass one passes all signals at frequencies between some frequency interval and reflects signals at frequencies outside this band. The band-stop filter reflects all signals over some frequency interval and transmits all signals outside it. Such filters may be combined in different ways to route a number of different signals in a radio-frequency carrier to different receivers or transmitters. In practice, idealized responses cannot be achieved, and some approximate representations must be sought. The approximation problem is indeed one of the main topics of this text. Four possible low-pass frequency responses are known as Butterworth, Chebyshev, inverse Chebyshev and elliptic filters respectively. The low-pass response which is equal ripple in both the passband and stopband is the elliptic function one. The Butterworth (or maximally flat), Chebyshev (equal ripple in passband) and inverse Chebyshev (equal ripple in stopband) are in fact all special cases of the elliptic prototype.

The first part of this text is essentially written as a one semester senior level introduction to the synthesis of one- and two-port reactance functions and filter networks. The second part with its emphasis on the synthesis of distributed circuits and the design of planar microwave circuits should also be of value as a postgraduate text and to young engineers in the microwave industry. The design of commercial filters is, of course, best carried out by having recourse to the appropriate professional literature. In order to permit (encourage) the instructor to proceed from realizability to the approximation and synthesis problem in one semester, the emphasis of the text is on the realization of one-port reactance networks and resistively terminated two-port reactance networks. The latter circuits include the important low-pass, band-pass, band-stop and high-pass filter networks. *RC*, *RL* and transformer coupled networks are therefore omitted but both distributed and lumped networks are included in the text. Since the theory of distributed quarter-wave long matching networks between different resistances and the lumped element broadband matching problem are closely related to the filter approximation problem, the former two topics are included for completeness. In keeping with modern usage the approximation problem is posed in terms of the scattering matrix. Worked examples are used throughout the text to provide the student with working experience with much of the principles discussed in the book. The standard and length of each chapter has been kept as uniform as possible in order not to emphasize one topic over another. It is hoped that the text will provide a good introduction to the student to the language of the modern synthesis and approximation problem of two-port reactance networks and to permit him or her to proceed without too much difficulty to modern practical filter practice.

The scattering matrix of a network with more than one port is described in Chapter 1. This matrix is admirably suited for the formulation of one-port immittances from a magnitude squared transfer ratio specification of a two-port reactance network terminated in a 1-ohm resistor. It also provides a classic testing procedure for the realizability of this class of networks. Chapters 2 and 3 discuss realizability and synthesis of one-port reactance functions. Realization of a function as a one-port network is ensured by satisfying Brune's tests. Such functions are known as positive real. Four possible canonical forms for the synthesis of a one-port reactance function are the first and second Cauer forms and the first and second Foster forms. Chapter 3 also includes the problems of zero shifting and partial pole removal. Networks with more than one port are defined in terms of impedance and admittance matrices in Chapter 4. Chapter 5 deals with the problem of whether a square immittance matrix is realizable as a two-port network. It is demonstrated that such matrices must be semi-positive real in order to ensure realizability. For a square symmetrical immittance matrix to be realizable as a two-port network it is both necessary and sufficient for its eigenvalues to be positive real. This problem is tackled in Chapter 6. The Darlington synthesis of a one-port *LCR* immittance as a two-port *LC* network terminated in a 1-ohm resistor is developed in Chapter 7. This statement forms the basis of modern filter theory. Chapters 8 and 9 introduce the student to the amplitude and phase approximation problems and that of phase equalization using all-pass networks. Chapter 10 outlines the synthesis of the low-pass network for which all the attenuation poles of the amplitude approximation reside at infinite frequency. These include the classic Butterworth and Chebyshev problems. Chapter 11 deals with the case where the attenuation poles lie at finite frequency. This problem corresponds by and large to modern practice. The mapping between the low-pass, band-pass, band-stop and high-pass filter circuits is dealt with in Chapter 12. The topologies of these circuits are given in terms of the element value of the all-pole low-pass prototype (g_i), the cutoff or midband frequencies (ω_0) and, in the cases of the pass and stopband circuits, the bandwidth of the filter (BW). The design of some lumped element circuits in terms of a filter specification is also outlined in this chapter. The broadband matching problem using lumped element reactances is separately considered in Chapter 13. *ABCD* parameters are introduced in Chapter 14. This notation is suitable for the analysis problem. Chapter 15 outlines the use of immittance inverters and other microwave techniques which permit a network with two kinds of elements to be replaced by one using only one kind. The uniform transmission line is dealt with in Chapter 16, and Chapter 17 discusses the use of low-pass and high-pass distributed circuits. The exact synthesis of distributed networks using the Richards variable (t) is introduced in Chapter 18. This chapter also provides an introduction to the unit element (UE). Chapter 19 deals with the approximation problem of a cascade arrangement of n unit elements and m series or shunt stubs. Chapter 20 describes the synthesis of stepped impedance transducers. A number of classic microwave circuits may be realized by terminating two of the four ports of two coupled parallel transmission lines by electric and magnetic walls in all combinations. Chapter 21 describes scattering and immittance matrices of this type of circuit and Chapter 22 derives the equivalent circuits of some possible solutions using an exact synthesis method. Aspects of microwave design using planar circuits are introduced in Chapter 23. It includes the design of band-pass filters using half-wave long cavities spaced by inductive posts and low-pass and band-pass filters using parallel line circuits. The geometries and properties of some TEM transmission lines are discussed in Chapter 24.

J. Helszajn

CHAPTER
1

THE
SCATTERING
MATRIX

INTRODUCTION

The scattering matrix to be dealt with in this chapter is admirably suited for the description of networks with more than one port. Such a matrix exists for every linear, passive, time-invariant network. The elements along its main diagonal are reflection coefficients and those along its off-diagonal are transmission coefficients. Furthermore, it lends itself well to the formulation of one-port immittances from a magnitude squared transfer ratio specification of a two-port reactance network terminated in a 1-Ω resistor, which is the main topic of this text. It is possible to deduce important general properties of networks containing a number of ports from this matrix by invoking such properties as symmetry, reciprocity and energy conservation. The latter condition is of special importance in circuit theory in that it indicates the permissible relationships between the entries of the scattering matrix.

The scattering parameters of symmetrical two-port networks can be readily deduced from their equivalent circuits by forming its two eigennetworks. These eigennetworks are obtained by bisecting the two-port network and open- and short-circuiting the exposed terminals. The reflection coefficients of these two one-port eigennetworks are just the two eigenvalues of the scattering matrix. Since the scattering coefficients are the sum and difference of the two possible eigenvalues this approach immediately yields the entries of the scattering matrix. This topic is dealt with in Chapter 6.

THE SCATTERING MATRIX

The scattering matrix of an n-port network is a square matrix of order n whose entries relate suitably chosen incident and reflected waves at the terminals of the network:

$$\bar{b} = \bar{S}\bar{a} \tag{1-1}$$

where $\bar{a}$ and $\bar{b}$ are incident and reflected vectors, which for a two-port network are given by

$$\bar{a} = \begin{bmatrix} a_1 \\ a_2 \end{bmatrix} \tag{1-2}$$

$$\bar{b} = \begin{bmatrix} b_1 \\ b_2 \end{bmatrix} \tag{1-3}$$

and the corresponding scattering matrix $\bar{S}$ is described by

$$\bar{S} = \begin{bmatrix} S_{11} & S_{12} \\ S_{21} & S_{22} \end{bmatrix} \tag{1-4}$$

The elements along the main diagonal of this matrix are the reflection coefficients at the ports of the network; the elements along the off-diagonal represent the transmission coefficient between the ports of the same network. The relationship between the incoming and outgoing waves for a two-port network are therefore described using this nomenclature by

$$b_1 = a_1 S_{11} + a_2 S_{12} \tag{1-5}$$

$$b_2 = a_1 S_{21} + a_2 S_{22} \tag{1-6}$$

A schematic diagram of this relationship is depicted in Fig. 1-1.

The scattering parameters of the two-port network are defined in terms of the incident and reflected waves by the two preceding equations as

$$S_{11} = \frac{b_1}{a_1}\bigg|_{a_2 = 0} \tag{1-7}$$

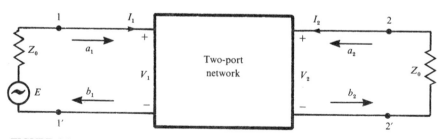

FIGURE 1-1
Schematic diagram showing definition of scattering waves.

$$S_{21} = \frac{b_2}{a_1}\bigg|_{a_2 = 0} \tag{1-8}$$

$$S_{12} = \frac{b_1}{a_2}\bigg|_{a_1 = 0} \tag{1-9}$$

$$S_{22} = \frac{b_2}{a_2}\bigg|_{a_1 = 0} \tag{1-10}$$

If it is assumed that a_i and b_i are normalized in such a way that $\frac{1}{2}a_i a_i^*$ is the available power at port i and $\frac{1}{2}b_i b_i^*$ is the emergent power at the same port, then

$$a_1 = \frac{1}{2}\left(\frac{V_1}{\sqrt{Z_0}} + \sqrt{Z_0}I_1\right) \tag{1-11}$$

$$b_1 = \frac{1}{2}\left(\frac{V_1}{\sqrt{Z_0}} - \sqrt{Z_0}I_1\right) \tag{1-12}$$

$$a_2 = \frac{1}{2}\left(\frac{V_2}{\sqrt{Z_0}} + \sqrt{Z_0}I_2\right) \tag{1-13}$$

$$b_2 = \frac{1}{2}\left(\frac{V_2}{\sqrt{Z_0}} - \sqrt{Z_0}I_2\right) \tag{1-14}$$

Adopting these definitions indicates that the a's and b's at any port are linear combinations of the voltage and current variables at the same port. Such linear combinations are in fact also met in the description of a uniform transmission line for which the solution to the transmission line equations is given in terms of forward and backward travelling waves $A \exp(-\gamma z)$ and $B \exp(\gamma z)$ by

$$V = A \exp(-\gamma z) + B \exp(\gamma z) \tag{1-15}$$

$$Z_0 I = A \exp(-\gamma z) - B \exp(\gamma z) \tag{1-16}$$

Writing the incident and reflected waves in terms of V and I:

$$A \exp(-\gamma z) = \tfrac{1}{2}(V + Z_0 I) \tag{1-17}$$

$$B \exp(\gamma z) = \tfrac{1}{2}(V - Z_0 I) \tag{1-18}$$

indicates that the travelling waves on such a distributed line are linear combinations of the voltage and current on the line as asserted. The power normalization adopted in the scattering formulation dealt with here is satisfied at the appropriate terminals provided the following substitutions are introduced:

$$a = \frac{A \exp(-\gamma z)}{\sqrt{Z_0}} \tag{1-19}$$

$$b = \frac{B \exp(\gamma z)}{\sqrt{Z_0}} \tag{1-20}$$

To show that $\frac{1}{2}a_1a_1^*$ is the available power at port 1 it is only necessary to form the voltage V_1 in Fig. 1-1 in terms of the generator voltage E_1 and the internal impedance Z_0:

$$V_1 = E_1 - Z_0I_1 \tag{1-21}$$

and to substitute this value of V_1 into the definition of a_1 in Eq. (1-11). The result is

$$a_1 = \frac{E_1}{2\sqrt{Z_0}} \tag{1-22}$$

The related quantity

$$\tfrac{1}{2}a_1a_1^* = \frac{|E_1|^2}{8Z_0} \tag{1-23}$$

is readily recognized as the available power of a generator of e.m.f. E_1 and internal impedance Z_0 as asserted.

To demonstrate that $\frac{1}{2}b_2b_2^*$ is the emergent power at port 2 it is necessary to combine Eqs (1-13) and (1-14) with $a_2 = 0$:

$$b_2 = \frac{V_2}{\sqrt{Z_0}} \tag{1-24}$$

The related quantity

$$\tfrac{1}{2}b_2b_2^* = \frac{|V_2|^2}{2Z_0} \tag{1-25}$$

is readily recognized as the power in the load at port 2.

The significance of the transmission parameters may be separately understood by evaluating S_{21} in Eq. (1-8) in terms of the original variables. This may be done by rewriting a_1 in Eq. (1-11) in terms of V_1 by making use of Eq. (1-21) and by rewriting V_2 in the definition of b_2 in Eq. (1-14) in terms of the load conditions at port 2:

$$V_2 = -I_2Z_0 \tag{1-26}$$

The required result

$$S_{21} = 2\frac{V_2}{E_1} \tag{1-27}$$

is easily recognized as the usual voltage transfer ratio of the network.

The meaning of the reflection parameter may be separately deduced by evaluating S_{11} in Eq. (1-7) in terms of the original variables. This gives

$$S_{11} = \frac{b_1}{a_1}\bigg|_{a_2=0} = \frac{V_1/\sqrt{Z_0} - I_1\sqrt{Z_0}}{V_1/\sqrt{Z_0} + I_1\sqrt{Z_0}} \tag{1-28}$$

The required result obtained by replacing V_1 by $Z_{\text{in}}I_1$,

$$S_{11} = \frac{Z_{\text{in}} - Z_0}{Z_{\text{in}} + Z_0} \tag{1-29}$$

is readily recognized as the usual bilinear transformation between the impedance and the reflection coefficient of a one-port network.

The significance of the scattering matrix may now be appreciated by recognizing that its entries represent the usual transfer ratios and reflection coefficients of a two-port network, the reflection coefficients being related to the one-port input immittances of the circuit. For a lossless network the entries of this matrix are related in a simple way by the unitary condition to be derived later in this chapter.

RELATIONSHIPS BETWEEN $\bar{S}$, $\bar{Z}$ AND $\bar{Y}$ MATRICES

Networks may be specified by either their scattering or immittance matrices. Other descriptions are the T matrix to be dealt with later in this chapter and the $ABCD$ description dealt with elsewhere in this text.

The correspondence between the scattering and impedance matrices may be derived by starting with the definition of the normalized scattering vectors in terms of the normalized voltage and current ones:

$$\bar{a} = \tfrac{1}{2}(\bar{V} + \bar{\imath}) \tag{1-30a}$$

$$\bar{b} = \tfrac{1}{2}(\bar{V} - \bar{\imath}) \tag{1-30b}$$

which have the form of Eqs (1-11) to (1-14).

Substituting $\bar{S}\bar{a}$ for $\bar{b}$ into the last equation indicates that

$$\bar{S}(\bar{V} + \bar{\imath}) = \bar{V} - \bar{\imath} \tag{1-31}$$

Replacing $\bar{V}$ by $\bar{Z}\bar{\imath}$ in this relationship,

$$\bar{S}(\bar{Z} + \bar{I}) = \bar{Z} - \bar{I} \tag{1-32}$$

and postmultiplying both sides by $(\bar{Z} + \bar{I})^{-1}$ gives the required result:

$$\bar{S} = \frac{\bar{Z} - \bar{I}}{\bar{Z} + \bar{I}} \tag{1-33}$$

The other one-port relationships are readily described by

$$\bar{Z} = \frac{\bar{I} + \bar{S}}{\bar{I} - \bar{S}} \tag{1-34}$$

$$\bar{S} = \frac{\bar{I} - \bar{Y}}{\bar{I} + \bar{Y}} \tag{1-35}$$

$$\bar{Y} = \frac{\bar{I} - \bar{S}}{\bar{I} + \bar{S}} \tag{1-36}$$

and
$$\bar{Z} = \bar{Y}^{-1} \tag{1-37}$$

The relationship between the entries of the $\bar{S}$, $\bar{Z}$ and $\bar{Y}$ matrices are summarized in Table 1-1.

By way of an example consider the evaluation of the scattering parameters of the network in Fig. 1-2 whose open-circuit parameters are described by

$$Z_{11}(s) = \frac{2s^2 + 1}{2s}$$

$$Z_{22}(s) = \frac{3s^2 + 1}{2s}$$

$$Z_{12}(s) = Z_{21}(s) = \frac{1}{2s}$$

The square impedance matrix of this network is positive real since it is associated with the physical circuit in Fig. 1-2.

TABLE 1-1

S parameters in terms of Y and Z parameters	Y and Z parameters in terms of S parameters
$S_{11} = \dfrac{(Z_{11} - 1)(Z_{22} + 1) - Z_{12}Z_{21}}{(Z_{11} + 1)(Z_{22} + 1) - Z_{12}Z_{21}}$	$Z_{11} = \dfrac{(1 + S_{11})(1 - S_{22}) + S_{12}S_{21}}{(1 - S_{11})(1 + S_{22}) - S_{12}S_{21}}$
$S_{12} = \dfrac{2Z_{12}}{(Z_{11} + 1)(Z_{22} + 1) - Z_{12}Z_{21}}$	$Z_{12} = \dfrac{2S_{12}}{(1 - S_{11})(1 - S_{22}) - S_{12}S_{21}}$
$S_{21} = \dfrac{2Z_{21}}{(Z_{11} + 1)(Z_{22} + 1) - Z_{12}Z_{21}}$	$Z_{21} = \dfrac{2S_{21}}{(1 - S_{11})(1 - S_{22}) - S_{12}S_{21}}$
$S_{22} = \dfrac{(Z_{11} + 1)(Z_{22} - 1) - Z_{12}Z_{21}}{(Z_{11} + 1)(Z_{22} + 1) - Z_{12}Z_{21}}$	$Z_{22} = \dfrac{1 + S_{22})(1 - S_{11}) + S_{12}S_{21}}{(1 - S_{11})(1 - S_{22}) - S_{12}S_{21}}$
$S_{11} = \dfrac{(1 - Y_{11})(1 + Y_{22}) + Y_{12}Y_{21}}{(1 + Y_{11})(1 + Y_{22}) - Y_{12}Y_{21}}$	$Y_{11} = \dfrac{(1 + S_{22})(1 - S_{11}) + S_{12}S_{21}}{(1 + S_{11})(1 + S_{22}) - S_{12}S_{21}}$
$S_{12} = \dfrac{-2Y_{12}}{(1 + Y_{11})(1 + Y_{22}) - Y_{22}Y_{21}}$	$Y_{12} = \dfrac{-2S_{12}}{(1 + S_{11})(1 + S_{22}) - S_{12}S_{21}}$
$S_{21} = \dfrac{-2Y_{21}}{(1 + Y_{11})(1 + Y_{22}) - Y_{12}Y_{21}}$	$Y_{21} = \dfrac{-2S_{21}}{(1 + S_{11})(1 + S_{22}) - S_{12}S_{21}}$
$S_{22} = \dfrac{1 + Y_{11})(1 - Y_{22}) + Y_{12}Y_{21}}{(1 + Y_{11})(1 + Y_{22}) - Y_{12}Y_{21}}$	$Y_{22} = \dfrac{1 + S_{11})(1 - S^2{}_2) + S_{12}S_{21}}{(1 + S_{22})(1 + S_{11}) - S_{12}S_{21}}$

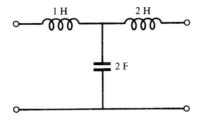

FIGURE 1-2
Schematic diagram of asymmetric network.

Making use of the results in Table 1-1 indicates that

$$S_{12}(s) = S_{21}(s) = \frac{1}{1.5s^3 + 2.5s^2 + 2.25s + 1}$$

$$S_{11}(s) = \frac{-(-1.5s^3 + 0.50s^2 - 0.25s)}{1.5s^3 + 2.5s^2 + 2.25s + 1}$$

$$S_{22}(s) = \frac{1.5s^3 + 0.50s^2 + 0.25s}{1.5s^3 + 2.5s^2 + 2.25s + 1}$$

THE UNITARY CONDITION

Since the entries of the scattering matrix have the nature of reflection and transmission parameters the amplitude of its entries are bounded by zero and unity. The permissible relationships between these entries will now be deduced. The derivation starts by recognizing that the power dissipated in an n-port network is the difference between the incident power at all ports and the reflected power at the same ports:

$$P_{\text{diss}} = \frac{1}{2}\left(\sum_i^n a_i a_i^* - \sum_i^n b_i b_i^* \right) \tag{1-38}$$

It may be readily demonstrated that

$$\sum_i^n a_i a_i^* = (\bar{a})^{\mathrm{T}}(\bar{a}*) \tag{1-39}$$

$$\sum_i^n b_i b_i^* = (\bar{a})^{\mathrm{T}}(\bar{S})^{\mathrm{T}}(\bar{S}*)(\bar{a}*) \tag{1-40}$$

so that the power dissipated in the circuit may be expressed in matrix form as

$$P_{\text{diss}} = \tfrac{1}{2}(\bar{a})^{\mathrm{T}}[\bar{I} - (\bar{S})^{\mathrm{T}}(\bar{S}*)](\bar{a}*) \tag{1-41}$$

where $\bar{I}$ is the unit matrix, $(\bar{S})^{\mathrm{T}}$ is the transpose of $\bar{S}$ and $(\bar{a}*)$ is the conjugate of $\bar{a}$.

The energy function defined by Eq. (1-41) is a Hermitian form in that the matrix

$$\bar{Q} = \bar{I} - (\bar{S})^{\mathrm{T}}(\bar{S}*) \tag{1-42}$$

is its own conjugate transpose:

$$[\bar{I} - (\bar{S})^{\mathrm{T}}(\bar{S}*)]^{*\mathrm{T}} = \bar{I} - (\bar{S})^{\mathrm{T}}(\bar{S}*) \tag{1-43}$$

All energy functions are in fact either Hermitian or quadratic forms.

Since the power dissipated in a network is always positive the Hermitian form in Eq. (1-41) is positive semi-definite:

$$\tfrac{1}{2}(\bar{a})^{\mathrm{T}}[\bar{I} - (\bar{S})^{\mathrm{T}}(\bar{S}*)](\bar{a}*) \geq 0 \tag{1-44}$$

For a reactance function the Hermitian form in (1-41) is satisfied with the equals sign. The condition for a lossless junction therefore becomes

$$\bar{I} - (\bar{S})^{\mathrm{T}}(\bar{S}*) = 0 \tag{1-45}$$

This last statement indicates that the scattering matrix of a dissipationless network is unitary. The unitary condition is widely used to establish the permissible relationships between the entries of the matrix $\bar{S}$ in a reactance network. Its significance may be examined by applying it to form the relationships between the scattering parameters of a two-port network:

$$S_{11}(s)S_{11}^*(s) + S_{21}(s)S_{21}^*(s) = 1 \tag{1-46a}$$

$$S_{22}(s)S_{22}^*(s) + S_{12}(s)S_{12}^*(s) = 1 \tag{1-46b}$$

$$S_{11}^*(s)S_{12}(s) + S_{21}^*(s)S_{22}(s) = 0 \tag{1-46c}$$

$$S_{11}(s)S_{12}^*(s) + S_{21}(s)S_{22}^*(s) = 0 \tag{1-46d}$$

The first two equations are expressions of the conservation of energy when there is an input wave at either port 1 or 2; they form part of the foundation of modern network synthesis. The third and fourth relationships represent the situations when the network is excited by its two possible eigenvectors. These important relationships are considered further later in this chapter and in more detail in Chapter 8, and are satisfied by the scattering parameters of the circuits in Figs 1-2 and 1-3.

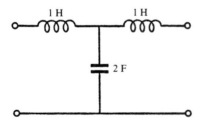

FIGURE 1-3
Schematic diagram of symmetrical network.

FELDTKELLER CONDITION

A relationship of some importance in the synthesis of modern filter circuits is the Feldtkeller equation. This gives a relationship between the numerator and denominator polynomials of the scattering parameters of a two-port reactance network. Its derivation starts by recognizing that $S_{11}(s)$ and $S_{22}(s)$ are related in such a network by the unitary condition

$$S_{11}(s)S_{11}^*(s) = S_{22}(s)S_{22}^*(s) \tag{1-47}$$

One solution that satisfies the preceding equation is

$$S_{11}(s) = \frac{h(s)}{g(s)} \tag{1-48}$$

$$S_{22}(s) = \pm\frac{h^*(s)}{g(s)} \tag{1-49}$$

In obtaining this solution use is made of the fact that the denominator polynomial of $S_{11}(s)$ or $S_{22}(s)$ must be strictly Hurwitz whereas the numerator polynomial need not be Hurwitz. Thus, in constructing a realizable solution for $S_{22}(s)$ from a knowledge of $S_{11}(s)$ it is not permissible to take the complex conjugate of $g(s)$ although it is permitted to take that of $h(s)$. The Hurwitz character of a polynomial is treated in some detail in Chapter 2. The problem in Fig. 1-2 suggests that the negative sign applies in Eq. (1-49).

The nature of $S_{21}(s)$ may also be now understood by again having recourse to the unitary condition. One possible solution is

$$S_{21}(s) = S_{12}(s) = \frac{f(s)}{g(s)} \tag{1-50}$$

This possibility satisfies

$$f(s)f^*(s) + h(s)h^*(s) = g(s)g^*(s) \tag{1-51}$$

and is known as the Feldtkeller condition.

Scrutiny of the relationships in Table 1-1 and anticipating the observation that the open-circuit parameters of two-port reactance networks are the ratios of odd to even or even to odd functions indicates that $f(s)$ for such a network is either an odd or an even polynomial. It is also apparent from the Feldtkeller conditions that the degree of $f(s)$ and $h(s)$ cannot exceed that of $g(s)$.

If the scattering matrix of a two-port reactance network is posed in terms of the polynomials $f(s)$, $g(s)$ and $h(s)$, then

$$\bar{S} = \frac{1}{g(s)}\begin{bmatrix} h(s) & f(s) \\ f(s) & \pm h^*(s) \end{bmatrix} \tag{1-52}$$

The task of relating the open- or short-circuit parameters to the polynomials associated with the scattering parameters described here is undertaken in Chapter 11.

SCATTERING TRANSFER PARAMETERS

In dealing with circuits in cascade, the scattering formalism is not the best description of the network. To overcome this difficulty, scattering transfer parameters are sometimes defined. This new matrix is obtained by rearranging the scattering relationships so that the input waves a_1 and b_1 are the dependent variables and the output waves a_2 and b_2 are the independent ones. In the original $\bar{S}$ matrix, the backward waves b_1 and b_2 are the dependent variables and the forward waves a_1 and a_2 are the independent variables. This new matrix is known as the $\bar{T}$ matrix. The standard $\bar{S}$ matrix relates a_1, a_2 to b_1, b_2 by

$$\begin{bmatrix} b_1 \\ b_2 \end{bmatrix} = \begin{bmatrix} S_{11} & S_{12} \\ S_{21} & S_{22} \end{bmatrix} \begin{bmatrix} a_1 \\ a_2 \end{bmatrix} \tag{1-53}$$

Rearranging this matrix so that a_1 and b_1 are the dependent variables and a_2 and b_2 are the independent ones gives

$$\begin{bmatrix} b_1 \\ a_1 \end{bmatrix} = \begin{bmatrix} T_{11} & T_{12} \\ T_{21} & T_{22} \end{bmatrix} \begin{bmatrix} a_2 \\ b_2 \end{bmatrix} \tag{1-54}$$

where

$$T_{11} = S_{12} - \frac{S_{11}S_{22}}{S_{21}} \tag{1-55a}$$

$$T_{12} = \frac{S_{11}}{S_{21}} \tag{1-55b}$$

$$T_{21} = \frac{S_{22}}{S_{21}} \tag{1-55c}$$

$$T_{22} = \frac{1}{S_{21}} \tag{1-55d}$$

A transfer matrix is also sometimes defined with the input waves as the independent variables and the output ones as the dependent variables.

The transfer matrices for the two individual networks with the nomenclature in Fig. 1-4 are given by

$$\begin{bmatrix} b_1 \\ a_1 \end{bmatrix} = \begin{bmatrix} T_{11} & T_{12} \\ T_{21} & T_{22} \end{bmatrix} \begin{bmatrix} a_2 \\ b_2 \end{bmatrix} \tag{1-56}$$

$$\begin{bmatrix} b_1' \\ a_1' \end{bmatrix} = \begin{bmatrix} T_{11}' & T_{12}' \\ T_{21}' & T_{22}' \end{bmatrix} \begin{bmatrix} a_2' \\ b_2' \end{bmatrix} \tag{1-57}$$

Making use of the fact that

$$\begin{bmatrix} b_1' \\ a_1' \end{bmatrix} = \begin{bmatrix} a_2 \\ b_2 \end{bmatrix} \tag{1-58}$$

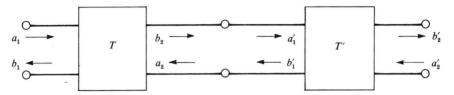

FIGURE 1-4
Cascade connection of two two-port networks.

gives

$$\begin{bmatrix} b_1 \\ a_1 \end{bmatrix} = \begin{bmatrix} T_{11} & T_{12} \\ T_{21} & T_{22} \end{bmatrix} \begin{bmatrix} T'_{11} & T'_{12} \\ T'_{21} & T'_{22} \end{bmatrix} \begin{bmatrix} a'_2 \\ b'_2 \end{bmatrix} \qquad (1\text{-}59)$$

The ratio b_1/a_1 gives S_{11} for the overall network.

Since matrix multiplication is not commutative, these $\bar{T}$ matrices must be multiplied in the proper order. Using the alternate definition for the $\bar{T}$ parameters mentioned earlier, this matrix multiplication must be shown in the reverse order.

MODERN FILTER SYNTHESIS

The scattering matrix description of a two-port circuit is of special importance in the modern synthesis of filters. The classic approach rests on the definition of a low-pass amplitude squared transfer function:

$$S_{21}(s)S_{21}^*(s)$$

and the use of the unitary condition to form the associated amplitude squared reflection coefficient:

$$S_{11}^*(s)S_{11}(s) = 1 - S_{21}(s)S_{21}^*(s)$$

and relies on the standard bilinear relationship between immittance and reflection to construct an *LCR* immittance function:

$$Y_{in}(s) = \frac{1 - S_{11}(s)}{1 + S_{11}(s)}$$

which is then synthesized as a two-port *LC* low-pass ladder network with a cutoff frequency of 1 rad/s terminated in a 1-ohm resistor in such a way as to display the attenuation poles of the transfer function. Band-pass, stopband and high-pass filters are then obtained using suitable frequency transformations. The required network is separated by impedance scaled from 1 to 50 Ω. This approach, due to Darlington, will always succeed provided $S_{11}(s)$ is formed from $S_{11}(s)S_{11}^*$ so that $Y_{in}(s)$ is associated with a one-port *LCR* network. Such immittances belong to a class of mathematical functions known as positive real functions. This is the main topic of Chapter 2.

PROBLEMS

1-1 Obtain the scattering parameters for the circuit below:

1-2 Check that the scattering parameters in Prob. 1-1 satisfy the unitary condition.

1-3 Verify Eqs (1-34) to (1-37).

1-4 Derive $S_{11}(s)$ and $S_{21}(s)$ of a symmetrical T network by using Eqs (1-7) to (1-10).

1-5 Repeat Prob. 1-4 for a π network.

1-6 Show that the input reflection coefficient of a two-port symmetrical network terminated in a load which is described by $b_2 = S_L a_2$ is

$$S'_{11} = S_{11} + \frac{S_{21}^2 S_L}{1 - S_{11} S_L}$$

BIBLIOGRAPHY

Belevitch, V., 'Elementary applications of the scattering formalism in network design', *IRE Trans. on Circuit Theory*, Vol. CT-3, pp. 97–104, 1956.

Carlin, H. J., 'The scattering matrix in network theory', *IRE Trans. on Circuit Theory*, Vol. CT-3, pp. 97–104, 1956.

CHAPTER
2

ONE-PORT IMMITTANCE FUNCTIONS

INTRODUCTION

Analysis typically involves the calculation of the response of a known circuit to a given excitation; synthesis or design consists of finding a new circuit that provides a desirable response to a given excitation. An analysis problem has normally a solution; by contrast, in synthesis a solution may not exist. This chapter indicates that whereas a one-port LCR immittance function is the ratio of two polynomials, the ratio of any two polynomials need not be realizable as a one-port network. To ensure realizability, it is necessary for it to be a positive real (p.r.) function. A general testing procedure due to Brune, applicable to one-port networks, is first stated without proof. This testing procedure is then specialized to the special class of reactive LC functions which forms the main topic of this text. One-port LC functions are shown to be the ratio of odd to even or even to odd Hurwitz polynomials with simple poles on the $j\omega$ axis with positive real residues.

A bounded real condition for the reflection coefficient is separately obtained by making use of the bilinear transformation between the reflection coefficient and the impedance of a network. This bounded real condition may be used to determine whether a one-port immittance is p.r. or not without the need to evaluate residues, as is required in the case of a reactance function.

ONE-PORT IMMITTANCES

The input immittance $I(s)$ (impedance or admittance) of a one-port passive network involves at most algebraic manipulation of terms such as R, sL and $1/sC$, and may always be reduced to be the ratio of two polynomials with positive real coefficients. The general form of such an immittance or rational function is

$$I(s) = \frac{P(s)}{Q(s)} = \frac{a_n s^n + a_{n-1} s^{n-1} + \cdots + a_1 s + a_0}{b_m s^m + b_{m-1} s^{m-1} + \cdots + b_1 s + b_0} \qquad (2\text{-}1)$$

The nature of $I(s)$ for which the one-port network in Fig. 2-1 is realizable in terms of LCR elements is the main topic of this section. This rational function may also be expressed as a product of factors:

$$I(s) = \frac{a_n(s - z_1)(s - z_2) \cdots (s - z_n)}{b_m(s - p_1)(s - p_2) \cdots (s - p_m)} \qquad (2\text{-}2)$$

The poles and zeros of $I(s)$ are known as the natural frequencies of the system. If the poles and zeros are all different then the p's and z's are simple; in some situations two or more poles or zeros may coincide to give a multiple pole or zero. This representation is important in that the location of the poles and zeros in the complex frequency plane furnishes a means of classifying networks.

Another useful form for Eq. (2-1) or (2-2) is obtained by forming a partial fraction expansion of Eq. (2-2):

$$I(s) = k_n s^n + k_{n-1} s^{n-1} + \cdots + k_1 s + k_0 + \frac{k_{-1}}{s} + \frac{k_{-2}}{s^2} + \cdots + \frac{k_{-m}}{s^m}$$

$$+ \frac{k_{p1}}{s - j\omega_{p1}} + \frac{k'_{p1}}{s + j\omega_{p1}} + \frac{k_{p2}}{s - j\omega_{p2}} + \frac{k'_{p2}}{s + j\omega_{p2}} + \cdots \qquad (2\text{-}3)$$

Terms like k_0, $k_1 s$ and k_{-1}/s are realizable in terms of LCR elements, whereas terms like $k_{n-1} s^{n-1}$, k_{-m}/s^m and $k_{p1}/(s - j\omega_{p1})$ are not. Thus a general immittance function represented by either Eq. (2-1), (2-2) or (2-3) is not necessarily realizable in terms of simple LCR elements. Functions that are realizable as LCR one-port

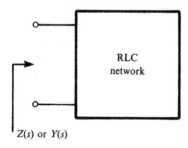

Z(s) or Y(s)

FIGURE 2-1
Schematic diagram of one-port network.

networks are known as positive real. A similar conclusion may be deduced from either Eq. (2-1) or (2-2).

POSITIVE REAL FUNCTIONS

A positive real function $I(s)$ is an analytic function of a complex variable $s = \sigma + j\omega$ which has the following properties:

$$I(s) \text{ is real for } s \text{ real.} \tag{2-4}$$

$$\text{Re}[I(s)] \geq 0 \text{ for } \sigma \geq 0. \tag{2-5}$$

These two conditions, due to Brune, may be interpreted in the complex s plane in the manner illustrated in Fig. 2-2. The first condition states that the real axis of the s plane maps onto the real axis of the $I(s)$ plane. The second condition indicates that the right half of the s plane maps onto the right half of the $I(s)$ plane.

Taking the impedance of a series LCR circuit as an example,

$$Z(s) = R + \left(sL + \frac{1}{sC} \right)$$

indicates that the first statement is satisfied and forming

$$\text{Re}[Z(s)]_{s=j\omega+\sigma} = R + \sigma L + \frac{\sigma C}{\sigma^2 + \omega^2}$$

indicates that the second test is also met.

The first of these tests is easy to implement since it is only necessary to ensure that the a's and b's in Eq. (2-1) are positive. However, to satisfy the second

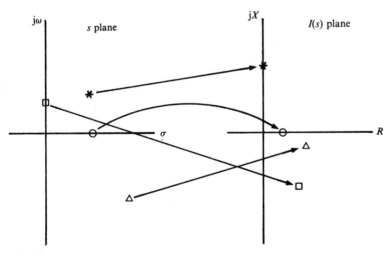

FIGURE 2-2
Mapping between s and $I(s)$ planes, for p.r. functions.

test it is necessary to evaluate every single point in the right half of the s-plane (RHP), which is not readily done. To overcome this difficulty the statement in (2-5) may be replaced by the following three equivalent necessary and sufficient conditions (stated again without proof):

$$I(s) \text{ is real for } s \text{ real.} \tag{2-6}$$

$$I(s) \text{ must have no RHP poles.} \tag{2-7}$$

$I(s)$ may only have simple poles on the $j\omega$ axis
with positive real residues. $\tag{2-8}$

$$\mathrm{Re}[I(j\omega)] \geq 0 \text{ for all } \omega. \tag{2-9}$$

The statement that $I(s)$ is real for s real may be appreciated by recognizing that network functions are obtained by solving simple algebraic equations which involve at most terms like sL and sC and their reciprocals. The condition $\mathrm{Re}[I(j\omega)] \geq 0$ for all ω may be understood by noting that passive LCR networks contain no energy sources and as such can only dissipate but not deliver energy. This must be true at any point on the $j\omega$ axis. The two conditions that $I(s)$ may have no RHP poles and may only have simple poles on the $j\omega$ axes with positive real residues may be appreciated by noting that such poles lead to growing waves in the time domain and are therefore prohibited.

Taking Ls as an example it is readily seen that Ls is real for s real, Ls has no poles on the RHP and Ls has a simple pole at $s = j\omega$. To evaluate the condition $\mathrm{Re}[I(\omega)] \geq 0$ for all ω it is necessary to replace s by $\sigma + j\omega$ and take the real part of $L(\sigma + j\omega)$, which readily satisfies the last test. It is apparent that $1/cS$ and R also satisfy conditions (2-6) to (2-9).

It will be shown in the next section that the above conditions may be replaced by

$$P(s) \text{ and } Q(s) \text{ are strictly Hurwitz.} \tag{2-10}$$

$$\mathrm{Re}[I(j\omega)] \geq 0 \text{ for all } \omega. \tag{2-11}$$

or

$$P(s) \text{ and } Q(s) \text{ are Hurwitz.} \tag{2-12}$$

$I(s)$ may only have simple poles on the $j\omega$ axis
with positive real residues. $\tag{2-13}$

$$\mathrm{Re}[I(j\omega)] \geq 0 \text{ for all } \omega. \tag{2-14}$$

HURWITZ POLYNOMIALS

The roots of a polynomial are determined by solving its characteristic equation

$$p(s) = a_n s^n + a_{n-1} s^{n-1} + \cdots + a_0 = 0 \tag{2-15}$$

or in factored form

$$p(s) = a_n(s - s_1)(s - s_2) \cdots (s - s_n) = 0 \tag{2-16}$$

where s_k are the roots of $p(s)$.

The factors of the characteristic equation consist of the following four types:

$$(s \pm \sigma_k) \tag{2-17a}$$

$$(s + \sigma_k \pm j\omega_k) \tag{2-17b}$$

$$(s - \sigma_k \pm j\omega_k) \tag{2-17c}$$

$$(s \pm j\omega_k) \tag{2-17d}$$

A polynomial is either Hurwitz, strictly Hurwitz or non-Hurwitz according to whether its roots are restricted to the LHP including the $j\omega$ axis, its roots are restricted to the LHP excluding the $j\omega$ axis or its roots do not correspond to a Hurwitz or strictly Hurwitz polynomial. A Hurwitz polynomial must also satisfy $p(s)$ is real for s real. Thus all complex roots must appear as conjugate pairs, otherwise the coefficients of $p(s)$ are not all real. The three possible situations are depicted in Fig. 2-3.

The only possible factors in a Hurwitz polynomial are therefore

$$(s + q_k) \tag{2-18}$$

$$(s + \sigma_k)^2 + \omega_k^2 \tag{2-19}$$

$$(s^2 + \omega_k^2) \tag{2-20}$$

It is observed that if $I(s)$ in Eq. (2-1) refers to $Z(s)$, $Q(s)$ must be a Hurwitz polynomial. However, since the zeros of $Z(s)$ are the poles of $Y(s)$, $P(s)$ must also be a Hurwitz polynomial. A p.r. function is therefore the ratio of two Hurwitz polynomials.

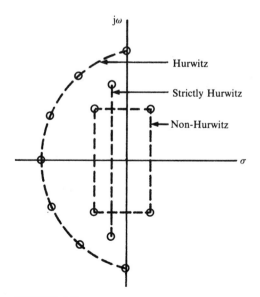

FIGURE 2-3
Zeros of Hurwitz, strictly Hurwitz and non-Hurwitz polynomials.

It is of note that an even polynomial involves only products of the form

$$(s^2 + \omega_k^2)$$

and that an odd polynomial involves products of the types

$$(s + q_k)$$

$$(s^2 + \omega_k^2)$$

with $q_k = 0$. It therefore follows that the roots of even and odd Hurwitz polynomials lie on the $j\omega$ axis with positive real residues.

Another property of these classes of polynomials is that no adjacent terms of order more than two may be missing in either an even or an odd Hurwitz polynomial. To demonstrate this property consider the following even polynomial with $a_2 = 0$:

$$m(s) = a_4 s^4 + a_0 = 0$$

The roots of this polynomial are located on a unit circle:

$$s_k = \left(\frac{a_0}{a_4}\right)^{1/4} e^{j(2k-1)\pi/4} \qquad \text{for } k = 0, 1, 2, 3$$

which do not lie on the $j\omega$ axis as required.

The fact that a polynomial possesses positive coefficients, although necessary, is by itself not sufficient, as the following polynomial illustrates:

$$q(s) = (s + 2)(s^2 - s + 4) = s^3 + s^2 + 2s + 8$$

The factors of this polynomial do not coincide with those in (2-18) to (2-20) although the characteristic equation possesses positive coefficients.

Whether a polynomial is Hurwitz or not merely involves finding the factors of the characteristic equation. Equations up to the fourth order are solvable in this way since there are analytical expressions to determine their roots. No such expressions exist for equations of higher order and numerical solutions must be used.

A polynomial may be tested to deduce whether it is Hurwitz or not by constructing a continued fraction expansion of the odd to even or even to odd parts of $Q(s)$ or $P(s)$. If all the quotients are positive then it is Hurwitz; otherwise it is not.

As an example of the evaluation of the Hurwitz test consider the following polynomial:

$$Q(s) = s^4 + s^3 + 6s^2 + 4s + 5$$

The odd and even parts of $Q(s)$ are

$$n(s) = s^3 + 4s$$

$$m(s) = s^4 + 6s^2 + 5$$

The ratio of the even to the odd parts of $Q(s)$ is

$$\psi(s) = \frac{s^4 + 6s^2 + 5}{s^3 + 4s}$$

The continued fraction expansion of $\psi(s)$ is

$$
s^3 + 4s \,)\, s^4 + 6s^2 + 5 \quad (s
$$
$$
\underline{s^4 + 4s^2 \quad\; 5}
$$
$$
2s^2 + 5 \,)\, s^3 + 4s \quad (\tfrac{s}{2}
$$
$$
\underline{s^3 + 2.5s}
$$
$$
1.5s \,)\, 2s^2 + 5 \,(\tfrac{4s}{3}
$$
$$
\underline{2s}
$$
$$
5 \,)\, 1.5s \,(\tfrac{3s}{10}
$$
$$
\underline{1.5s}
$$
$$
0
$$

Since all the quotients are positive this polynomial is Hurwitz. This expansion is sometimes written as

$$\psi(s) = s + \cfrac{1}{\cfrac{s}{2} + \cfrac{1}{\cfrac{4s}{3} + \cfrac{1}{\cfrac{3s}{10} + 0}}}$$

Proceeding in a similar fashion indicates that the following polynomial is not Hurwitz:

$$Q(s) = s^3 + s^2 + 2s + 8$$

since all the quotients of $Q(s)$ are not positive.

If the continued fraction expansion is terminated prematurely through a common factor $W(s)$ it is necessary to write $Q(s)$ as $W(s)Q_1(s)$ and ensure that each polynomial is Hurwitz. One such example is

$$Q(s) = s^3 + 2s^2 + 3s + 6$$

for which

$$W(s) = s^3 + 3s$$

$$Q_1(s) = 1 + \frac{2}{s}$$

ONE-PORT *LC* REACTANCE FUNCTIONS

If an immittance function is a reactance one, $\text{Re}[I(j\omega)] = 0$, the necessary and sufficient conditions in (2-11) to (2-14) then become

$I(s)$ is the ratio of odd to even or even to odd
Hurwitz polynomials.   (2-21)

$I(s)$ has only simple poles on the $j\omega$ axis with
positive real residues.   (2-22)

An even polynomial Ev $F(s)$ has the property that

$$\text{Ev } F(-s) = \text{Ev } F(s) \qquad (2\text{-}23a)$$

and an odd one that

$$\text{Od } F(-s) = -\text{Od } F(s) \qquad (2\text{-}23b)$$

In general

$$F(s) = \text{Ev } F(s) + \text{Od } F(-s) \qquad (2\text{-}23c)$$

To demonstrate that a reactance function is the ratio of an odd to even or even to odd Hurwitz polynomial as asserted in (2-21) it is necessary to make use of the fact that the power dissipated in a reactance network equals zero. This may be done by first writing the numerator and denominator polynomials in terms of their even and odd parts:

$$Z(s) = \frac{m_1(s) + n_1(s)}{m_2(s) + n_2(s)} \qquad (2\text{-}24)$$

where $m_1(s)$, $n_1(s)$ and $m_2(s)$, $n_2(s)$ are the even and odd parts of the numerator and denominator polynomials of $Z(s)$ respectively. The even and odd parts of $Z(s)$ are now formed by writing $Z(s)$ as

$$Z(s) = \frac{m_1(s) + n_1(s)}{m_2(s) + n_2(s)} \frac{m_2(s) - n_2(s)}{m_2(s) - n_2(s)} \qquad (2\text{-}25a)$$

which gives

$$Z(s) = \frac{[m_1(s)m_2(s) - n_1(s)n_2(s)] + [m_2(s)n_1(s) - m_1(s)n_1(s)]}{m_2^2(s) - n_2^2(s)} \qquad (2\text{-}25b)$$

Noting that the products of two even or two odd polynomials is an even polynomial and that the product of an even and an odd polynomial is an odd polynomial gives

$$\text{Ev}[Z(s)] = \frac{m_1(s)m_2(s) - n_1(s)n_2(s)}{m_2^2(s) - n_2^2(s)} \qquad (2\text{-}25c)$$

$$\text{Od}[Z(s)] = \frac{m_2(s)n_1(s) - m_1(s)n_2(s)}{m_2^2(s) - n_2^2(s)} \qquad (2\text{-}25d)$$

Replacing s by $j\omega$ indicates that

$$\text{Re}[Z(j\omega)] = \text{Ev}[Z(s)]_{s=j\omega} \qquad (2\text{-}26a)$$

$$j\,\text{Im}[Z(j\omega)] = \text{Od}[Z(s)]_{s=j\omega} \qquad (2\text{-}26b)$$

For a reactance function

$$\text{Re}[Z(j\omega)] = 0 \qquad (2\text{-}27)$$

and

$$m_1(s)m_2(s) - n_1(s)n_2(s) = 0 \qquad (2\text{-}28)$$

The two possible solutions to this equation are

$$m_1(s) = n_2(s) = 0 \qquad (2\text{-}29)$$

or

$$m_2(s) = n_1(s) = 0 \qquad (2\text{-}30)$$

and $Z(s)$ is the ratio of an even to odd Hurwitz polynomial:

$$Z(s) = \frac{m_1(s)}{n_2(s)} \qquad (2\text{-}31)$$

or it is the ratio of an odd to even Hurwitz polynomial:

$$Z(s) = \frac{n_1(s)}{m_2(s)} \qquad (2\text{-}32)$$

These two conditions indicate that a reactance function is either the ratio of odd to even or even to odd Hurwitz polynomials. However, while either of these conditions is necessary it is not sufficient since $Z(s)$ must also satisfy condition (2-22).

The fact that a reactance function has only simple poles on the $j\omega$ axis may be understood by recalling that the roots of even and odd Hurwitz polynomials lie on the $j\omega$ axis with positive real residues. It may also be deduced by noting that if $Z(s)$ is a reactance function given by the ratio of odd to even or even to odd polynomials as in Eq. (2-31) or (2-32) then it is an odd function of s and

$$Z(s) = -Z(-s) \qquad (2\text{-}33)$$

If $Z(s)$ has no poles or zeros on the RHP, $Z(-s)$ has no poles or zeros on the LHP. Since there are no poles of $Z(s)$ on either the LHP or the RHP, the poles and zeros of reactance functions lie on the $j\omega$ axis as already noted.

Condition (2-22) also states that $Z(s)$ or $Y(s)$ must have no multiple poles at the origin and at infinity. Thus the highest degree terms of $P(s)$ and $Q(s)$ must differ by unity. Likewise, the lowest degree terms of $P(s)$ and $Q(s)$ must differ by unity. Reactance functions must therefore have either a pole or zero at the origin and at infinity. Figure 2-4 depicts the four possible pole zero arrangements whose features lead to the first and second Cauer realizations of a one-port network to be discussed in Chapter 3.

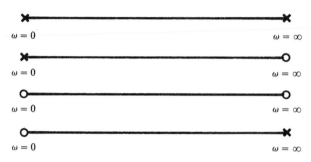

FIGURE 2-4
Pole zero diagrams for LC reactance or susceptance functions.

To demonstrate the first requirement consider the following LC function by taking the ratio of two Hurwitz polynomials as in (2-21):

$$Z(s) = \frac{a_5 s^5 + a_3 s^3 + a_1 s}{b_4 s^4 + b_2 s^2 + b_0} \tag{2-34}$$

This reactance function has a simple zero at the origin and a simple pole at infinity. It therefore is p.r. since it meets both (2-21) and (2-22).
Putting $b_4 = 0$ yields

$$Z(s) = \frac{a_5 s^5 + a_3 s^3 + a_1 s}{b_2 s^2 + b_0} \tag{2-35}$$

This function is still the ratio of two Hurwitz polynomials in keeping with (2-21), but is no longer p.r. since it has multiple poles at infinity contrary to the condition in (2-22).
Forming a new function $Y(s)$ instead of $Z(s)$ and putting $a_1 = 0$ gives

$$Y(s) = \frac{b_4 s^4 + b^2 s^2 + b_0}{a_5 s^5 + a_3 s^3} \tag{2-36}$$

This function is again not p.r. since it has multiple poles at the origin, which is not compatible with (2-22).
The notation in (2-42) to (2-45) reflects the important fact that the poles and zeros of reactance functions interlace on the $j\omega$ axis. To demonstrate this property it is necessary to extract the realizable terms in the partial fraction expansion in Eq. (2-3) and replace s by $j\omega$:

$$X(\omega) = k_\infty \omega + \frac{2k_1 \omega}{-\omega^2 + \omega_1^2} + \cdots \tag{2-37}$$

Differentiating this quantity with respect to ω leads to

$$\frac{dX(\omega)}{d\omega} = k_\infty + \frac{2k_1(\omega^2 + \omega_1^2)}{(-\omega^2 + \omega_1^2)^2} + \cdots \tag{2-38}$$

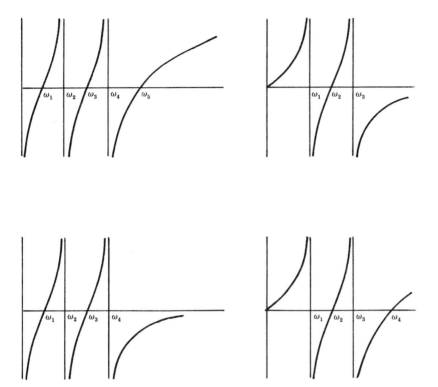

FIGURE 2-5
Typical reactance or susceptance curves for LC functions.

Since all residues are positive,

$$\frac{dX(\omega)}{d\omega} \geq 0 \qquad (2\text{-}39)$$

This inequality implies that the slope of a reactance function is always positive for real frequencies and that the poles and zeros of reactance functions alternate on the $j\omega$ axis.

Figure 2-5 summarizes the four possible reactance functions.

PROPERTIES OF REACTANCE AND SUSCEPTANCE FUNCTIONS

Combining the properties of Hurwitz polynomials with those of reactance or susceptance functions gives the necessary and sufficient properties of such functions as:

1. $Z(s)$ and $Y(s)$ are the ratio of odd to even or even to odd Hurwitz polynomials.
2. No adjacent terms of more than two may be missing in either the numerator or denominator polynomials (even or odd Hurwitz polynomials).

3. The poles and zeros of $Z(s)$ and $Y(s)$ are interlaced on the $j\omega$ axis with positive real residues.
4. The highest degree terms of the numerator and denominator polynomials of $Z(s)$ or $Y(s)$ differ by unity.
5. The lowest degree terms of the numerator and denominator polynomials of $Z(s)$ or $Y(s)$ differ by unity.
6. $Z(s)$ or $Y(s)$ has either a pole or zero at the origin and at infinity.

The general form of a realizable LC reactance function in polynomial form is therefore

$$Z(s) = \frac{s(a_n s^n + a_{n-2} s^{n-2} + \cdots + a_0)}{(b_m s^m + b_{m-2} s^{m-2} + \cdots + b_0)} \quad \text{for } m = n \text{ or } m = n+2 \quad (2\text{-}40)$$

or

$$Z(s) = \frac{(a_n s^n + a_{n-2} s^{n-2} + \cdots + a_0)}{s(b_m s^m + b_{m-2} s^{m-2} + \cdots + b_0)} \quad \text{for } m = n \text{ or } m = n-2 \quad (2\text{-}41)$$

The origin is either a pole or a zero and there is either a pole or zero at infinity depending on whether the numerator or denominator has the higher degree. This is indicated in Fig. 2-4.

The four terms defined by Eqs (2-40) and (2-41) may be factored as in Eq. (2-2):

$$Z(s) = \frac{s(s^2 + z_2^2)(s^2 + z_4^2) \cdots}{(s^2 + p_1^2)(s^2 + p_3^2) \cdots} \quad (2\text{-}42)$$

$$Z(s) = \frac{(s^2 + z_1^2)(s^2 + z_3^2) \cdots}{s(s^2 + p_2^2)(s^2 + p_4^2) \cdots} \quad (2\text{-}43)$$

The poles and zeros occur on the $j\omega$ axis either as conjugate pairs as in (2-20) or as a simple pole or zero at the origin and at infinity.

$Z(s)$ in Eqs (2-40) and (2-41) may also be written in the form indicated in Eq. (2-3):

$$Z(s) = k_\infty s + \frac{2k_1 s}{s^2 + p_1^2} + \frac{2k_3 s}{s^2 + p_3^2} + \cdots \quad (2\text{-}44)$$

and

$$Z(s) = \frac{k_0}{s} + \frac{2k_2 s}{s^2 + p_2^2} + \frac{2k_4 s}{s^2 + p_4^2} + \cdots \quad (2\text{-}45)$$

respectively.

The residues k_0, k_1, etc., must be positive real and are obtained in the usual way:

$$k_i = \frac{s^2 + p_i^2}{2s} Z(s)|_{s^2 = p_i^2} \quad (2\text{-}46a)$$

$$k_0 = sZ(s)|_{s=0} \quad (2\text{-}46b)$$

$$k_\infty = \lim \frac{Z(s)}{s}\bigg|_{s \to \infty} \quad (2\text{-}46c)$$

Dual equations to (2-40) to (2-46) apply for $Y(s)$.

REALIZABILITY PROCEDURE FOR IMMITTANCE FUNCTIONS

The general testing procedure for *LCR* functions summarized in conditions (2-12) to (2-14) will now be used to verify the p.r. character of one immittance function:

$$Z(s) = \frac{s^2 + 1}{s^3 + 4s}$$

The first condition defined by (2-12) is met by inspection since $P(s)$ and $Q(s)$ are even and odd polynomials which are Hurwitz by definition.

The second condition given by (2-13) may be evaluated by factoring $Z(s)$ and finding its residues:

$$Z(s) = \frac{s^2 + 1}{s(s + j2)(s - j2)}$$

$s = 0$ residue:

$$k_0 = sZ(s)\big|_{s=0} = \tfrac{1}{4}$$

$s = -j2$ residue:

$$k_1 = (s + j2)|Z(s)|_{s=-j2} = \tfrac{3}{8}$$

$s = j2$ residue:

$$k_2 = (s - j2)Z(s)\big|_{s=j2} = \tfrac{3}{8}$$

The third and last test (2-14) may be evaluated by replacing s by $j\omega$ and taking the real part of $Z(s)$:

$$Z(s)\big|_{s=j\omega} = j\frac{(\omega^2 - 1)}{(-\omega^3 + 4\omega)}$$

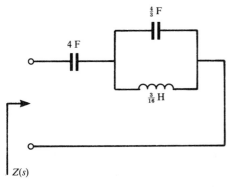

FIGURE 2-6
Equivalent circuit for $(s^2 + 1)/(s^3 + 4s)$.

Thus

$$Re[Z(j\omega)] = 0$$

Since $Z(s)$ satisfies (2-12) to (2-14) it is indeed p.r. Further, since $Z(s)$ satisfies (2-21) and (2-22) it is a reactance function.
One possible equivalent circuit for $Z(s)$ is depicted in Fig. 2-6.

THE BOUNDED REAL CONDITIONS

An alternate testing procedure for a one-port immittance function which avoids the necessity of computing residues may be deduced by examining the bilinear transformation between $Z(s)$ and the reflection coefficient $\Gamma(s)$:

$$\Gamma(s) = \frac{Z(s) - 1}{Z(s) + 1} \qquad (2\text{-}47)$$

Making use of conditions (2-6) to (2-9) indicates that

$\Gamma(s)$ is real for s real. $\qquad (2\text{-}48)$

$\Gamma(s)$ must have no right half or $j\omega$ axis poles. $\qquad (2\text{-}49)$

$0 \leq |\Gamma(j\omega)| \leq |$ for all ω. $\qquad (2\text{-}50)$

The first condition is satisfied by ensuring that the coefficients of the polynomials of $\Gamma(s)$ are real and need not be further laboured.
To demonstrate the second requirement it is merely necessary to write $Z(s)$ in terms of its odd and even parts:

$$Z(s) = \frac{m_1(s) + n_1(s)}{m_2(s) + n_2(s)} \qquad (2\text{-}51)$$

and to form $\Gamma(s)$ by making use of the bilinear transformation between $\Gamma(s)$ and $Z(s)$:

$$\Gamma(s) = \frac{m_2(s) + n_2(s) - m_1(s) - n_1(s)}{m_2(s) + n_2(s) + m_1(s) + n_1(s)} \qquad (2\text{-}52)$$

Scrutiny of the denominator polynomial of this quantity indicates that for both LC and LCR networks it cannot be odd or even. Such a polynomial is therefore strictly Hurwitz. Since this type of polynomial does not have roots on the $j\omega$ axis, this testing procedure avoids the need to evaluate residues. The numerator polynomial, being the difference between two Hurwitz polynomials, need not be Hurwitz.

PROBLEMS

2-1 Determine which of the following polynomials are Hurwitz:

$$s^2 + 3s + 1$$
$$s^3 + 3s^2 - 6s + 10$$
$$s^3 + 4s^2 + 5s + 6$$

2-2 Show that $\text{Re}|Z(j\omega)| = \text{Ev}[Z(s)]$, $j\,\text{Im}|Z(j\omega)| = \text{Od}[Z(s)]$.

2-3 Show that the poles and zeros of $Z(s) = s(s^2 + \omega_3^2)/[(s^2 + \omega_2^2)(s^2 + \omega_4^2)]$ interlace on the $j\omega$ axis.

2-4 Demonstrate Eqs (2-31) and (2-32) by starting with the condition $\text{Re}|Z(j\omega)| = 0$.

2-5 Show that condition (2-14) may be written as $m_1(s)m_2(s) - n_1(s)n_2(s)|_{s=j\omega} \geq 0$ for all ω.

2-6 Verify that $k_1/(s - j\omega_1)$ is not p.r.

2-7 Verify that $k_1/(s - j\omega_1) + k_2/(s + j\omega_1)$ is not p.r. unless $k_1 = k_2$, when it can be written as $2k_1s/(s^2 + \omega_1^2)$.

2-8 Show that sL and $1/sC$ satisfy (2-4) and (2-5).

2-9 Show that

$$Z(s) = \frac{s+1}{s^2 + s + 1}$$

is p.r., whereas

$$Z(s) = \frac{s+2}{s^2 + s + 2}$$

is not.

2-10 Verify that $s^3 + s^2 + 2s + 8$ is not Hurwitz.

2-11 Investigate whether

$$\psi(s) = \frac{s^4 + 6s^2 + 5}{s^3 + 4s}$$

is a reactance function.

BIBLIOGRAPHY

Brune, O., 'On the synthesis of a two terminal network whose driving point impedance is a prescribed function of frequency', *J. Math. Phys.*, Vol. 10, pp. 191–236, 1931.

Cauer, W., 'Ein interpolations problem mit funktionen mit positiven realteil', *Mathematics Zeitschrift*, Vol. 38, pp. 1–44, 1933.

Guillemin, E. A., *The Mathematics of Circuit Analysis*, John Wiley, New York.

Hurwitz, A., 'Ueber die Bedingungen unter welchen eine Gleichung nur Wurzeln mit Negativen Reelen Teilen Besitzt', *Math. Ann.*, Vol. 46, p. 273, 1895.

Temes, G. C. and Lapratra, J. W., *Introduction to Circuit Synthesis and Design*, McGraw-Hill, New York, 1977.

CHAPTER
3

SYNTHESIS OF ONE-PORT REACTANCE CIRCUITS

INTRODUCTION

Four simple canonical forms for the realization of one-port reactance functions are the first and second Foster forms and the first and second Cauer forms. These four topologies are canonical in that they can always be realized, and they are also minimal in that they realize the driving point immittance with the least number of elements. The four canonical forms described in this chapter indicate that there is more than one way to synthesize a one-port driving point immittance.

The two Foster forms involve a sequential removal of the poles of an immittance function by first putting it into a partial fraction form. If the pole is at the origin or at infinity the order of the function is reduced by unity. If it is at $s = \pm j\omega_i$ the order of the function is reduced by two. The two Cauer canonical forms involve the removal of poles exclusively at the origin or at infinity. It is of note that the zeros of the immittance function are not involved in the construction of one-port networks.

In many instances it may be desirable to extract a pole at a finite frequency from a reactance function that does not coincide with its pole zero diagram. One way this may be achieved is by employing a zero shifting technique. This principle is separately outlined in this chapter.

FIRST FOSTER
CANONICAL FORM

The first Foster form realization of a one-port immittance function is obtained by expanding $Z(s)$ by partial fractions and identifying terms in the summation with impedances of simple networks:

$$Z(s) = Z_1(s) + Z_2(s) + Z_3(s) + \cdots \tag{3-1}$$

The general form for Eq. (3-1) is obtained by combining Eqs (2-44) and (2-45) in Chapter 2:

$$Z(s) = \frac{K_0}{s} + \sum_{i=1}^{n} \frac{2K_i s}{s^2 + \omega_i^2} + K_\infty s \tag{3-2}$$

The residues are given by

$$K_0 = sZ(s)|_{s=0} \tag{3-3}$$

$$2K_i = \frac{s^2 + \omega_i^2}{s} Z(s)|_{s^2 = -\omega_i^2} \tag{3-4}$$

$$K_\infty = \mathrm{Im} \left. \frac{Z(s)}{s} \right|_{s \to \infty} \tag{3-5}$$

The first term in the partial fraction expansion is recognized as the reactance of a series capacitance C_0 having a value

$$C_0 = \frac{1}{K_0} \quad \mathrm{F} \tag{3-6}$$

The second term is recognized as the reactance of a parallel tuned circuit with capacitance and inductance C_i and L_i:

$$C_i = \frac{1}{2K_i} \quad \mathrm{F} \tag{3-7}$$

$$L_i = \frac{2K_i}{\omega_i^2} \quad \mathrm{H} \tag{3-8}$$

respectively. The last term in this expansion represents the reactance of an inductance with a value

$$L_\infty = K_\infty \quad \mathrm{H} \tag{3-9}$$

If the function in Eq. (3-2) has a pole at $s = 0$ and at $s = j\infty$ its order is reduced by one if either of these poles is extracted from $Z(s)$. If the function $Z(s)$ has a pair of conjugate poles at $s = \pm j\omega_i$ the order is reduced by two if these poles are extracted in the form of a tank circuit. Figure 3-1 gives the schematic diagram of the first Foster form realization of an impedance function.

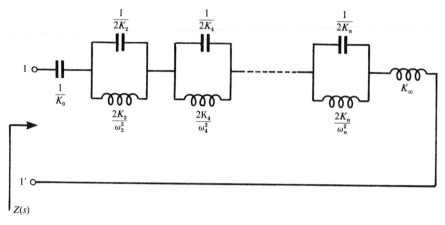

FIGURE 3-1
First Foster realization of a one-port impedance function.

To demonstrate the realization of a one-port reactance function in the first Foster form consider the following reactance function:

$$Z(s) = \frac{(s^2 + 1)(s^2 + 9)}{s(s^2 + 4)}$$

$Z(s)$ is a reactance function in that it satisfies the properties of an LC function listed in Chapter 2. Its poles and zeros are interlaced in the manner shown in Fig. 3-2.

The residues of the poles in the partial fraction expansion of $Z(s)$ are given by Eqs (3-3) to (3-5) as

$$K_0 = sZ(s)\big|_{s=0} = \tfrac{9}{4}$$

$$2K_2 = \frac{s^2 + 4}{s} Z(s)\big|_{s^2 = -4} = \tfrac{15}{4}$$

$$K_\infty = \operatorname{Im} \frac{Z(s)}{s}\bigg|_{s \to \infty} = 1$$

Each residue is positive in keeping with the p.r. condition in Chapter 2. The partial

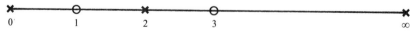

FIGURE 3-2
Pole zero diagram of
$$Z(s) = \frac{(s^2 + 1)(s^2 + 9)}{s(s^2 + 4)}$$

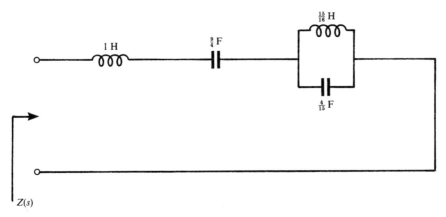

FIGURE 3-3
First Foster one-port network of

$$Z(s) = \frac{(s^2 + 1)(s^2 + 9)}{s(s^2 + 4)}$$

fraction expansion of $Z(s)$ is therefore

$$Z(s) = \frac{9/4}{s} + \frac{15s/4}{s^2 + 4} + s$$

Figure 3-3 shows the one-port equivalent circuit of $Z(s)$. It is observed that this equivalent circuit may be obtained by a sequential extraction of the poles of $Z(s)$ in Fig. 3-2.

If $Z(s)$ is p.r., $1/Z(s)$ is also p.r. Defining such a new reactance function yields

$$Z(s) = \frac{s(s^2 + 4)}{(s^2 + 1)(s^2 + 9)}$$

The pole zero diagram for this reactance function is depicted in Fig. 3-4. Its residues are

$$2K_1 = \frac{s^2 + 1}{s} Z(s)\Big|_{s^2 = -1} = \frac{3}{8}$$

$$2K_3 = \frac{s^2 + 9}{s} Z(s)\Big|_{s^2 = -9} = \frac{5}{8}$$

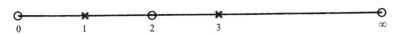

FIGURE 3-4
Pole zero diagram of

$$Z(s) = \frac{s(s^2 + 4)}{(s^2 + 1)(s^2 + 9)}$$

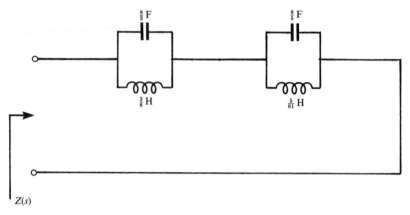

FIGURE 3-5
First Foster one-port network of

$$Z(s) = \frac{s(s^2 + 4)}{(s^2 + 1)(s^2 + 9)}$$

Thus

$$Z(s) = \frac{3s/8}{s^2 + 1} + \frac{5s/8}{s^2 + 9}$$

It is again observed that the poles of $Z(s)$ in Fig. 3-4 have been extracted in forming the equivalent one-port circuit in Fig. 3-5.

SECOND FOSTER CANONICAL FORM

The second Foster form realization of a one-port immittance function is the dual of the first one except that it involves a partial fraction expansion of an admittance function instead of an impedance one:

$$Y(s) = Y_1(s) + Y_2(s) + Y_3(s) + \cdots \qquad (3\text{-}10)$$

The general form for the above equation is deduced by duality from Eq. (3-2) as

$$Y(s) = \frac{K_0}{s} + \sum_{i=1}^{n} \frac{2K_i s}{s^2 + \omega_i^2} + \cdots + K_\infty s \qquad (3\text{-}11)$$

where

$$K_0 = sY(s)\big|_{s=0} \qquad (3\text{-}12)$$

$$2K_i = \frac{s^2 + \omega_i^2}{s} \, Y(s)\big|_{s^2 = -\omega_i^2} \qquad (3\text{-}13)$$

$$K_\infty = \lim \frac{Y(s)}{s}\bigg|_{s \to \infty} \tag{3-14}$$

The term K_0/s is recognized as the susceptance of a shunt inductance L_0 having a value

$$L_0 = \frac{1}{K_0} \quad \text{H} \tag{3-15}$$

The second term represents the reactance of a series tank circuit with elements C_i and L_i in shunt with the network:

$$C_i = \frac{2K_i}{\omega_i^2} \quad \text{F} \tag{3-16}$$

$$L_i = \frac{1}{2K_i} \quad \text{H} \tag{3-17}$$

respectively. The term $K_\infty s$ is recognized as the susceptance of a shunt capacitance given by

$$C_\infty = K_\infty \quad \text{F} \tag{3-18}$$

The second Foster realization of a one-port immittance function has the form depicted in Fig. 3-6. It is observed that it is the exact dual of that in Fig. 3-1.

The order of the function in Eq. (3-11) is reduced by one if it has a pole at $s = 0$ or $s = j\infty$ removed in the form of shunt inductance or capacitance. It is reduced by two if it has a pair of conjugate poles at $s = \pm j\omega_i$ extracted in the form of a shunt tank circuit in shunt with the input terminals.

To illustrate the synthesis of a one-port admittance function in the second Foster form, consider the susceptance function obtained by taking the reciprocal of $Z(s)$ used in the first Foster form problem.

$$Y(s) = \frac{s(s^2 + 4)}{(s^2 + 1)(s^2 + 9)}$$

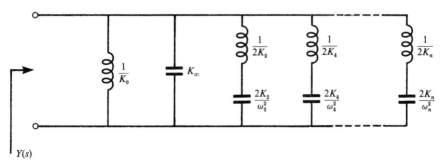

FIGURE 3-6
Second Foster realization of a one-port admittance function.

FIGURE 3-7
Pole zero diagram of

$$Y(s) = \frac{s(s^2 + 4)}{(s^2 + 1)(s^2 + 9)}$$

The pole zero diagram for this susceptance function is depicted in Fig. 3-7. The residues of the poles of $Y(s)$ in Eq. (3-11) are given with the help of Eqs (3-12) to (3-14) by

$$2K_1 = \frac{s^2 + 1}{s} \left. Z(s) \right|_{s^2 = -1} = \frac{3}{8}$$

$$2K_3 = \frac{s^2 + 9}{s} \left. Z(s) \right|_{s^2 = -9} = \frac{5}{8}$$

The partial fraction expansion of $Y(s)$ has the following form:

$$Y(s) = \frac{3s/8}{s^2 + 1} + \frac{5s/8}{s^2 + 9}$$

The one-port circuit formed in this way is illustrated in Fig. 3-8. The tank circuits in this network are recognized as the poles of the susceptance function.

As a final example consider the synthesis of the admittance function below in the second Foster form

$$Y(s) = \frac{(s^2 + 1)(s^2 + 9)}{s(s^2 + 4)}$$

The pole zero diagram for this function is shown in Fig. 3-9.

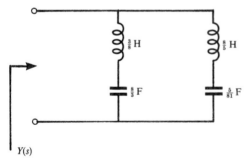

$Y(s)$

FIGURE 3-8
Second Foster one-port network of

$$Y(s) = \frac{s(s^2 + 4)}{(s^2 + 1)(s^2 + 9)}$$

FIGURE 3-9
Pole zero diagram of
$$Y(s) = \frac{(s^2 + 1)(s^2 + 9)}{s(s^2 + 4)}$$

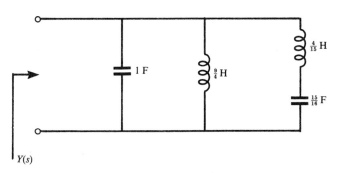

$Y(s)$

FIGURE 3-10
Second Foster one-port network of
$$Y(s) = \frac{(s^2 + 1)(s^2 + 9)}{s(s^2 + 4)}$$

The partial fraction expansion for this susceptance function is readily determined as

$$Y(s) = \frac{4/9}{s} + \frac{15s/4}{s^2 + 4}$$

The one-port circuit for this admittance is illustrated in Fig. 3-10.

FIRST CAUER FORM (REMOVAL OF POLES AT INFINITY)

A useful property of a reactance or susceptance function is that it has either a pole or zero at $s = j\infty$. An immittance function has a pole at infinity if the degree of the numerator polynomial $P(s)$ is of degree one higher than that of the denominator polynomial $Q(s)$. If $Z(s)$ or $Y(s)$ have no poles at infinity their reciprocals do. One ladder network due to Cauer is obtained by a repeated removal of poles at infinity. If such a pole is removed from an reactance function there is now a zero there. However, the susceptance function obtained by inverting $Z(s)$ has now a pole there which can be again extracted. This process is repeated until the degree of the polynomial is reduced to zero.

To illustrate this synthesis technique consider the reactance in the example in the previous section:

$$Z(s) = \frac{s^4 + 10s^2 + 9}{s^3 + 4s}$$

$Z(s)$ has a pole at infinity since the degree of $P(s)$ is one larger than that of $Q(s)$. The residue of this pole can be obtained by dividing $P(s)$ by $Q(s)$:

$$Z(s) = s + \frac{6s^2 + 9}{s^3 + 4s}$$

$$= Z_1(s) + Z_2(s)$$

Thus

$$Z_1(s) = s$$

$$Z_2(s) = \frac{6s^2 + 9}{s^3 + 4s}$$

Extracting the pole at infinity from $Z(s)$ yields the equivalent circuit in Fig. 3-11a.

$Z_2(s)$ has now a zero at infinity instead of a pole there. However, $Y_2(s)$ has a pole there. Forming $Y_2(s)$ gives

$$Y_2(s) = \frac{1}{Z_2(s)} = \frac{s^3 + 4s}{6s^2 + 4s}$$

The residue of the pole of $Y_2(s)$ is now evaluated by dividing $P(s)$ by $Q(s)$:

$$Y_2(s) = \frac{s}{6} + \frac{5s/2}{6s^2 + 9}$$

or

$$Y_2(s) = Y_3(s) + Y_4(s)$$

where

$$Y_3(s) = \frac{s}{6}$$

$$Y_4(s) = \frac{5s/2}{6s^2 + 9}$$

Extracting the pole of $Y_2(s)$ at infinity produces the equivalent circuit in Fig. 3-11b.

$Y_4(s)$ has now a zero at infinity. Forming $Z_4(s) = 1/Y_4(s)$ locates a pole there:

$$Z_4(s) = \frac{1}{Y_4(s)} = \frac{6s^2 + 9}{5s/2}$$

The residue of the new pole at infinity may now be evaluated by dividing $P(s)$ by $Q(s)$:

$$Z_4(s) = \frac{12s}{5} + \frac{18}{5s}$$

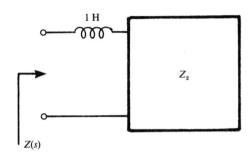

(a) After removal of first pole at infinity

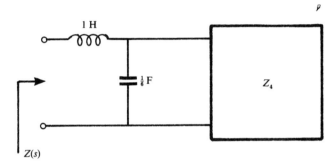

(b) After removal of second pole at infinity

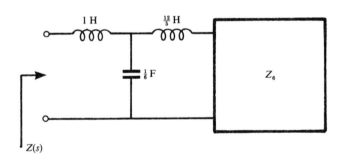

(c) After removal of third pole at infinity

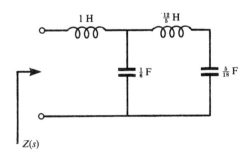

(d) After removal of fourth pole at infinity

FIGURE 3-11

First Cauer form of one-port network of

$$Z(s) = \frac{s^4 + 10s^2 + 9}{s^3 + 4s}$$

or
$$Z_4(s) = Z_5(s) + Z_6(s)$$

where
$$Z_5(s) = \frac{12s}{5}$$

$$Z_6(s) = \frac{18}{5s}$$

Figure 3-11c depicts the equivalent circuit of $Z(s)$ with this pole removed. $Z_6(s)$ has a zero at infinity but $Y_6(s)$ has a pole there which may again be extracted. Forming $Y_6(s)$ by inverting $Z_6(s)$ gives

$$Y_6(s) = \frac{5s}{18}$$

The residue of the pole of $Y_6(s)$ is 5/18 by definition. Thus the final equivalent circuit for $Z(s)$ has the form shown in Fig. 3-11d.

A similar procedure indicates that the equivalent circuit in Fig. 3-12 corresponds to the susceptance function below:

$$Y(s) = \frac{s^4 + 10s^2 + 9}{s^3 + 4s}$$

The realization of a one-port reactance or susceptance function in the first Cauer form involves the extraction of a pole at infinity by long division, inverting the remainder and dividing again to remove the next pole there (and so on). It is therefore concluded that it is possible to synthesize an LC ladder network by a continued fraction expansion of the reactance or susceptance function. This method may be illustrated by repeating the synthesis of $Z(s)$ in Fig. 3-11:

$$Z(s) = \frac{s^4 + 10s^2 + 9}{s^3 + 4s}$$

Forming a continued fraction expansion of $Z(s)$ indicates that

$$
\begin{array}{r}
S^3 + 4s)\,\overline{s^4 + 10s^2 + 9}\,(s \to z \\
\underline{s^4 + 4s^2} \\
6s^2 + 9)\,\overline{s^3 + 4s}\,(s/6 \to y \\
\underline{s^3 + \dfrac{3s}{2}} \\
\dfrac{5s}{2})\,\overline{6s^2 + 9}\,(12s/5 \to z \\
\underline{6s} \\
9)\,\overline{\dfrac{5s}{2}}\,(5s/18 \to y \\
\underline{\dfrac{5s}{2}} \\
0
\end{array}
$$

$$(3\text{-}19)$$

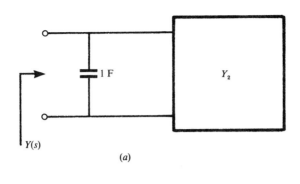

(a)

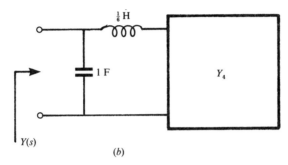

(b)

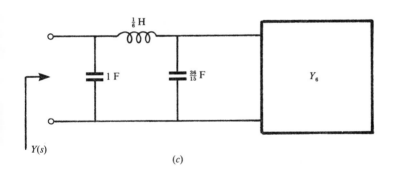

(c)

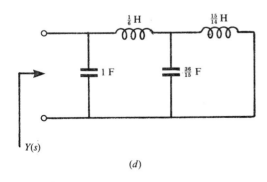

(d)

FIGURE 3-12

First Cauer form of one-port network of

$$Y(s) = \frac{s^4 + 10s^2 + 9}{s^3 + 4s}$$

The first element of the ladder network is an impedance s, the next one is a shunt admittance $s/6$, the third element of the ladder network is a series impedance $12s/6$ and the last element of the network is a shunt admittance $5s/18$. These elements of the continued fraction expansion are in keeping with those of the ladder network in Fig. 3-11.

If $Z(s)$ is replaced by $Y(s)$ in the above example the first element of the ladder network would be a shunt element, as indicated in Fig. 3-12.

SECOND CAUER FORM
(REMOVAL OF POLES AT THE ORIGIN)

Another useful property of an LC function is that it has always a pole or zero at the origin. If $Z(s)$ or $Y(s)$ has no pole there its reciprocal does. The second Cauer ladder synthesis of an immittance function consists of the successive extraction of such poles. If a pole is removed at the origin there is now a zero there. However, inverting the function introduces a new pole there which may once more be extracted. This process is repeated until the degree of the function is reduced to zero.

Synthesis of a ladder network in the second Cauer form proceeds as in the case of the first Cauer form except that it is necessary to arrange both $P(s)$ and $Q(s)$ in ascending order before division to identify the residue of the pole. This will now be illustrated for the reactance function used in the earlier example:

$$Z(s) = \frac{s^4 + 10s^2 + 9}{s^3 + 4s}$$

Rearranging $P(s)$ and $Q(s)$ in ascending order and forming a continued fraction expansion of $Z(s)$ yields

$$4s + s^3)9 + 10s^2 + s^4(9/4s \rightarrow z$$
$$9 + \frac{9s^2}{4}$$

$$\frac{31s^2}{4} + s^4)4s + s^3(16/31s \rightarrow y$$
$$4s + \frac{16s^3}{31}$$

$$\frac{15s^3}{31})\frac{31s^2}{4} + s^4(961/15s \rightarrow z$$
$$\frac{31s^2}{31}$$

$$s^4)\frac{15s^3}{31} \quad (15/31s \rightarrow y$$
$$\frac{15s^3}{31}$$

$$0 \qquad (3\text{-}20)$$

The ladder network obtained in this way is depicted in Fig. 3-13.

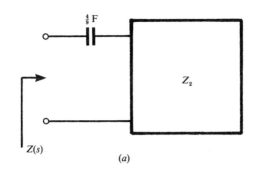

(a)

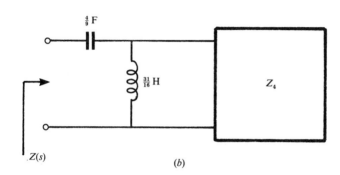

(b)

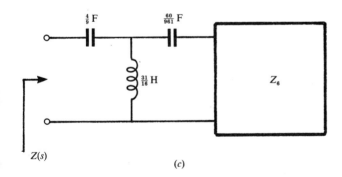

(c)

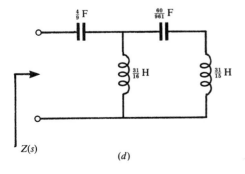

(d)

FIGURE 3-13
Second Cauer form of one-port network of

$$Z(s) = \frac{s^4 + 10s^2 + 9}{s^3 + 4s}$$

ZERO SHIFTING AND POLE REMOVAL

The poles of the input immittance of a two-port reactance network terminated in a resistive termination correspond to the transmission zero (attenuation poles) of the circuit. Much of modern filter synthesis therefore involves extracting the poles of an *LCR* rational function in such a way that the ensuing circuit has the proper topology and exhibits the required attenuation poles of the transfer function. If all the poles either reside at the origin or at infinity then a first or second Cauer synthesis procedure will directly display them without further ado. If the attenuation poles of the two-port network lie at finite frequencies, as in the case of a transfer function that has equal ripple in the pass- and stopbands, then some other procedure is necessary. Scrutiny of this problem suggests that whereas it is not possible to directly extract a pole at a specified frequency, it may in fact be done indirectly by zero shifting followed by inversion in order to interchange the positions of the poles and zeros of the immittance function. This technique may be understood by noting that the extraction of a pole in one or other of the two Cauer forms moves a zero to the origin or infinity to replace whichever pole has been extracted, since one or the other must always appear there in a rational reactance function. In fact, all the zeros of the rational function are shifted by the removal of a pole, although the other poles are left unperturbed. A procedure based on zero shifting and inversion is one common technique utilized in practice for the removal of poles at finite frequencies.

ZERO SHIFTING BY POLE REMOVAL

The technique of zero shifting and inversion to enable the extraction of poles at finite frequencies will now be demonstrated by way of an example for the reactance function associated with the pole zero diagram in Fig. 3-9:

$$Y(s) = \frac{s^4 + 10s^2 + 9}{s(s^2 + 4)}$$

This susceptance function has a pole at the origin and at infinity which may be removed in a first and second Cauer manner; the remainder functions are

$$Y_1(s) = Y(s) - s$$

$$Y_1(s) = Y(s) - \frac{9}{4s}$$

respectively. Evaluating these two quantities gives

$$Y_1(s) = \frac{6s^2 + 9}{s(s^2 + 4)}$$

$$Y_1(s) = \frac{s(s^2 + 31/4)}{s^2 + 4}$$

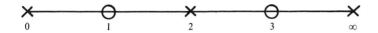

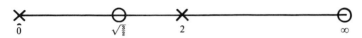

FIGURE 3-14
Pole zero diagrams for

$$Y(s) = \frac{s^4 + 10s^2 + 9}{s(s^2 + 4)}$$

and for same after removal of pole at infinity and at origin.

In the first remainder a zero has been shifted to infinity to replace the pole that has been removed there; in the second one a zero has been moved to the origin to replace the pole that has been extracted. In both instances the other poles of the function have been left unperturbed as asserted. Since the numerator polynomials are also altered all the zeros have in fact been perturbed. Scrutiny of some typical pole zero diagrams suggests that the zeros move in every instance towards the pole being removed; if it is at infinity the zero shifts to infinity and if it is removed at the origin it moves to the origin. Figures 3-14 and 3-15 depict the two situations here. Indeed, the two Cauer forms have realized, in this manner, all the poles of the reactance function at either the origin or at infinity. Since the poles and zeros of reactance functions are interlaced on the $j\omega$ axis it is not possible to shift a zero past an adjacent pole; it is therefore always necessary to partially remove the pole nearest to the zero that must be shifted.

ZERO SHIFTING BY PARTIAL POLE REMOVAL

The extraction of a pole at a specified frequency with the aid of zero shifting or partial pole removal will now be illustrated by way of an example

$$Y(s) = \frac{s^4 + 10s^2 + 9}{s(s^2 + 4)}$$

This admittance has a pole at infinity but has no pole or zero at $s^2 = -10$ (say). A pole may, however, be extracted from it by first placing a zero there by a partial removal of the nearest pole which happens in this case to lie at infinity such that

$$Y(s) - ks|_{s^2 = -10} = 0$$

This equation may now be solved for the residue k of the pole which must be partially extracted:

$$k = \frac{s^4 + 10s^2 + 9}{s^2(s^2 + 4)}\bigg|_{s^2 = -10} = \tfrac{3}{20}$$

Extracting this partial pole at infinity in the form of a shunt capacitor from $Y(s)$ gives

$$Y_1(s) = \frac{17s^4 + 188s^2 + 180}{20s(s^2 + 4)}$$

$Y_1(s)$ has still a pole at s equal to infinity but it now also has a zero at $s^2 = -10$, as is readily verified. Forming $Z_1(s)$ gives a pole there which may be removed:

$$Z_1(s) = \frac{1}{Y_1(s)} = \frac{20s(s^2 + 4)}{17s^4 + 188s^2 + 180}$$

The pole of $Z_1(s)$ at $s^2 = -10$ may now be extracted by expanding it in partial fractions:

$$Z_1(s) = \frac{15s/19}{s^2 + 10} + \frac{250s/38}{17s^2 + 18}$$

A pair of conjugate poles may now be removed at $s^2 = -10$ in the form of a parallel LC circuit in series with the network giving a remainder reactance

$$Z_2(s) = \frac{250s/38}{17s^2 + 18}$$

This reactance has a zero at infinity but its reciprocal has a pole there:

$$Y_2(s) = \frac{1}{Z_2(s)} = \frac{646s}{250} + \frac{684}{250s}$$

The remaining susceptance may now be realized by inspection by removing a pole at both the origin and at infinity in the form of a shunt capacitor and series inductor.

Figure 3-15 illustrates the development of the circuit. This circuit has one pole at $s^2 = -10$ and a pair of poles at infinity. It is noted that the network in Fig. 3-15 is no longer canonical in that it requires one more element for its realization than was necessary to synthesize the same rational function using either the two Foster or Cauer forms. Indeed, whenever a pole at either the origin or at infinity is partially removed, one additional element is required. If an internal pole is partially removed, two additional elements are required. Thus, where possible, partial pole removals usually only involve poles at the origin or at infinity.

If it is desired to remove a pole at $s^2 = -\tfrac{1}{2}$ (say) from the rational function in Eq. (3-1) a partial removal of a pole at the origin is required and the first step takes the following form:

$$Y_1(s) = Y(s) - \frac{k}{s}\bigg|_{s^s = -\frac{1}{2}} = 0$$

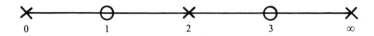

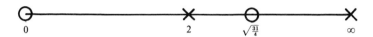

FIGURE 3-15
Pole zero diagrams for
$$Y(s) = \frac{s^4 + 10s^2 + 9}{s(s^2 + 4)}$$
showing zero shifting to remove pole at $s^2 = -10$.

As another example, the following susceptance associated with a degree 3 elliptic filter is to be realized with a pair of complex poles at $s^2 = -1.742\,29^2$:

$$Y(s) = \frac{s(2s^2 + 3.686\,33)}{2.272\,67s^2 + 3.007\,52}$$

The first step in the realization of this circuit is to move a zero to $s^2 = -1.742\,29^2$ by partially removing a pole at infinity:

$$Y_1(s) = Y(s) - ks|_{s^2 = -3.035\,58}$$

The residue is

$$k = \frac{2s^2 + 3.686\,33}{2.272\,67s^2 + 3.007\,52}\bigg|_{s^2 = -3.035\,58} = 0.6129$$

and the first element is a shunt capacitor

$$C_1 = 0.6129 \text{ H}$$

and the remainder susceptance is

$$Y_1(s) = \frac{s(s^2 + 3.035\,58)}{2.272\,67s^2 + 3.007\,52}$$

The required complex poles may now be extracted by forming $Z_1(s)$ and either expanding it in partial fractions or evaluating the required residue. Employing the latter approach gives

$$2k = \frac{s^2 + 3.035\,58}{s} Z_1(s)|_{s^2 = -3.035\,58} = 2.111\,77$$

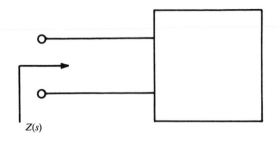

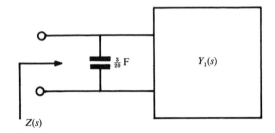

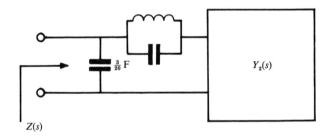

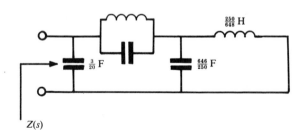

FIGURE 3-16
Schematic diagram for
$$Y(s) = \frac{2s^3 + 3.68633s}{2.27267s^2 + 3.00752}$$

The reactance associated with the pair of complex conjugate poles at $s^2 = -3.035\,58$,

$$Z_2(s) = \frac{2.111\,77s}{s^2 + 3.035\,58}$$

may be realized as a parallel LC network in series with the circuit with elements

$$C_2 = 0.4735 \text{ F}$$

$$L_2 = 0.6955 \text{ H}$$

The remainder reactance is

$$Z_3(s) = Z_1(s) - Z_2(s)$$

or $\qquad\qquad Z_3(s) = \dfrac{1.631\,84}{s}$

$Y_3(s)$ has a pole at infinity:

$$Y_3(s) = 0.6129s$$

which may be realized as a shunt element given by

$$C_3 = 0.6129 \text{ F}$$

The required circuit is given in Fig. 11-11b in Chapter 11.

PROBLEMS

3-1 Obtain the two Foster canonical realizations for each of the following two immittance functions:

$$\frac{s^3 + 8s}{s^4 + 10s^2 + 9}$$

$$\frac{(s^2 + 1)(s^2 + 3)(s^2 + 5)}{s(s^2 + 2)(s^2 + 4)}$$

3-2 Synthesize the two Cauer forms for each of the functions in Prob. 3-1.

3-3 Deduce the upper and lower bounds on the coefficient Q in order for the following function to be a positive real one:

$$\frac{s^3 + Qs}{s^4 + 10s^2 + 9}$$

BIBLIOGRAPHY

Cauer, W., 'The realization of impedance with prescribed frequency dependence', *Arch. Electrotech.*, Vol. 15, pp. 355–388, 1926.

Foster, R. M., 'A reactance theorem', *Bell System Tech. J.*, No. 3, pp. 259–267, 1924.

CHAPTER
4

IMMITTANCE MATRICES

INTRODUCTION

Networks with more than one port are usually described in terms of impedance, admittance, scattering or *ABCD* transfer matrices. This chapter deals with impedance and admittance (immittance) matrices. The problem of whether a square immittance matrix is realizable as a two-port network will be dealt with in Chapters 5 and 7. The chapter also includes the synthesis of immittance matrices in the form of series and parallel connections of two-port networks.

It is sometimes possible that a network has no impedance or admittance matrix, some of its elements being infinitely large. In such a situation, an immittance matrix may usually be obtained by shifting the reference terminals. One network that has no immittance matrix description (impedance or admittance) is the ideal transformer.

An important point to observe is that whereas an equivalent T or π network may correctly characterize voltage and current at the ports of the circuit, it need not be realizable in those two forms. Some of the branches may not be p.r. This important problem is discussed further in Chapters 5 and 7.

IMPEDANCE MATRIX

The complete relation between voltages and currents of an *n*-port network is defined through an impedance matrix given by

$$\bar{V} = \bar{Z}\bar{I} \qquad (4\text{-}1)$$

In the case of a two-port network the impedance matrix $\bar{Z}$ has the following form:

$$\bar{Z} = \begin{bmatrix} Z_{11} & Z_{12} \\ Z_{21} & Z_{22} \end{bmatrix} \qquad (4\text{-}2)$$

Voltage and current matrices for a two-port network are

$$\bar{V} = \begin{bmatrix} V_1 \\ V_2 \end{bmatrix} \qquad (4\text{-}3)$$

$$\bar{I} = \begin{bmatrix} I_1 \\ I_2 \end{bmatrix} \qquad (4\text{-}4)$$

Figure 4-1 defines voltages and currents in a two-port network. The entries of the $\bar{Z}$ matrix are known as open-circuit impedances. They are defined from Eqs (4-1) to (4-4) by

$$Z_{11} = \left. \frac{V_1}{I_1} \right|_{I_2 = 0} \qquad (4\text{-}5)$$

$$Z_{12} = \left. \frac{V_1}{I_2} \right|_{I_1 = 0} \qquad (4\text{-}6)$$

$$Z_{22} = \left. \frac{V_2}{I_2} \right|_{I_1 = 0} \qquad (4\text{-}7)$$

$$Z_{21} = \left. \frac{V_2}{I_1} \right|_{I_2 = 0} \qquad (4\text{-}8)$$

For a reciprocal network

$$Z_{12} = Z_{21} \qquad (4\text{-}9)$$

whereas for a symmetrical one

$$Z_{11} = Z_{22} \qquad (4\text{-}10)$$

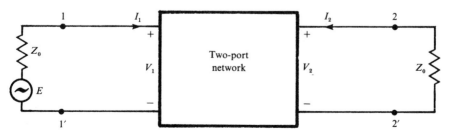

FIGURE 4-1
Definition of voltage and current for a two-port network.

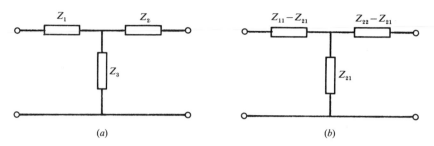

FIGURE 4-2
T network in terms of (a) Z_1, Z_2 and Z_3 and (b) Z_{11}, Z_{22} and Z_{21}.

One possible equivalent circuit for Eq. (4-1) is the T network depicted in Fig. 4-2a. The values of Z_1, Z_2 and Z_3 in this circuit may be related to the entries of the $\bar{Z}$ matrix by applying the open-circuit definitions in Eqs (4-5) to (4-8) to it:

$$Z_{11} = Z_1 + Z_3 \qquad (4\text{-}11)$$

$$Z_{22} = Z_2 + Z_3 \qquad (4\text{-}12)$$

$$Z_{12} = Z_{21} = Z_3 \qquad (4\text{-}13)$$

The required branch impedance is obtained by rearranging the preceding equations. The result is

$$Z_1 = Z_{11} - Z_{21} \qquad (4\text{-}14)$$

$$Z_2 = Z_{22} - Z_{21} \qquad (4\text{-}15)$$

$$Z_3 = Z_{21} = Z_{12} \qquad (4\text{-}16)$$

Figure 4.2b depicts the equivalent circuit of the two-port network in terms of its open-circuit parameters.

The input impedance of a network terminated in an impedance Z_L is readily formulated in terms of its open-circuit parameters. This may be done by expanding Eq. (4-1) and introducing the following boundary conditions at the load:

$$V_2 = -Z_L I_2 \qquad (4\text{-}17)$$

Expanding Eq. (4-1) and eliminating V_2 in terms of the preceding equation gives

$$V_1 = Z_{11} I_1 + Z_{12} I_2 \qquad (4\text{-}18)$$

$$-I_2 Z_L = Z_{21} I_1 + Z_{22} I_2 \qquad (4\text{-}19)$$

Solving these equations for V_1/I_1 gives the required result:

$$Z_{in} = \frac{V_1}{I_1} = Z_{11} - \frac{Z_{12} Z_{21}}{Z_L + Z_{22}} \qquad (4\text{-}20)$$

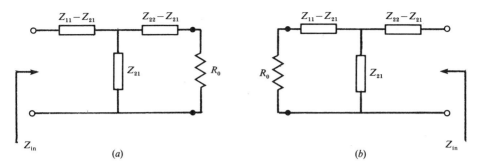

FIGURE 4-3
Definition of input impedance of T network terminated at (*a*) port 2 and (*b*) port 1.

Terminating port 1 instead of port 2 by Z_L leads to

$$Z_{in} = \frac{V_2}{I_2} = Z_{22} - \frac{Z_{12}Z_{21}}{Z_L + Z_{11}} \qquad (4\text{-}21)$$

These two situations are summarized in Fig. 4-3*a* and *b*.

THE ADMITTANCE MATRIX

It is also possible to define the square admittance matrix $\bar{Y}$ for a network provided the inverse of the impedance matrix $\bar{Z}$ exists:

$$\bar{Y} = \bar{Z}^{-1} \qquad (4\text{-}22)$$

This matrix relates currents and voltages at the *n*-ports of the junction by

$$\bar{I} = \bar{Y}\bar{V} \qquad (4\text{-}23)$$

For a two-port network the admittance matrix $\bar{Y}$ is a square matrix given by

$$\bar{Y} = \begin{bmatrix} Y_{11} & Y_{12} \\ Y_{21} & Y_{22} \end{bmatrix} \qquad (4\text{-}24)$$

and the voltage and current matrices $\bar{V}$ and $\bar{I}$ are column matrices defined by Eqs (4-3) and (4-4).
The admittance parameters are defined by Eq. (4-23) as

$$Y_{11} = \frac{I_1}{V_1}\bigg|_{V_2 = 0} \qquad (4\text{-}25)$$

$$Y_{12} = \frac{I_2}{V_2}\bigg|_{V_1 = 0} \qquad (4\text{-}26)$$

$$Y_{21} = \frac{I_2}{V_1}\bigg|_{V_2 = 0} \qquad (4\text{-}27)$$

$$Y_{22} = \frac{I_2}{V_2}\bigg|_{V_1=0} \tag{4-28}$$

The above quantities are often referred to as the short-circuit parameters of the network.

Reciprocity and symmetry imply

$$Y_{11} = Y_{22} \tag{4-29}$$

$$Y_{12} = Y_{21} \tag{4-30}$$

A π equivalent circuit is often utilized to represent the short-circuit parameters. This circuit is illustrated in Fig. 4-4a. The equivalence between the short-circuit admittances Y_{11}, Y_{22} and Y_{21} and the branch admittances Y_1, Y_2 and Y_3 is arrived at by applying the short-circuit admittance definitions to the equivalent circuit in Fig. 4-4a:

$$Y_{11} = Y_1 + Y_3 \tag{4-31}$$

$$Y_{22} = Y_2 + Y_3 \tag{4-32}$$

$$Y_{12} = Y_{21} = Y_3 \tag{4-33}$$

Rearranging the above equations gives

$$Y_1 = Y_{11} - Y_{21} \tag{4-34}$$

$$Y_2 = Y_{22} - Y_{21} \tag{4-35}$$

$$Y_3 = Y_{21} \tag{4-36}$$

The π equivalent circuit for this notation is depicted in Fig. 4-4b.

The input admittances of the π network in Fig. 4-5a and b terminated in Y_L at ports 2 and 1 are expressed as

$$Y_{in} = \frac{V_1}{I_1} = Y_{11} + \frac{Y_{12}Y_{21}}{Y_L + Y_{22}} \tag{4-37}$$

$$Y_{in} = \frac{I_2}{V_2} = Y_{22} + \frac{Y_{12}Y_{21}}{Y_L + Y_{11}} \tag{4-38}$$

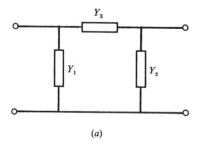
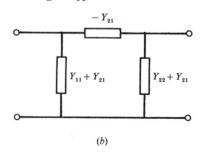

(a)   (b)

FIGURE 4-4
π Network in terms of (a) Y_1, Y_2 and Y_3 and (b) Y_{11}, Y_{22} and Y_{21}.

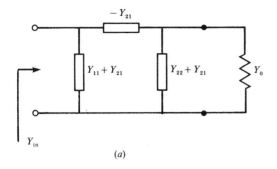

(a)

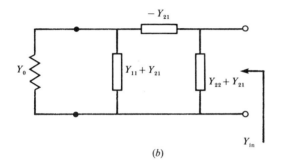

(b)

FIGURE 4-5
Definition of input admittance of π network terminated at (a) port 2 and (b) port 1.

Although the open- and short-circuit parameters always represent the voltage–current relations at ports 1 and 2, the branch immittances themselves need not be p.r. Thus the equivalent circuits in Figs 4-3 and 4-5 are not necessarily realizable by simple *LCR* elements.

$\bar{Y}(s)$ in Eq. (4-22) can be expressed in terms of the entries of $\bar{Z}(s)$ by forming $\bar{Z}(s)^{-1}$. The result is

$$\bar{Y}(s) = \bar{Z}(s)^{-1} = \frac{\text{adj } \bar{Z}(s)}{|Z(s)|}$$

For a 2×2 matrix this last equation can be expanded as

$$\bar{Y}(s) = \frac{1}{Z_{11}(s)Z_{22}(s) - Z_{12}(s)Z_{21}(s)} \begin{bmatrix} Z_{22}(s) & -Z_{21}(s) \\ -Z_{12}(s) & Z_{11}(s) \end{bmatrix}$$

SERIES AND PARALLEL REALIZATIONS OF IMMITTANCE MATRICES

It will be shown in Chapter 5 that the open-circuit reactance parameters of a two-port reciprocal network can be expanded in partial fraction form as

$$Z_{11}(s) = \frac{k_{11}^0}{s} + \frac{2k_{11}^1 s}{s^2 + \omega^2} + \cdots + k_{11}^\infty s \qquad (4\text{-}39)$$

$$Z_{22}(s) = \frac{k_{22}^0}{s} + \frac{2k_{22}^1 s}{s^2 + \omega^2} + \cdots + k_{22}^\infty s \qquad (4\text{-}40)$$

$$Z_{12}(s) = Z_{21}(s) = \frac{k_{21}^0}{s} + \frac{2k_{21}^1 s}{s^2 + \omega_1^2} + \cdots + k_{21}^\infty s \qquad (4\text{-}41)$$

subject to the residue condition discussed in Chapter 5.

A partial fraction expansion of such a matrix is

$$Z(s) = \frac{1}{s}\begin{bmatrix} k_{11}^0 & k_{21}^0 \\ k_{21}^0 & k_{22}^0 \end{bmatrix} + \frac{2s}{s^2 + \omega_1^2}\begin{bmatrix} k_{11}^1 & k_{21}^1 \\ k_{21}^1 & k_{22}^1 \end{bmatrix}$$

$$+ \cdots + s\begin{bmatrix} k_{11}^\infty & k_{21}^\infty \\ k_{21}^\infty & k_{22}^\infty \end{bmatrix} \qquad (4\text{-}42)$$

One circuit representation for this function due to Cauer is depicted in Fig. 4-6.

The realization of this circuit may be understood by considering the series connection of two two-port networks in Fig. 4-7.

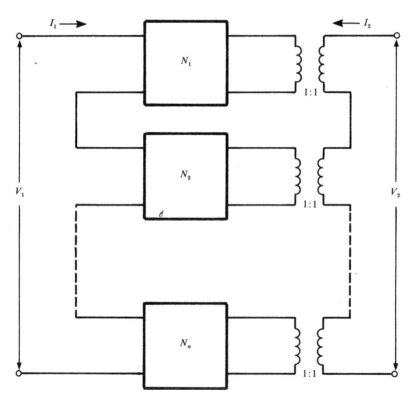

FIGURE 4-6
Cauer form series realization of two-port network.

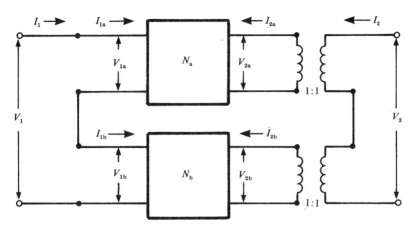

FIGURE 4-7
Series connection of two two-port networks.

The voltage–current relationships for networks a and b in Fig. 4-7 are given in terms of their open-circuit parameters by

$$V_{2a} = Z_{11a}(s)I_{1a} + Z_{21a}(s)I_{2a} \qquad (4\text{-}43)$$

$$V_{2a} = Z_{21a}(s)I_{1a} + Z_{22a}(s)I_{2a} \qquad (4\text{-}44)$$

and

$$V_{1b} = Z_{11b}(s)I_{1b} + Z_{21b}(s)I_{2b} \qquad (4\text{-}45)$$

$$V_{2b} = Z_{21b}(s)I_{1b} + Z_{22b}(s)I_{2b} \qquad (4\text{-}46)$$

The ideal transformers at the output port of the networks makes possible the following boundary conditions:

$$V_1 = V_{1a} + V_{2a} \qquad (4\text{-}47)$$

$$V_2 = V_{1b} + V_{2b} \qquad (4\text{-}48)$$

and

$$I_1 = I_{1a} = I_{1b} \qquad (4\text{-}49)$$

$$I_2 = I_{2a} = I_{2b} \qquad (4\text{-}50)$$

The open-circuit parameters defined by this arrangement are

$$V_1 = [Z_{11a}(s) + Z_{11b}(s)]I_1 + [Z_{21a}(s) + Z_{21b}(s)]I_2 \qquad (4\text{-}51)$$

$$V_2 = [Z_{21a}(s) + Z_{21b}(s)]I_1 + [Z_{22a}(s) + Z_{22b}(s)]I_2 \qquad (4\text{-}52)$$

These equations indicate that the open-circuit parameters of the two networks add for the circuit arrangement in Fig. 4-7.

A dual development, in the case of a square admittance matrix, indicates

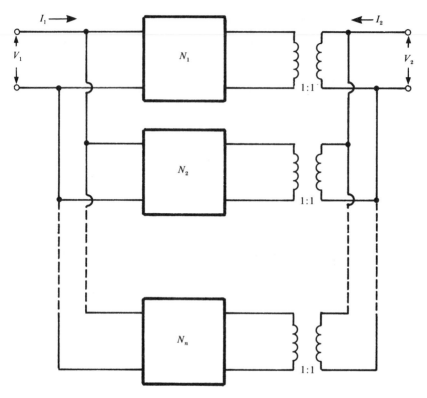

FIGURE 4-8
Cauer form parallel realization of two-port network.

the possibility of expanding the matrix in the form depicted in Fig. 4-8. In this realization, each term in the partial fraction expansion is connected in parallel.

The realization of this canonical form may be understood by considering the connection of networks a and b in Fig. 4-9. The ideal transformers in this circuit make possible the following boundary conditions:

$$V_1 = V_{1a} = V_{1b} \tag{4-53}$$

$$V_2 = V_{2a} = V_{2b} \tag{4-54}$$

$$I_1 = I_{1a} + I_{1b} \tag{4-55}$$

$$I_2 = I_{2a} + I_{2b} \tag{4-56}$$

Combining these boundary conditions with the current–voltage relationships for networks a and b readily yields the required result:

$$I_1 = [Y_{11a}(s) + Y_{11b}(s)]V_1 + [Y_{21a}(s) + Y_{21b}(s)]V_2 \tag{4-57}$$

$$I_2 = [Y_{21a}(s) + Y_{21b}(s)]V_1 + [Y_{22a}(s) + Y_{22b}(s)]V_2 \tag{4-58}$$

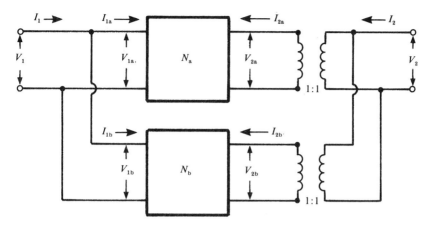

FIGURE 4-9
Parallel connection of two two-port networks.

PROBLEMS

4-1 Demonstrate that there is no $\bar{Z}$ matrix for a series network and no $\bar{Y}$ matrix for a shunt one.

4-2 Find the open-circuit parameters of the T network in the illustration below:

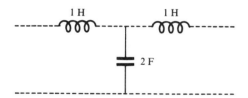

4-3 Show that the input impedance for the circuit in Prob. 4-2 terminated in a 1-Ω resistor is

$$Z(s) = \frac{2s^2 + 2s^2 + 2s + 1}{2s^2 + 2s + 1}$$

4-4 Obtain short-circuit parameters of the π network below:

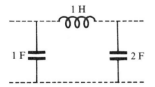

4-5 Form the input admittance of the network in Prob. 4-4 terminated in an admittance Y_L at port 2. Repeat with Y_L at port 1.

4-6 Verify Eqs (4-37) and (4-38) from first principles.

4-7 The open- and short-circuit parameters of a section of uniform transmission line $\theta/2$ long are $Z_{o.c.} = Z_0 \coth \theta/2$ and $Z_{s.c.} = Z_0 \tanh \theta/2$. Find the open-circuit parameters of a transmission line θ radians long with characteristic impedance Z_0.

4-8 Obtain T and π equivalent circuits for the transmission line in Prob. 4-7.

4-9 Deduce the equivalent circuit for the following impedance matrix:

$$\begin{bmatrix} \dfrac{2s^2+1}{2s} & \dfrac{1}{2s} \\[3mm] \dfrac{1}{2s} & \dfrac{2s^2+1}{2s} \end{bmatrix}$$

4-10 Obtain the inverse matrix of the following impedance matrix:

$$\bar{Z} = \begin{bmatrix} Z_{11} & Z_{12} \\ Z_{21} & Z_{22} \end{bmatrix}$$

4-11 Evaluate the inverse matrix of the admittance matrix below:

$$\bar{Y} = \begin{bmatrix} Y_{11} & Y_{12} \\ Y_{21} & Y_{22} \end{bmatrix}$$

4-12 Invert the matrix in Prob. 4-10.

CHAPTER
5

POSITIVE
REAL
IMPEDANCE
MATRICES

INTRODUCTION

Networks with more than one port are usually described by square immittance or scattering matrices. Before attempting the synthesis of such a network it is also desirable to develop realizability conditions to ensure that it is realizable as an *LCR* network. The diagonal elements in a square immittance matrix represents one-port immittances and must therefore be p.r. functions by definition. The off-diagonal elements represent transfer parameters which it will be shown need not be p.r. functions. The nature of the off-diagonal elements of immittance matrices may be investigated by forming the quadratic form of the square matrix. The quadratic form is a scalar quantity which it will be demonstrated must satisfy Brune's one-port realizability conditions in Chapter 2. The quadratic form of the square immittance matrix must therefore be real for s real, and the quadratic form of the real parts of the square matrix must be equal or larger than zero (positive semi-definite) for σ larger or equal to zero. Since definiteness of a quadratic form is an inherent property of a square matrix, it is natural to refer to it as a positive semi-definite matrix. An important property of a two-port *LC* function is that the off-diagonal element of its immittance description is an odd function of the frequency variable.

POSITIVE DEFINITE AND
POSITIVE SEMI-DEFINITE
QUADRATIC FORMS

It will be demonstrated in the next section that in order for a square immittance matrix to be realizable as a two-port network the quadratic form associated with it must satisfy the two tests given by Brune in connection with that of a one-port immittance function. The quadratic form associated with the matrix $\bar{A}(s)$ is a scalar quantity [say $a(s)$] defined by

$$a(s) = \bar{x}^{\mathrm{T}} \bar{A}(s) \bar{x} \tag{5-1}$$

where $\bar{x}$ is an unknown column vector

$$\bar{x} = \begin{bmatrix} x_1 \\ x_2 \end{bmatrix} \tag{5-2}$$

and $\bar{x}^{\mathrm{T}}$ is its transpose

$$\bar{x}^{\mathrm{T}} = [x_1 x_2] \tag{5-3}$$

$\bar{A}(s)$ is the matrix of the quadratic form

$$\bar{A}(s) = \begin{bmatrix} a_{11}(s) & a_{12}(s) \\ a_{21}(s) & a_{22}(s) \end{bmatrix} \tag{5-4}$$

If $\bar{x}$ is an eigenvector and $\bar{A}(s)$ is a symmetric matrix, $a(s)$ is an eigenvalue [see Chapter 6, Eq. (6-1)].

Expanding $a(s)$ in Eq. (5-1) indicates that it is a scalar quantity as asserted:

$$a(s) = x_1^2 a_{11}(s) + x_2^2 a_{22}(s) + x_1 x_2 a_{12}(s) + x_1 x_2 a_{21}(s) \tag{5-5}$$

If the quadratic form is greater than but not equal to zero it is said to be positive definite (p.d.). If it is equal to or greater than zero it is said to be positive semi-definite (p.s.d.). A necessary and sufficient condition for a quadratic form to be p.s.d. is that the principal minors of the matrix $\bar{A}$ of the quadratic form arranged in ascending order from the upper left corner be equal to or greater than zero. For the square matrix considered here this becomes

$$a_{11}(s) \geqq 0 \tag{5-6}$$

$$a_{22}(s) \geqq 0 \tag{5-7}$$

$$a_{11}(s) a_{22}(s) - a_{12}(s) a_{21}(s) \geqq 0 \tag{5-8}$$

As discussed in the chapter on matrices, this condition may also be met by requiring that the eigenvalues of A should be p.d. or p.s.d.

REALIZABILITY CONDITIONS
FOR TWO-PORT
LCR NETWORKS

Before proceeding with the synthesis of a square impedance or admittance matrix as a two-port network it is essential to develop equivalent necessary and sufficient conditions for it to meet those given by Brune in the case of a one-port immittance function. Such a testing procedure may be derived with the aid of the network arrangement in Fig. 5-1.

In this circuit both ports of the network have voltages applied to them through ideal transformers n_1 and n_2, and the task is to formulate the one-port impedance of the arrangement. The $\bar{Z}$ matrix of the original network is defined by

$$\begin{bmatrix} E_1 \\ E_2 \end{bmatrix} = \begin{bmatrix} Z_{11}(s) & Z_{12}(s) \\ Z_{21}(s) & Z_{22}(s) \end{bmatrix} \begin{bmatrix} I_1 \\ I_2 \end{bmatrix} \tag{5-9}$$

The relation between the original voltage and current variables and those given by the new generator are

$$I_1 = n_1 I \tag{5-10}$$

$$I_2 = -n_2 I \tag{5-11}$$

$$n_1 E_1 - n_2 E_2 = E \tag{5-12}$$

In matrix notation the relationships become

$$E = (\bar{n})^{\mathrm{T}} \bar{E} \tag{5-13}$$

$$I = (\bar{n})^{-1} \bar{I} \tag{5-14}$$

Combining the preceding equations and noting that $(\bar{A} \cdot \bar{B})^{-1} = (\bar{B})^{-1} \cdot (\bar{A})^{-1}$ gives

$$z(s) = \frac{E}{I} = [n_1 \quad -n_2] \begin{bmatrix} Z_{11}(s) & Z_{12}(s) \\ Z_{21}(s) & Z_{22}(s) \end{bmatrix} \begin{bmatrix} n_1 \\ -n_2 \end{bmatrix} \tag{5-15}$$

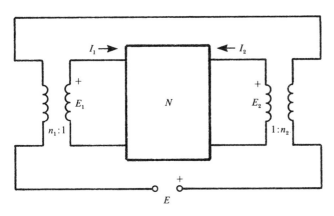

FIGURE 5-1
Schematic diagram showing definition of quadratic form.

This equation is recognized as having the nature of the quadratic form in Eq. (5-1). In order for $z(s)$ in Eq. (5-15) to represent the impedance of a one-port network it must satisfy the p.r. conditions in (2-4) and (2-5) in Chapter 2. The necessary and sufficient conditions for a square impedance or admittance matrix to be realizable as a two-port network is therefore that its quadratic form is a p.r. function. This statement may be written as

The quadratic form of the matrix $\bar{Z}(s)$ is real
for s real. (5-16)

The quadratic form of the matrix $[\operatorname{Re} \bar{Z}(s)]$ must
be positive semi-definite for $\sigma \geqq 0$. (5-17)

A consequence of these two tests is that terms such as $Z_{21}(s)$, $Z_{12}(s)$, $Y_{21}(s)$ and $Y_{12}(s)$ need not be p.r. functions. Terms like $Z_{11}(s)$, $Z_{22}(s)$, $Y_{11}(s)$ and $Y_{22}(s)$ are, however, p.r. ones, since these quantities represent one-port immittances. These two statements may be readily demonstrated by investigating Eq. (5-15) with $n_1 = 1/\sqrt{2}$, $n_2 = 1/\sqrt{2}$ and $n_1 = 1/\sqrt{2}$, $n_2 = -1/\sqrt{2}$:

$$2z_1(s) = Z_{11}(s) + Z_{22}(s) + Z_{12}(s) + Z_{21}(s) \tag{5-18}$$

$$2z_2(s) = Z_{11}(s) + Z_{22}(s) - Z_{12}(s) - Z_{21}(s) \tag{5-19}$$

Taking linear combinations of the above two equations in the case of a network which is both reciprocal and symmetrical indicates that

$$Z_{12}(s) = Z_{21}(s) = \frac{z_1(s) - z_2(s)}{2} \tag{5-20}$$

$$Z_{11}(s) = Z_{22}(s) = \frac{z_1(s) + z_2(s)}{2} \tag{5-21}$$

Since the difference between two p.r. functions is not necessarily p.r., terms like $Z_{21}(s)$ need not be p.r. However, the sum of two p.r. functions is p.r., and therefore $Z_{11}(s)$ and $Z_{22}(s)$ are p.r. quantities. The pole zero diagrams of terms like $Z_{11}(s)$ are therefore in keeping with the discussion in Chapter 2, but quantities such as $Z_{21}(s)$ must be considered separately.

It is again helpful to replace the conditions in (5-16) and (5-17) by equivalent necessary and sufficient ones as in (2-12) to (2-14) in Chapter 2:

$P(s)$ and $Q(s)$ of the quadratic form of the matrix $\bar{Z}(s)$
are Hurwitz. (5-22)

The quadratic form of the matrix $\bar{Z}(s)$ may only have simple poles
on the jω axis with positive real residues. (5-23)

The real parts of the quadratic form of the matrix $\bar{Z}(j\omega)$ must be
$\geqq 0$ for all ω. (5-24)

It will presently be demonstrated that (5-23) and (5-24) may be replaced by so-called residue and real part conditions. The necessary and sufficient conditions for a square matrix to be realizable as a two-port network may therefore be restated as:

$P(s)$ and $Q(s)$ of the quadratic form of the matrix
$\bar{Z}(s)$ are Hurwitz. (5-25)

$\bar{Z}(s)$ may only have simple poles on the $j\omega$ axis which must satisfy the residue condition $k_{11}^i \geqq 0$, $k_{22}^i \geqq 0$, $k_{11}^i k_{22}^i - k_{21}^{i2} \geqq 0$. (5-26)

$\bar{Z}(s)$ must satisfy the real part condition $\text{Re}[Z_{11}(j\omega)] \geqq 0$,
$\text{Re}[Z_{22}(j\omega)] \geqq 0$, $\text{Re}[Z_{11}(j\omega)]\,\text{Re}[Z_{22}(j\omega)] - \text{Re}[Z_{21}^2(j\omega)] \geqq 0$. (5-27)

RESIDUES CONDITION

The condition in (5-23) that the quadratic form of the matrix $\bar{Z}(s)$ must only have simple poles on the $j\omega$ axis with positive real residues requires that the matrix formed by the residues of the open-circuit parameters is associated with a p.s.d. quadratic form. This condition is deduced by assuming that the open-circuit parameters of the impedance matrix have a common pair of conjugate poles as $s = \pm j\omega_i$:

$$Z_{11}(s) = \frac{2k_{11}^i s}{s^2 + \omega_i^2} + \text{other terms} \qquad (5\text{-}28)$$

$$Z_{22}(s) = \frac{2k_{22}^i s}{s^2 + \omega_i^2} + \text{other terms} \qquad (5\text{-}29)$$

$$Z_{21}(s) = \frac{2k_{21}^i s}{s^2 + \omega_i^2} + \text{other terms} \qquad (5\text{-}30)$$

Equation (5-5) or (5-15) indicates that $Z(s)$ has also a pair of poles at $s = \pm j\omega_i$:

$$Z(s) = \frac{2k^i s}{s^2 + \omega_i^2} + \text{other terms} \qquad (5\text{-}31)$$

Introducing the above quantities into Eq. (5-15) and evaluating it in the vicinity of $s^2 = -\omega_i^2$ indicates that

$$k^i = n_1^2 k_{11}^i + n_2^2 k_{22}^i + 2n_1 n_2 k_{21}^i \qquad (5\text{-}32)$$

The residue itself is therefore a quadratic form whose matrix is the matrix of the residue. Since the residue of the poles on the $j\omega$ axis must be equal or greater than zero the quadratic form itself must be positive semi-definite. Making use of the relationships in Eqs (5-6) to (5-8) indicates that

$$k_{11}^i \geqq 0 \qquad (5\text{-}33)$$

$$k_{22}^i \geqq 0 \qquad (5\text{-}34)$$

$$k_{11}^i k_{22}^i - k_{21}^{i2} \geqq 0 \qquad (5\text{-}35)$$

Equations (5-33) to (5-35) are known as the residue condition. It gives a necessary relationship between the residues of the poles on the $j\omega$ axis of the open-circuit parameters $Z_{11}(s)$, $Z_{22}(s)$ and $Z_{21}(s)$.

Equation (5-35) also indicates that where as k_{21}^i must be real it may be negative since only $(k_{21}^i)^2$ appears in this equation. $Z_{21}(s)$ need therefore not be a p.r. function. This property is already understood in connection with Eq. (5-20).

It is always possible to partially remove a pole from $Z_{11}(s)$ or $Z_{22}(s)$ such that

$$(k_{11} - k_{11}')k_{22} - k_{21}^2 = 0 \tag{5-36a}$$

or

$$(k_{22} - k_{22}')k_{11} - k_{21}^2 = 0 \tag{5-36b}$$

Such a network is said to be compact. k_{11}' and k_{22}' are known as the private poles of $Z_{11}(s)$ and $Z_{22}(s)$.

REAL PART
CONDITION

Since the quadratic form of the real parts of $Z(j\omega)$ in (5-24) must be positive semi-definite for all ω the entries of the matrix of the quadratic form satisfy Eqs (5-6) to (5-8).

$$\text{Re}[Z_{11}(j\omega)] \geqq 0 \quad \text{for all } \omega \tag{5-37}$$

$$\text{Re}[Z_{22}(j\omega)] \geqq 0 \quad \text{for all } \omega \tag{5-38}$$

$$\text{Re}[Z_{11}(j\omega)] \, \text{Re}[Z_{22}(j\omega)] - \text{Re}[Z_{21}^2(j\omega)] \geqq 0 \quad \text{for all } \omega \tag{5-39}$$

The preceding equations are known as the real part condition. Making use of the fact that the real parts of $Z_{11}(j\omega)$ and $Z_{22}(j\omega)$ are zero for a reactance function indicates that

$$\text{Re}[Z_{21}(j\omega)] = 0 \tag{5-40}$$

The open-circuit parameter $Z_{21}(s)$ of a reactance network is therefore an odd rational function of s—an important property that is of note in the development of filters with attenuation poles at finite frequencies.

REALIZABILITY CONDITIONS
FOR TWO-PORT
LC NETWORKS

If the two-port network is a reactance network the real part condition in (5-27) implies that the quadratic form of the matrix $\bar{Z}(s)$ is odd [see also Eq. (5-40)]. The necessary and sufficient conditions for a square matrix to be realizable as an *LC* network are given from (5-25) to (5-27) by

The quadratic form of the matrix $\bar{Z}(s)$ is the ratio of an odd to even or even to odd Hurwitz polynomial.   (5-41)

$\bar{Z}(s)$ has only simple poles on the $j\omega$ axis which satisfy the residue condition $k_{11}^i \geq 0$, $k_{11}^i k_{22}^i \geq 0$, $k_{22}^i - (k_{21}^i)^2 \geq 0$. (5-42)

The quadratic form of the matrix $\bar{Z}(s)$ associated with a two-port reactance network satisfies the necessary properties of one-port LC functions listed in Chapter 2.

POLES OF $Z_{21}(s)$
OF TWO-PORT
REACTANCE NETWORKS

Since $Z_{21}(s)$ involves terms such as in Eq. (5-30) its general form is

$$Z_{21}(s) = \frac{k_{21}^0}{s} + \sum \frac{2k_{21}^i s}{s^2 + \omega_i^2} + k_{21}^\infty s$$ (5-43)

The poles of $Z_{21}(s)$ must be simple since from Eq. (5-31) they also belong to $Z_{11}(s)$ and $Z_{22}(s)$ [but $Z_{11}(s)$ and $Z_{22}(s)$ may have private or semi-private poles]. The residues of $Z_{21}(s)$ must satisfy the residue condition in (5-42). They must therefore be real but may be positive, negative or zero.

$Z_{21}(s)$ may also be written as the quotient of two polynomials $P(s)$ and $Q(s)$:

$$Z_{21}(s) = \frac{a_n s^n + a_{n-1} s^{n-1} + \cdots + a_1 s + a_0}{b_m s^m + b_{m-1} s^{m-1} + \cdots + b_1 s + b_0}$$ (5-44)

n can exceed m by at most one, while m may be larger than n by more than one. The denominator polynomial is odd or even Hurwitz but the numerator polynomial need not be Hurwitz. This is discussed further in the next section which deals with the zeros of $Z_{21}(s)$. For a reactance function (5-41) requires that $Z_{21}(s)$ is an odd function of s.

ZEROS OF $Z_{21}(s)$
OF TWO-PORT
REACTANCE NETWORKS

Since the residues of the poles in Eq. (5-43) may be negative or zero the numerator polynomial $P(s)$ of $Z_{21}(s)$ in Eq. (5-44) does not need to be Hurwitz; some of its coefficients may be missing. $Z_{21}(s)$ is therefore not required to be p.r. as already demonstrated in connection with Eq. (5-34). Its zeros are thus not restricted to the LHP. Imaginary zeros, except at the origin and at infinity occur in conjugate pairs since the coefficients of $P(s)$ are real. Complex zeros have quadrantal symmetry because $Z_{21}(s)$ is an odd function of s. The different zero locations found in practice with reactance functions may be tabulated as follows:

1. The zeros of $Z_{21}(s)$ are all at infinity.
2. The zeros of $Z_{21}(s)$ are all at the origin.

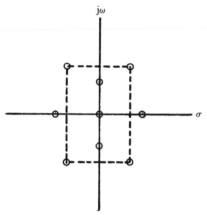

FIGURE 5-2
Realizable zeros of transmission for two-port network.

3. The zeros of $Z_{21}(s)$ are at the origin and at infinity.
4. The zeros of $Z_{21}(s)$ are at arbitrary points on the $j\omega$ axis in conjugate pairs.
5. The zeros of $Z_{21}(s)$ are complex and have quadrantal symmetry.

Figure 5-2 depicts the possible types of zero diagrams. If the zeros are in the LHP or on the $j\omega$ axis the network is a minimum phase one. If the zeros of $Z_{21}(s)$ are on the RHP the network is non-minimum phase. RHP zeros in LC networks require more than one path between the load and generator for their realization.

For an LCR ladder network the zeros of $Z_{21}(s)$ all lie on the LHP. This stems from the fact that since the branches themselves are p.r. the poles and zeros of the network cannot lie in the RHP. If the ladder network is an LC one the poles and zeros of the branches are interlaced on the $j\omega$ axis and the zeros of $Z_{21}(s)$ must therefore also lie there. Figure 5-3 depicts examples of some ladder branches associated with zeros of transmission.

If all the zeros of $Z_{21}(s)$ lie at infinity the resultant network is known as a low-pass one. Such a ladder network has only series inductors and shunt capacitors in the manner illustrated in Fig. 5-4. This is the main transfer function considered in this text. Other filter prototypes will rely on frequency transformations from the low-pass prototype for their synthesis. The presence of m zeros at infinity in $Z_{21}(s)$ implies

$$a_n = a_{n-1} = \cdots = a_1 = 0$$

in Eq. (5-44). Thus

$$Z_{21}(s) = \frac{a_0}{b_m s^m + b_{m-1} s^{m-1} + \cdots + b_1 s + b_0} \qquad (5\text{-}45)$$

This transfer function is obtained by a term-by-term cancellation in Eq. (5-20).

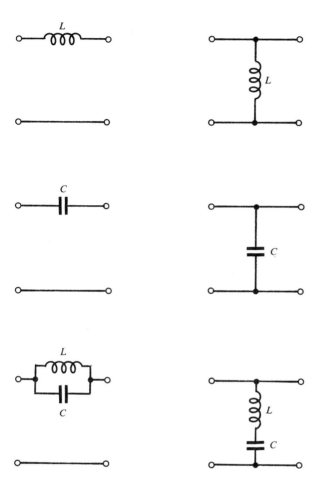

FIGURE 5-3
Ladder branches associated with zeros of transmission.

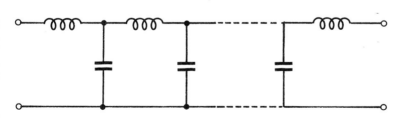

FIGURE 5-4
Ladder network with zeros of transmission at infinity.

REALIZABILITY TESTING

The necessary and sufficient conditions for a square matrix to be realizable as a two-port reactance network will now be illustrated by way of an example:

$$Z_{11}(s) = \frac{s^2 + 2}{s}$$

$$Z_{22}(s) = \frac{4s^2 + 2}{s}$$

$$Z_{12}(s) = Z_{21}(s) = \frac{2s^2 + 2}{s}$$

$Z_{11}(s)$ and $Z_{22}(s)$ are p.r. since they meet the condition in Chapter 2. Evaluating the residues at the origin leads to

$$k_{11}^0 = sZ_{11}(s)|_{s=0} = 2$$

$$k_{22}^0 = sZ_{22}(s)|_{s=0} = 2$$

$$k_{21}^0 = sZ_{21}(s)|_{s=0} = 2$$

Substituting these values into the residue equation indicates that the pole at the origin is compact and satisfies the residue condition. Forming the residues at infinity gives

$$k_{11}^\infty = \text{Im} \frac{Z_{11}(s)}{s} = 1$$

$$k_{22}^\infty = \text{Im} \frac{Z_{22}(s)}{s} = 4$$

$$k_{21}^\infty = \text{Im} \frac{Z_{21}(s)}{s} = 2$$

This pole is also compact and again meets the residue condition. It is also observed that the quadratic form of the square matrix with entries $Z_{11}(s)$, $Z_{22}(s)$ and $Z_{12}(s) = Z_{21}(s)$ satisfies (5-41). Since the conditions in (5-42) are met the impedance matrix of the quadratic form is realizable as a two-port network.

It is readily shown that the following open-circuit parameters are incompatible with the residue equation:

$$Z_{11}(s) = \frac{s^2 + 2}{s}$$

$$Z_{22}(s) = \frac{4s + 2}{s}$$

$$Z_{12}(s) = Z_{21}(s) = \frac{5s^2 + 2}{s}$$

These entries are therefore not realizable as a two-port network.

It is left as an exercise for the reader to show that the open-circuit parameters below are realizable as a two-port *LC* network:

$$Z_{11}(s) = Z_{22}(s) = \frac{2s^3 + 5s}{s^4 + 10s^2 + 9}$$

$$Z_{12}(s) = Z_{21}(s) = \frac{s}{s^4 + 10s^2 + 9}$$

THE BOUNDED
REAL CONDITION

The bilinear transformation between reflection and immittance on a one-port network has been employed in Chapter 2 to deduce a bounded real condition on the reflection coefficient of the circuit which avoids the need to evaluate residues. This testing procedure will now be extended to all the scattering parameters of a two-port circuit. The bounded real conditions on the reflection coefficient $[\Gamma(s)]$ enunciated in Chapter 2 are reproduced below:

$\Gamma(s)$ is real for s real. $\hspace{4cm}$ (5-46)

$\Gamma(s)$ must have no right half or $j\omega$ axis poles. $\hspace{1.5cm}$ (5-47)

$0 \leq |\Gamma(j\omega)| \leq 1$ for all ω. $\hspace{3.5cm}$ (5-48)

The additional condition imposed on the transmission coefficient $[\tau(s)]$ in the case of a two-port reactance network is defined using the unitary condition by

$$1 \geq |\tau(j\omega)| \geq 0 \text{ for all } \omega \hspace{2cm} (5\text{-}49)$$

The bounded real condition will now be used to verify that the scattering parameters below are realizable as a two-port reactance network:

$$S_{11}(s) = S_{22}(s) = \frac{s^3}{s^3 + 2s^2 + 2s + 2}$$

$$S_{12}(s) = S_{21}(s) = \frac{1}{s^3 + 2s^2 + 2s + 2}$$

$S_{11}(s)$ readily meets the condition in (5-46) because it is real for s real. Likewise, it is readily verified that the denominator polynomial is Hurwitz. Forming the roots of this polynomial also indicates that $S_{11}(s)$ has no poles on the $j\omega$ axis. $S_{11}(s)$ therefore satisfies the requirement in (5-47). To ensure that (5-48) is met the amplitude squared function of the reflection coefficient must be evaluated. This gives

$$S_{11}(s)S_{11}(-s)\big|_{s=j\omega} = \frac{\omega^6}{1 + \omega^6}$$

$S_{11}(s)$ is therefore also compatible with the last requirement.

As $S_{11}(s)$ satisfies all the required tests it is bounded real and the impedance associated with it is a p.r. function.

Forming $S_{21}(s)S_{21}(-s)$ from a knowledge of $S_{11}(s)S_{11}(-s)$ by having recourse to the unitary conditions indicates that

$$S_{21}(s)S_{21}(-s)|_{s=j\omega} = \frac{1}{1+\omega^6}$$

is compatible with the condition in (5-49).

The scattering matrix in the example is therefore bounded real, defines a realizable two-port reactance network and is compatible with the unitary condition.

ENERGY FUNCTIONS

Whether a square $n \times n$ impedance matrix can be realized as a passive n-port network has been the main topic of this chapter. The fact that such matrices must be positive semi-definite or positive defnite matrices may also be understood by considering the complex power in such circuits. In an n-port network with m meshes, the loop equations may be written as

$$\bar{Z}_m \bar{I}_m = \bar{V}_m$$

where $\bar{I}_m$ is the mesh current vector, $\bar{V}_m$ is the mesh voltage vector and $\bar{Z}_m$ is the mesh impedance matrix

$$\bar{Z}_m = \bar{R}_m + s\bar{L}_m + \frac{\bar{D}_m}{s}$$

where $\bar{D}_m$ is the reciprocal of the loop capacitances.

The complex power in ths circuit is

$$(\bar{I}_m^*)^T \bar{V}_m$$

Writing $\bar{V}_m$ in terms of $\bar{I}_m$ leads to

$$(\bar{I}_m^*)^T \bar{V}_m = (\bar{I}_m^*)^T \bar{Z}_m \bar{I}_m$$

If there are no voltage sources in the hidden meshes, it may be shown that

$$(\bar{I}_m^*)^T \bar{Z}_m \bar{I}_m = (\bar{I}^*)^T \bar{Z}\bar{I}$$

where $\bar{I}$, $\bar{V}$ and $\bar{Z}$ are the port voltages and currents and $\bar{Z}$ is the open-circuit impedance matrix.

Combining the preceding three equations indicates that the complex power in the circuit may be expressed in terms of the open-circuit variables of the network provided there are no hidden voltage sources within the network:

$$(\bar{I}_m^*)^T \bar{V}_m = (\bar{I}^*)^T \bar{Z}\bar{I}$$

Writing $(\bar{I}_m^*)^T \bar{V}_m$ in the preceding equation in terms of the mesh impedance matrix gives

$$(\bar{I}^*)^T \bar{Z}\bar{I} = (\bar{I}_m^*)^T \bar{R}_m \bar{I}_m + (\bar{I}_m^*)^T (s\bar{L}_m)\bar{I}_m + (\bar{I}_m^*)^T \left(\frac{\bar{D}_m}{s}\right)\bar{L}_m$$

Replacing s by $j\omega$ in this equation indicates that the complex power in the circuit, the power dissipated in the circuit and the energy stored within the circuit are all quadratic forms. Since the power dissipated in the circuit and the average stored energy cannot be negative, the open-circuit impedance $\bar{Z}$ must be positive semi-definite or positive definite as asserted. This problem will be studied further in Chapter 6.

PROBLEMS

5-1 Obtain the quadratic form of a square admittance matrix.

5-2 Find a realizable $Z_{22}(s)$ for the given impedance parameters

$$Z_{21}(s) = \frac{1}{s(s^2 + 2)}$$

$$Z_{11}(s) = \frac{(s^2 + 1)(s^2 + 3)}{s(s^2 + 2)}$$

5-3 Investigate whether the following admittance parameters are realizable as a symmetrical network:

$$Y_{11}(s) = \frac{2s^3 + 5s}{s^4 + 10s^2 + 9}$$

$$Y_{21}(s) = \frac{2}{s^4 + 10s^2 + 9}$$

5-4 Show that if

$$Z(s) = \frac{m_1(s) + n_1(s)}{m_2(s) + n_2(s)}$$

is Hurwitz then $m_1(s) + n_1(s) + m_2(s) + n_2(s)$ is also Hurwitz.

5-5 A three-port network is defined by an impedance matrix. Find its quadratic form.

BIBLIOGRAPHY

Bode, H. W., 'A general theory of electric wave filters', *J. Math. Phys.*, Vol. 13, pp. 275–362, 1934.

CHAPTER

6

IMMITTANCE AND SCATTERING MATRICES OF SYMMETRICAL NETWORKS

INTRODUCTION

An algebraic equation that often appears in network theory in connection with a square real symmetric or Hermitian matrix is the eigenvalue equation. This equation relates the entries of a matrix of order n to a set of n scalar eigenvalues and n eigenvectors. If the inputs at the n ports of the network are proportional to an eigenvector the input immittance or reflection coefficient at any port corresponds to an eigenvalue.

The one-port circuit at any port corresponding to each eigenvector is known as an eigennetwork. The application of the two eigenvectors to a symmetrical two-port network indicates that the two eigennetworks may be obtained by bisecting the network and open- and short-circuiting the exposed terminals. However, if the eigenvalues are known, the eigennetworks may also be directly synthesized from them in the manner discussed in Chapter 3.

A circuit of particular importance in the development of all-pass networks, to be treated in Chapter 9, is the symmetrical lattice network. An important property of this network is that its branch immittances coincide with the two eigenvalues of its immittance matrix. Thus, a two-port network is always realizable as a lattice network provided its eigenvalues are p.r. Since the branch immittance of T or π networks involve linear combinations of their eigenvalues, some circuits may not be realizable in those two configurations.

A symmetrical scattering matrix can be diagonalized in a similar way to the immittance ones by noting that the eigenvectors of $\bar{S}$ are also those of $\bar{Z}$ and $\bar{Y}$. This is so because the matrices $\bar{Z}$ nd $\bar{Y}$ commute with $\bar{S}$ and the symmetry operators. Another property of the $\bar{S}$, $\bar{Z}$ and $\bar{Y}$ matrices is that their eigenvalues are related in a simple way. If the eigenvalues of a symmetrical scattering matrix are bounded real the matrix represents a realizable two-port network.

THE IMPEDANCE
MATRIX EIGENVALUES

The importance of the eigenvalue problem in network theory, to be described now, lies in the fact that the entries of a symmetrical or Hermitian square matrix may always be expressed in terms of the root (eigenvalues) of its characteristic equation. The formulation of the eigenvalue problem starts by examining the solutions of the voltage–current relationship:

$$\bar{V} = \bar{Z}\bar{I} \qquad (6\text{-}1a)$$

for the cases for which the n-independent variables are related by a scalar quantity of the form

$$\bar{I} = \bar{U}_i \qquad (6\text{-}1b)$$

$$V = z_i\bar{U}_i \qquad (6\text{-}1c)$$

The eigenvalue equation associated with the open-circuit parameters is then given by

$$z_i\bar{U}_i = \bar{Z}\bar{U}_i \qquad (6\text{-}1d)$$

where $\bar{U}_i$ is an eigenvector and z_i is an eigenvalue; U_i represents a possible excitation in the network with the fields at the terminal planes proportional to the elements of the eigenvector and z_i represents an impedance measured at any terminal plane. The definition of the eigenvalue equation is illustrated in Fig. 6-1. Equation (6-1d) has a non-vanishing value for $\bar{U}_i$ provided

$$\det|\bar{Z} - z_i\bar{I}| = 0 \qquad (6\text{-}2)$$

where $\bar{I}$ is a unit vector.

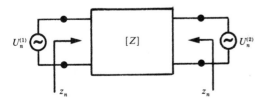

FIGURE 6-1
Schematic diagram of eigenvalue equation.

Equation (6-2) is known as the characteristic equation. The determinant given by this equation is a polynomial of degree n. Its roots are the n eigenvalues, of $\bar{Z}$, some of which may be equal (degenerate). These eigenvalues can be obtained once the coefficients of the immittance matrix are stated.

Since the eigenvalues of immittance matrices are one-port immittances these must be p.r. in order for the matrix to be realizable as a two-port network. Thus, the necessary and sufficient conditions for a square matrix to be realizable as a two-port network is that its eigenvalues are p.r. The sufficiency of this statement is demonstrated later in this chapter in the case of a two-port network, and is extended to networks with more than two ports in Chapter 5.

The characteristic equation associated with the matrix of a two-port symmetric network is

$$\begin{vmatrix} Z_{11} - z_i & Z_{21} \\ Z_{21} & Z_{11} - z_i \end{vmatrix} = 0 \tag{6-3}$$

provided the network is both reciprocal and symmetrical.

Expanding this determinant yields

$$(Z_{11} - z_i)^2 - Z_{21}^2 = 0 \tag{6-4}$$

The two roots of the characteristic equation are

$$z_1 = Z_{11} + Z_{21} \tag{6-5}$$

$$z_2 = Z_{11} - Z_{21} \tag{6-6}$$

Thus the eigenvalues are linear combinations of the entries of the impedance matrix.

The open-circuit parameters may also be written in terms of the eigenvalues as

$$Z_{11} = \frac{z_1 + z_2}{2} \tag{6-7}$$

$$Z_{21} = \frac{z_1 - z_2}{2} \tag{6-8}$$

This suggests that if either set of variables is known the other may be formed. Symmetrical networks may therefore be established in terms of either set of variables.

It is observed that since Z_{11} is the sum of two p.r. functions it is therefore also p.r., but as Z_{21} is the difference between two p.r. functions it need not be p.r.

EIGENVECTORS

A junction eigenvector is a unique set of incident waves determined by the symmetry of the network for which the input immittance at any port is the corresponding eigenvalue of the immittance matrix. Since the eigenvectors are completely determined by the junction symmetry, a symmetrical perturbation of the junction alters its eigenvalues but leaves the eigenvectors unchanged. For a two-port network the two eigenvectors may be obtained by forming the eigenvalue equation defined by Eq. (6-1d) one at a time. For the eigenvalue z_1 the eigenvalue equation becomes

$$\begin{bmatrix} Z_{11} & Z_{21} \\ Z_{21} & Z_{11} \end{bmatrix} \begin{bmatrix} U_1^{(1)} \\ U_1^{(2)} \end{bmatrix} = (Z_{11} + Z_{21}) \begin{bmatrix} U_1^{(1)} \\ U_1^{(2)} \end{bmatrix} \tag{6-9}$$

Expanding this equation gives

$$Z_{11} U_1^{(1)} + Z_{21} U_1^{(2)} = (Z_{11} + Z_{21}) U_1^{(1)} \tag{6-10}$$

$$Z_{21} U_1^{(1)} + Z_{11} U_1^{(2)} = (Z_{11} + Z_{21}) U_1^{(1)} \tag{6-11}$$

These two equations are satisfied provided

$$U_1^{(1)} = U_1^{(2)} = \frac{1}{\sqrt{2}} \tag{6-12}$$

This solution is shown schematically in Fig. 6-2a.

For the eigenvalue z_2 the eigenvalue equation is

$$\begin{bmatrix} Z_{11} & Z_{21} \\ Z_{21} & Z_{11} \end{bmatrix} \begin{bmatrix} U_2^{(1)} \\ U_2^{(2)} \end{bmatrix} = (Z_{11} - Z_{21}) \begin{bmatrix} U_2^{(1)} \\ U_2^{(2)} \end{bmatrix} \tag{6-13}$$

Expanding this equation yields

$$Z_{11} U_2^{(1)} + Z_{21} U_2^{(2)} = (Z_{11} - Z_{21}) U_2^{(1)} \tag{6-14}$$

$$Z_{21} U_2^{(1)} + Z_{11} U_2^{(2)} = (Z_{11} - Z_{21}) U_2^{(2)} \tag{6-15}$$

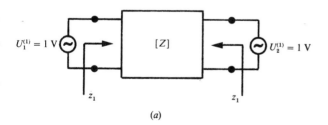

(a)

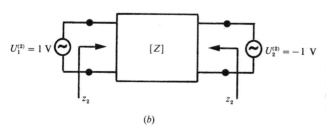

(b)

FIGURE 6-2
Schematic diagram of
(a) in-phase eigensolution
and (b) out-of-phase
eigensolution.

The two equations are consistent provided

$$U_2^{(1)} = -U_2^{(2)} = \frac{1}{\sqrt{2}}$$

(6-16)

This solution is shown schematically in Fig. 6-2b.

It is observed that the eigenvectors in Eqs (6-12) and (6-16) either produce an open circuit or a short circuit at the plane of symmetry of the network.

DIAGONALIZATION OF IMPEDANCE MATRIX

If the eigenvalues are known it is possible to form the coefficients of the matrix $\bar{Z}$:

$$\bar{Z} = \bar{U}\bar{\lambda}\bar{U}^{-1}$$

(6-17)

This equation is obtained by showing that the matrix $\bar{Z}$ is similar to a diagonal matrix with the eigenvalues of $\bar{Z}$ along its main diagonal, provided the columns of the matrix $\bar{U}$ are formed from the eigenvectors of the matrix $\bar{Z}$. If the matrix is real symmetric or Hermitian the matrix $\bar{U}$ is unitary:

$$\bar{U}^{-1} = (\bar{U}^*)^\mathrm{T}$$

(6-18)

where $(\bar{U}^*)^\mathrm{T}$ is the conjugate transpose of $\bar{U}$.

The diagonalization procedure defined by Eq. (6-17) will now be developed for a two-port network. This gives the relation between the eigenvalues and the coefficients of the immittance.

The matrix $\bar{U}$ having the eigenvectors of $\bar{Z}$ as its columns is

$$\bar{U} = \frac{1}{\sqrt{2}} = \begin{bmatrix} 1 & 1 \\ 1 & -1 \end{bmatrix}$$

(6-19)

and the diagonal matrix $\bar{\lambda}$ is

$$\bar{\lambda} = \begin{bmatrix} z_1 & 0 \\ 0 & z_2 \end{bmatrix}$$

(6-20)

Expanding the matrix $\bar{Z}$ in terms of its eigenvalues gives

$$\begin{bmatrix} Z_{11} & Z_{21} \\ Z_{21} & Z_{11} \end{bmatrix} = \frac{1}{2} \begin{bmatrix} 1 & 1 \\ 1 & -1 \end{bmatrix} \begin{bmatrix} z_1 & 0 \\ 0 & z_2 \end{bmatrix} \begin{bmatrix} 1 & 1 \\ 1 & -1 \end{bmatrix}$$

(6-21)

Thus

$$Z_{11} = \frac{z_1 + z_2}{2}$$

(6-22)

$$Z_{21} = \frac{z_1 - z_2}{2}$$

(6-23)

This is the result obtained earlier in Eqs (6-7) and (6-8).

EIGENNETWORKS

One method of finding the eigenvalues of a two-port network is obtained by observing that these coincide with the open- and short-circuit immittances of the half-sections obtained by bisecting the network and terminating the exposed terminals with open and short circuits.

If in-phase voltages are applied to a symmetrical network no current can flow across the plane of symmetry, so that for this type of excitation the exposed terminals may be left open-circuited without affecting the conditions at the input terminals. Similarly, if out-of-phase voltages are applied to the network the terminals crossing the plane of symmetry have the same potential, so that these may be short-circuited without affecting the terminal conditions.

In the case of the T circuit in Fig. 6-3a this leads to the two half-sections

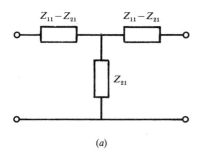

(a)

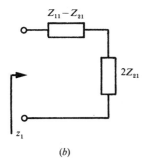

(b)

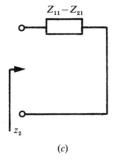

(c)

FIGURE 6-3
(a) Schematic diagram of symmetrical T network. (b) In-phase eigennetwork for T network. (c) Out-of-phase eigennetwork for T network.

depicted in Fig. 6-3b and c for which the input impedances are

$$Z_{o.c.} = Z_{11} + Z_{21} \tag{6-24}$$

$$Z_{s.c.} = Z_{11} - Z_{21} \tag{6-25}$$

Writing Z_{11} and Z_{21} in terms of their eigenvalues immediately gives

$$Z_{o.c.} = z_1 \tag{6-26}$$

$$Z_{s.c.} = z_2 \tag{6-27}$$

This result implies that the two half-sections are just the two eigennetworks associated with the impedance matrix.

For the π circuit in Fig. 6-4a the two half-sections become those in Fig. 6-4b and c for which the input admittances are

$$Y_{o.c.} = Y_{11} - Y_{21} \tag{6-28}$$

$$Y_{s.c.} = Y_{11} + Y_{21} \tag{6-29}$$

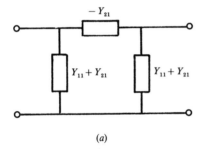

(a)

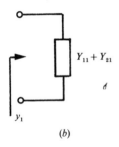

(b)

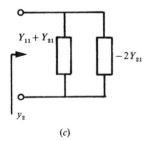

(c)

FIGURE 6-4
(a) Schematic diagram for symmetrical π network.
(b) In-phase eigennetwork for π network. (c)
Out-of-phase eigennetwork for π network.

In terms of the admittance eigenvalues these become

$$Y_{o.c.} = y_2 \qquad (6\text{-}30)$$
$$Y_{s.c.} = y_1 \qquad (6\text{-}31)$$

Thus the open- and short-circuit admittances of the half-sections are again just the two eigenvalues of the network.

EIGENVALUES OF SYMMETRICAL LATTICE NETWORKS

The two lattice networks considered in this section are depicted in Figs 6-5 and 6-6. An important property of the symmetric lattice network is that its branch immittances coincide with the two eigenvalues of its immittance matrix. Thus, an immittance matrix is always realizable as a symmetrical lattice network provided its eigenvalues are p.r.

Although it is possible to demonstrate the above statement by forming the eigennetworks of the lattice by bisecting it, there is some difficulty in deciding how to short-circuit the exposed terminals in the case where some of the exposed terminals cross each other. In the case of the lattice in Fig. 6-5, the required relationships can be obtained by redrawing the network in the form of the balanced bridge circuit in Fig. 6-7.

The open-circuit definitions of a symmetrical circuit are defined in Chapter 4 by

$$Z_{11} = Z_{22} = \left.\frac{V_1}{I_1}\right|_{I_2=0} \qquad (6\text{-}32)$$

$$Z_{21} = Z_{12} = \left.\frac{V_2}{I_1}\right|_{I_2=0} \qquad (6\text{-}33)$$

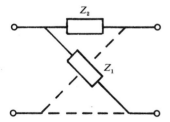

FIGURE 6-5
Symmetrical lattice network with branch impedances.

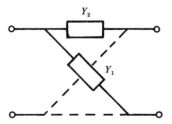

FIGURE 6-6
Symmetrical lattice network with branch admittances.

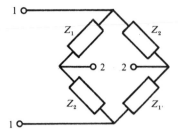

FIGURE 6-7
Bridge equivalent circuit for lattice
network with branch impedances.

Equation (6-32) immediately yields

$$Z_{11} = Z_{22} = \frac{Z_1 + Z_2}{2} \qquad (6\text{-}34)$$

In the case of Eq. (6-33), V_2 is obtained by recognizing that the bridge network operates as a voltage divider. Thus

$$V_2 = \frac{Z_1 V_1}{Z_1 + Z_2} - \frac{Z_2 V_1}{Z_1 + Z_2} \qquad (6\text{-}35)$$

Substituting Eq. (6-35) into (6-33) indicates that

$$Z_{21} = Z_{12} = \frac{Z_1 - Z_2}{Z_1 + Z_2} \frac{V_1}{I_1} \qquad (6\text{-}36)$$

Eliminating V_1/I_1 in the preceding equation in terms of Z_{11} gives

$$Z_{21} = Z_{12} = \frac{Z_1 - Z_2}{2} \qquad (6\text{-}37)$$

The forms of Z_{11} and Z_{21} indicate that Z_1 and Z_2 are just the two eigenvalues of the impedance matrix defined in Eqs (6-22) and (6-23):

$$Z_1 = Z_{11} + Z_{21} \qquad (6\text{-}38)$$

$$Z_2 = Z_{11} - Z_{21} \qquad (6\text{-}39)$$

It is readily demonstrated that the branch admittances of the lattice network in Fig. 6-6 are the eigenvalues of the short-circuit admittance parameters:

$$Y_1 = Y_{11} + Y_{21} \qquad (6\text{-}40)$$

$$Y_2 = Y_{11} - Y_{21} \qquad (6\text{-}41)$$

This derivation is left as an exercise for the reader.

An interesting property of a lattice network is that a constant immittance may always be symmetrically extracted from each branch of a lattice network without altering its overall description. This equivalence is depicted in Fig. 6-8*a* and *b*.

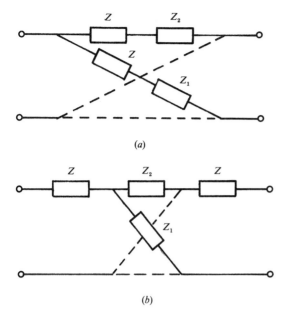

(a)

(b)

FIGURE 6-8
(a) Symmetrical lattice network with common branch impedances.
(b) Equivalent lattice circuit for symmetrical lattice with common branch impedances.

For the network in Fig. 6-8a the open-circuit parameters may be written from Eqs (6-7) and (6-8) as

$$Z_{11} = Z + \frac{(Z_1 - Z) + (Z_2 - Z)}{2} = \frac{Z_1 + Z_2}{2} \qquad (6\text{-}42)$$

$$Z_{21} = \frac{(Z_1 - Z) - (Z_2 - Z)}{2} = \frac{Z_1 - Z_2}{2} \qquad (6\text{-}43)$$

The open-circuit parameters for the circuit in Fig. 6-8b are given by definition by Eqs (6-7) and (6-8).

The equivalence between the networks in Fig. 6-9a and b follow directly.

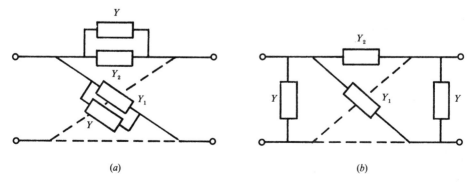

(a)

(b)

FIGURE 6-9
(a) Symmetrical lattice network with common branch admittance. (b) Equivalent lattice circuit for lattice with common branch admittances.

EQUIVALENCE BETWEEN SYMMETRICAL T, π AND LATTICE NETWORKS

Whereas it is not certain whether a square matrix is realizable as a T or π network (their shunt or series branches need not be p.r.), it is always guaranteed that it is realizable as a lattice one. The equivalence between the former two and the lattice network may be readily established by formulating their eigenvalues one at a time. This procedure will now be illustrated by way of an example by constructing the equivalence between the T network in Fig. 6-10a and that of the lattice one in Fig. 6-5.

The equivalence between the two circuits may be either obtained by finding the eigenvalues of the circuit in Fig. 6-10a from its open-circuit parameters or by obtaining them from its eigennetworks. Using the latter approach the eigenvalues

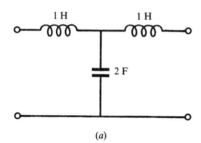

(a)

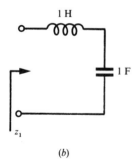

(b)

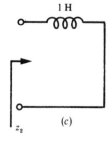

(c)

FIGURE 6-10
(a) Symmetrical T network with $Z_1 = Z_2 = s$, $Z_3 = 1/2s$. (b) In-phase eigennetwork for circuit in Fig. 6-10a. (c) Out-of-phase eigennetwork for circuit in Fig. 6-10a.

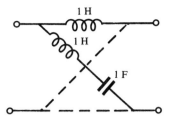

FIGURE 6-11
Symmetrical lattice network with $z_1 = s + 1/s$, $z_2 = s$.

are the input impedances of the two eigennetworks in Fig. 6-10b and c:

$$Z_1 = s + \frac{1}{s}$$

$$Z_2 = s$$

Since the branch impedances of the lattice network are its eigenvalues, its equivalent circuit is that depicted in Fig. 6-11. It is observed that the lattice representation requires more elements for its realization than is the case for the equivalent T network. Figure 6-12a, b and c illustrates the equivalence between the T and lattice networks based on the transformation illustrated in Fig. 6-8.

The equivalence between the π and lattice networks follow in a dual fashion to that of the T network.

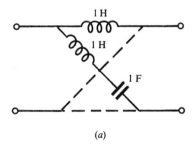

(a)

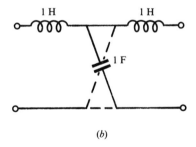

(b)

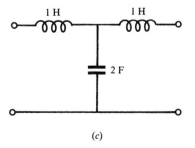

(c)

FIGURE 6-12
(a) Equivalent lattice network for T circuit in Fig. 6-10a.
(b) Lattice network for circuit in Fig. 6-12a
after removal of common branch impedances.
(c) Equivalent T network for lattice network in Fig. 6-12a.

CONSTANT RESISTANCE LATTICE NETWORKS

Another interesting property of a symmetrical lattice network is that its input impedance is equal to the terminating impedance R_0 provided

$$Z_1 Z_2 = R_0^2 \tag{6-44}$$

This property may be demonstrated by specializing Eq. (4-20) in Chapter 4 for a symmetrical network

$$Z_{in} = Z_{11} - \frac{Z_{21}^2}{R_0 + Z_{11}} \tag{6-45}$$

Writing the open-circuit parameters in terms of the eigenvalues in Eqs (6-7) and (6-8) gives

$$Z_{in} = \frac{Z_1 + Z_2}{2} - \frac{[(Z_1 - Z_2)/2]^2}{R_0 + (Z_1 + Z_2)/2} \tag{6-46}$$

Combining Eqs (6-44) and (6-46) immediately yields the required result:

$$Z_{in} = R_0 \tag{6-47}$$

One important consequence of Eq. (6-47) is that since such a network is always matched a number of them can be readily cascaded in the manner depicted in Fig. 6-13.

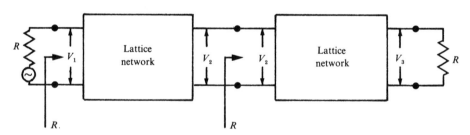

FIGURE 6-13
Cascade ladder circuit of lattice networks.

For a lattice network with admittance branches the dual relation to Eq. (6-44) is

$$Y_1 Y_2 = Y_0^2 \tag{6-48}$$

REALIZABILITY CONDITIONS FOR SYMMETRICAL TWO-PORT *LCR* NETWORKS

To ensure realizability of a symmetrical square impedance or admittance matrix as a two-port network it is only necessary to ensure that its eigenvalues are p.r.

functions. This statement may be understood by recognizing that the eigenvalues of two-port immittance matrices are one-port immittances that coincide with the branches of a lattice network.

By way of an example consider the following impedance matrix:

$$Z_{11}(s) = Z_{22}(s) = \frac{2s^3 + 5s}{s^4 + 10s^2 + 9}$$

$$Z_{12}(s) = Z_{21}(s) = \frac{s}{s^4 + 10s^2 + 9}$$

Making use of Eqs (6-5) and (6-6) gives

$$z_1(s) = \frac{2s^3 + 6s}{s^4 + 10s^2 + 9} = \frac{2s(s^2 + 3)^{\rho}}{(s^2 + 1)(s^2 + 9)}$$

$$z_2(s) = \frac{2s^3 + 4s}{s^4 + 10s^2 + 9} = \frac{2s(s^2 + 2)}{(s^2 + 1)(s^2 + 9)}$$

$z_1(s)$ and $z_2(s)$ are the ratio of odd to even polynomials and thus satisfy condition (2-21) in Chapter 2.

The residues of $z_1(s)$ are

$$k_1 = \frac{s^2 + 1}{2s} Z_1(s)\Big|_{s^2 = -1} = \tfrac{1}{4}$$

$$k_3 = \frac{s^2 + 9}{2s} Z_1(s)\Big|_{s^2 = -9} = \tfrac{3}{4}$$

and those of $z_2(s)$ are

$$k_1 = \frac{s^2 + 1}{2s} Z_2(s)\Big|_{s^2 = -1} = \tfrac{1}{8}$$

$$k_3 = \frac{s^2 + 9}{2s} Z_2(s)\Big|_{s^2 = -9} = \tfrac{7}{8}$$

Since the poles of $z_1(s)$ and $z_2(s)$ are simple with positive real residues condition (2-22) in Chapter 2 is also satisfied. As both (2-21) and (2-22) in Chapter 2 are satisfied, the impedance matrix is realizable as a two-port LC network. It is readily verified that the equivalent necessary and sufficient conditions in (5-41) and (5-42) in Chapter 5 are also satisfied by $Z_{11}(s)$ and $Z_{21}(s)$.

SCATTERING VARIABLES OF SYMMETRICAL NETWORKS

If an immittance eigenvalue or matrix description of a symmetrical network is known then it is possible to derive the corresponding scattering description without the need to invert matrices. This may be understood by noting that the eigenvalues are the usual reflection, impedance and admittance variables associated with one-port networks for which the standard one-port relationships apply:

$$s_n = \frac{z_n - Z_0}{z_n + Z_0} \tag{6-49}$$

$$\frac{z_n}{Z_0} = \frac{1 + s_n}{1 - s_n} \tag{6-50}$$

$$s_n = \frac{Y_0 - y_n}{Y_0 + y_n} \tag{6-51}$$

$$\frac{y_n}{Y_0} = \frac{1 - s_n}{1 + s_n} \tag{6-52}$$

and
$$z_n = \frac{1}{y_n} \tag{6-53}$$

The scattering matrix of a two-port symmetrical network may now be constructed by making use of the relationships between its entries and its eigenvalues:

$$S_{11} = S_{22} = \frac{s_1 + s_2}{2} \tag{6-54}$$

$$S_{12} = S_{21} = \frac{s_1 - s_2}{2} \tag{6-55}$$

and a knowledge of either eigenvalue relationship in Eq. (6-49) or (6-51).

Scrutiny of the above relationships indicates that in order to examine whether a symmetrical scattering matrix represents a realizable two-port network, it is both necessary and sufficient to evaluate its reflection eigenvalues, ensure that these are bounded real or verify that the corresponding immittance eigenvalues are p.r. functions.

The scattering coefficients of a symmetrical network characterized by the following open-circuit parameters will be now evaluated by way of an example:

$$Z_{11}(s) = Z_{22}(s) = \frac{2s^2 + 1}{2s}$$

$$Z_{12}(s) = Z_{21}(s) = \frac{1}{2s}$$

These parameters are realizable as a two-port network provided their eigenvalues are p.r.

The characteristic equation for this matrix is

$$\begin{vmatrix} Z_{11}(s) - z_n & Z_{21}(s) \\ Z_{21}(s) & Z_{11}(s) - z_n \end{vmatrix} = 0$$

Thus

$$Z_1(s) = s + \frac{1}{s}$$

$$Z_2(s) = \frac{1}{s}$$

As the eigenvalues are p.r., the matrix represents a realizable two-port network.

The scattering variables of the network may now be directly formed in terms of the open-circuit parameters by forming the appropriate eigenvalues:

$$s_1(s) = \frac{z_1(s) - 1}{z_1(s) + 1} = \frac{s^2 - s + 1}{s^2 + s + 1}$$

$$s_2(s) = \frac{z_2(s) - 1}{z_2(s) + 1} = \frac{1 - s}{1 + s}$$

The scattering parameters are now constructed using Eqs (6-54) and (6-55):

$$S_{11}(s) = S_{22}(s) = \frac{s^3}{s^3 + 2s^2 + 2s + 1}$$

$$S_{12}(s) = S_{21}(s) = \frac{1}{s^3 + 2s^2 + 2s + 1}$$

Since the open-circuit parameters represent a reactance network the scattering variables satisfy the unitary condition:

$$S_{11}(s)S_{11}(-s) + S_{21}(s)S_{21}(-s) = 1$$

as may be readily verified.

It can be verified that the eigenvalues of the problem in Fig. 6-2 are also bounded real (with the equality sign):

$$S_1(s)S_1^*(s)\big|_{s=j\omega} = \frac{s^2 - s + 1}{s^2 + s + 1}\frac{s^2 + s + 1}{s^2 - s + 1}\bigg|_{s=j\omega} = 1$$

$$S_2(s)S_2^*(s)\big|_{s=j\omega} = \frac{1 - s}{1 + s}\frac{1 + s}{1 - s}\bigg|_{s=j\omega} = 1$$

IMAGE IMMITTANCES

A common method of connecting networks is on the image immittance basis. The image immittances are a pair of quantities w_1 and w_2, such that, when these are connected to ports 1 and 2 respectively, the input immittance at port 1 is w_1 and that at port 2 is w_2.

The image admittance of ports 1 and 2 are defined in terms of the short-circuit parameters by

$$Y_{I1}^2 = \frac{Y_{11}}{Y_{22}}(Y_{11}Y_{22} - Y_{21}^2)$$

$$Y_{I2}^2 = \frac{Y_{22}}{Y_{11}}(Y_{11}Y_{22} - Y_{21}^2)$$

and for a symmetrical network by

$$Y_I^2 = Y_{I1}^2 = Y_{I2}^2 = Y_{11}^2 - Y_{21}^2$$

Factoring the right-hand side of the latter relationship suggests that this quantity may also be written as

$$Y_I^2 = (Y_{11} + Y_{21})(Y_{11} - Y_{21})$$

Scrutiny of this result indicates that the image immittance of a symmetric network is the product of the immittance eigenvalues:

$$Y_I^2 = (Y_{s.c.} Y_{o.c.})$$

PROBLEMS

6-1 Investigate whether the following immittance matrices are realizable as two-port networks:

$$\begin{bmatrix} \dfrac{2s^2+1}{2s} & \dfrac{1}{2s} \\[2ex] \dfrac{1}{2s} & \dfrac{2s^2+1}{2s} \end{bmatrix}$$

$$\begin{bmatrix} \dfrac{2s^2+1}{s} & \dfrac{1}{2s} \\[2ex] \dfrac{1}{2s} & \dfrac{2s^2+1}{s} \end{bmatrix}$$

6-2 Derive the equivalent circuit of a symmetrical π network in terms of its eigenvalues.

6-3 Repeat Prob. 6-2 for a T network.

6-4 Verify that the eigenvalues of the circuit below are p.r.

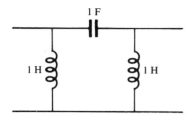

6-5 Construct the eigennetworks of the circuit below:

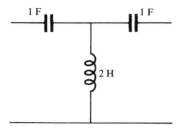

6-6 Obtain the short-circuit parameters of the network in Probs 6-4 and 6-5.

6-7 Synthesize the eigennetworks for the symmetrical network described by

$$Z_{11}(s) = Z_{22}(s) = \frac{2s^2 + 1}{2s}$$

$$Z_{12}(s) = Z_{21}(s) = \frac{1}{2s}$$

6-8 Show that the input admittance of a lattice network with eigenvalues Y_1 and Y_2 is $Y_{in} = Y_0$ provided $Y_1 Y_2 = Y_0^2$.

DARLINGTON SYNTHESIS OF TWO-PORT *LCR* NETWORKS

INTRODUCTION

An important property of a two-port *LCR* network is that it may always be realized as a two-port *LC* network terminated in a 1-Ω resistor. This theorem, due to Darlington, is the basis of modern network theory. It may be demonstrated by deriving a one-to-one correspondence between the input impedance of a two-port reactance network terminated in a 1-Ω resistance (written in terms of its two-port open-circuit parameters) and the input impedance of a two-port *LCR* network (written as the ratio of two Hurwitz polynomials). The derivation is completed by verifying that the square matrix obtained by writing the open-circuit impedance parameters in terms of the odd and even parts of the two Hurwitz polynomials is a p.r. matrix.

In the special case when the attenuation poles of the transfer function of the two-port network are all at infinity or at the origin, the Darlington method leads a simple ladder network without ideal transformers terminated in a 1-Ω resistor. Such a ladder network is obtained by forming a Cauer expansion (removal of poles at infinity or at the origin) of the input immittance of the network.

DARLINGTON THEOREM

Darlington's theorem may be established by writing the numerator and denominator polynomials of a one-port *LCR* immittance function in terms of odd to even or even to odd rational functions and making an equivalence between them and the open-circuit parameters of a two-port reactance network terminated in a 1-Ω resistor. The proof is completed by demonstrating that the open-circuit parameters are realizable as a two-port reactance network.

If $Z(s)$ is a *LCR* p.r. function it may be expressed as the ratio of two Hurwitz polynomials:

$$Z(s) = \frac{P(s)}{Q(s)} \tag{7-1}$$

Writing $P(s)$ and $Q(s)$ in terms of their odd and even parts yields

$$Z(s) = \frac{m_1(s) + n_1(s)}{m_2(s) + n_2(s)} \tag{7-2}$$

The odd and even parts of $P(s)$ and $Q(s)$ are Hurwitz by definition.

This relationship can be written in terms of one-port reactance functions by recalling that such parameters are the ratio of odd to even or even to odd Hurwitz polynomials. The two possible solutions are

$$Z(s) = \frac{m_1(s)}{n_2(s)} \frac{1 + n_1(s)/m_1(s)}{1 + m_2(s)/n_2(s)} \tag{7-3}$$

or

$$Z(s) = \frac{n_1(s)}{m_2(s)} \frac{1 + m_1(s)/n_1(s)}{1 + n_2(s)/m_2(s)} \tag{7-4}$$

The input impedance of the two-port reactance network terminated in a 1-Ω resistance in Fig. 7-1 in terms of the open-circuit parameters in Fig. 7-2 is

$$Z(s) = Z_{11}(s) - \frac{Z_{21}^2(s)}{1 + Z_{22}(s)} \tag{7-5}$$

as is readily verified starting with

$$[V] = [Z][I] \tag{7-6}$$

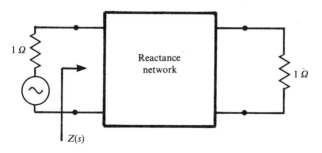

FIGURE 7-1
Two-port reactance network.

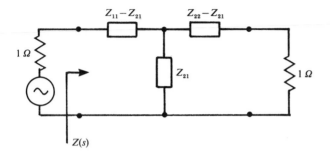

FIGURE 7-2
Two-port reactance network in terms of open-circuit parameters.

The immittance of the two-port reactance network may be put in the form of that in Eq. (7-3) or (7-4) by first writing it as the ratio of two polynomials:

$$Z(s) = \frac{Z_{11}(s) + Z_{11}(s)Z_{22}(s) - Z_{21}^2(s)}{1 + Z_{22}(s)} \qquad (7\text{-}7)$$

and then extracting a common term $Z_{11}(s)$ from the numerator polynomial:

$$Z(s) = Z_{11}(s)\frac{1 + \{[Z_{11}(s)Z_{22}(s) - Z_{21}^2(s)]/Z_{11}(s)\}}{1 + Z_{22}(s)} \qquad (7\text{-}8)$$

Comparing this immittance function with that in Eq. (7-3) or (7-4) indicates that a one-to-one correspondence may be possible between each situation provided the quantity in the brace in the numerator polynomial of the preceding equation can be shown to be a reactance function. This is fortunately the case since

$$\frac{Z_{11}(s)Z_{22}(s) - Z_{21}^2(s)}{Z_{11}(s)} = \frac{1}{Y_{22}(s)} \qquad (7\text{-}9)$$

as is readily verified by making use of the relationship between $[Z]$ and $[Y]$:

$$[Y] = [Z]^{-1} \qquad (7\text{-}10)$$

The input impedance of a two-port reactance network may therefore be written in terms of one-port reactance functions as

$$Z(s) = Z_{11}(s)\frac{1 + 1/Y_{22}(s)}{1 + Z_{22}(s)} \qquad (7\text{-}11)$$

Equation (7-3) or (7-4) is consistent with Eq. (7-11) provided

$$Z_{11}(s) = \frac{m_1(s)}{n_2(s)} \qquad (7\text{-}12a)$$

$$Z_{22}(s) = \frac{m_2(s)}{n_2(s)} \qquad (7\text{-}12b)$$

$$\frac{1}{Y_{22}(s)} = \frac{n_1(s)}{m_1(s)} \qquad (7\text{-}12c)$$

or

$$Z_{11}(s) = \frac{n_1(s)}{m_2(s)} \qquad (7\text{-}13a)$$

$$Z_{22}(s) = \frac{n_2(s)}{m_2(s)} \qquad (7\text{-}13b)$$

$$\frac{1}{Y_{22}(s)} = \frac{m_1(s)}{n_1(s)} \qquad (7\text{-}13c)$$

Writing Eq. (7-9) in terms of Eqs (7-12) or (7-13) gives

$$Z_{21}(s) = \frac{\sqrt{m_1(s)m_2(s) - n_1(s)n_2(s)}}{n_2(s)} \qquad (7\text{-}14)$$

or

$$Z_{21}(s) = \frac{\sqrt{n_1(s)n_2(s) - m_1(s)m_2(s)}}{m_2(s)} \qquad (7\text{-}15)$$

It is now necessary to ensure that $Z_{11}(s)$, $Z_{22}(s)$ and $Z_{21}(s)$ in Eqs (7-12a), (7-12b) and (7-14) or (7-13a), (7-13b) and (7-15) are realizable as a two-port reactance network. This requirement is always met, providing the following two tests are met.

The quadratic form of the matrix $\bar{Z}(s)$ is the ratio of odd to even or even to odd Hurwitz polynomials.

$\bar{Z}(s)$ has only simple poles on the $j\omega$ axis which satisfy the residue condition $k_{11}^i > 0$, $k_{22}^i > 0$, $k_{11}^i k_{22}^i - k_{21}^{i2} > 0$.

The first condition requires that $Z_{21}(s)$ as well as the one-port reactances $Z_{11}(s)$ and $Z_{22}(s)$ should be an odd reactance function. This condition determines whether Eq. (7-14) or (7-15) applies. To cater for the possibility that (7-14) or (7-15) may not be a perfect square, $Z(s)$ is augmented by a Hurwitz polynomial $m_0(s) + n_0(s)$:

$$Z(s) = \frac{m_0(s) + n_0(s)}{m_0(s) + n_0(s)} \frac{m_1(s) + n_1(s)}{m_2(s) + n_2(s)} \qquad (7\text{-}16a)$$

This gives

$$Z(s) = \frac{[m_1(s)m_0(s) + n_1(s)n_0(s)] + [m_1(s)n_0(s) + n_1(s)m_0(s)]}{[m_2(s)m_0(s) + n_2(s)n_0(s)] + [m_2(s)n_0(s) + n_2(s)m_0(s)]} \qquad (7\text{-}16b)$$

or

$$Z(s) = \frac{M_1(s) + N_1(s)}{M_2(s) + N_2(s)} \qquad (7\text{-}16c)$$

Although $Z(s)$ is unchanged, its odd and even parts are now different. The derivation of the open-circuit parameters of the *LCR* impedance function follows the development leading to Eqs $(7\text{-}12a)$, $(7\text{-}12b)$ and $(7\text{-}14)$. The result is

$$Z_{11}(s) = \frac{M_1(s)}{N_2(s)} \tag{7-17}$$

$$Z_{22}(s) = \frac{M_2(s)}{N_2(s)} \tag{7-18}$$

$$Z_{21}(s) = \frac{\sqrt{M_1(s)M_2(s) - N_1(s)N_2(s)}}{N_2(s)} \tag{7-19}$$

The other possibility for the open-circuit parameters is readily given by

$$Z_{11}(s) = \frac{N_1(s)}{M_2(s)} \tag{7-20}$$

$$Z_{22}(s) = \frac{N_2(s)}{M_2(s)} \tag{7-21}$$

$$Z_{21}(s) = \frac{\sqrt{N_1(s)N_2(s) - M_1(s)M_2(s)}}{M_2(s)} \tag{7-22}$$

The factor inside the square root sign in $(7\text{-}19)$ is given in terms of the original variables by

$$M_1(s)M_2(s) - N_1(s)N_2(s) = [m_0^2(s) - n_0^2(s)][m_1(s)m_2(s) - n_1(s)n_2(s)] \tag{7-23}$$

Factoring the quantity $m_1(s)m_2(s) - n_1(s)n_2(s)$ into odd and even factors leads to

$$m_1(s)m_2(s) - n_1(s)n_2(s) = \prod (s^2 - s_i^2)^{n_i} \prod (s^2 - s_j^2)^{n_j} \tag{7-24}$$

The terms involving s_i are even factors and those involving s_j are odd factors. To ensure that $Z_{21}(s)$ is a rational function $m_0^2(s) - n_0^2(s)$ is set equal to the factors in Eq. $(7\text{-}24)$ that are of odd multiplicity:

$$m_0^2(s) - n_0^2(s) = \prod (s^2 - s_j^2)^{n_j} \tag{7-25}$$

To ensure that $m_0(s) + n_0(s)$ is a Hurwitz polynomial it is constructed by the LHP factors in $(7\text{-}25)$:

$$m_0(s) + n_0(s) = \prod (s + s_j)^{n_j} \tag{7-26}$$

This choice of $m_0(s) + n_0(s)$ always guarantees a rational function for $Z_{21}(s)$ in Eq. $(7\text{-}19)$. A similar statement applies to the possibility in Eq. $(7\text{-}22)$.

The first realizability condition also requires that $Z_{11}(s)$, $Z_{22}(s)$ and $Y_{22}(s)$ be odd rational functions of s. $Z_{22}(s)$ and $Y_{22}(s)$ certainly meet this requirement since they are the ratios of odd to even or even to odd parts of Hurwitz polynomials $M_1(s) + N_1(s)$ or $M_2(s) + N_2(s)$. However, it is not immediately obvious that this condition is met by $Z_{11}(s)$ since it has not been demonstrated that $M_1(s) + N_2(s)$ or $M_2(s) + N_1(s)$ is a Hurwitz polynomial. To establish the Hurwitz character of these latter polynomials a new *LCR* impedance function $Z'(s)$ is formed such that

$$Z'(s) = \frac{M_1(s) + N_2(s)}{M_2(s) + N_1(s)} \qquad (7\text{-}27)$$

In order for this function to be p.r. it must satisfy conditions (2-10) and (2-11) in Chapter 2.

The first condition is met provided

$$M_1(s) + N_2(s) + M_2(s) + N_1(s) \qquad (7\text{-}28)$$

is Hurwitz.

The second condition is satisfied (see Prob. 2-5 in Chapter 2) provided

$$M_1(s)M_2(s) - N_1(s)N_2(s)\big|_{s=j\omega} \geqq 0 \qquad (7\text{-}29)$$

Applying the above two tests to the original impedance $Z(s)$ in Eq. (7-16) gives the same two relationships. Since $Z(s)$ is known to be p.r., $Z'(s)$ must also be p.r.; $M_1(s)/N_2(s)$ and $N_1(s)/M_2(s)$ in Eqs (7-17) and (7-20) are therefore odd rational functions of s required.

If it can now be demonstrated that the residue condition is satisfied by $Z_{11}(s)$, $Z_{22}(s)$ and $Z_{21}(s)$, then $Z(s)$ will be realizable as a two-port reactance network terminated in a 1-Ω resistor as asserted.

It is observed that the poles of Eqs (7-17), (7-18) and (7-19) are those of $N_2(s)$ or $M_2(s)$ (with the possibility of a pole at infinity if the degree of $P(s)$ is of degree one higher than $Q(s)$).

Residues may be evaluated with the help of Eqs (2-14a–c) in Chapter 2, but a more convenient formula for the present purpose (except at infinity) is

$$k = \frac{P(s)}{dQ(s)/ds}\bigg|_{Q(s)=0} \qquad (7\text{-}30)$$

Employing the latter expression to calculate the residue $Z_{11}(s)$ in Eq. (7-17) yields

$$k_{11} = \frac{M_1(s)}{N'_2(s)}\bigg|_{N_2(s)=0} \qquad (7\text{-}31)$$

Evaluating the residues of $Z_{22}(s)$ and $Z_{21}(s)$ in Eqs (7-18) and (7-19) gives

$$k_{22} = \frac{M_2(s)}{N'_2(s)}\bigg|_{N_2(s)=0} \qquad (7\text{-}32)$$

$$k_{21} = \frac{\sqrt{M_1(s)M_2(s)}}{N'_2(s)}\bigg|_{N_2(s)=0} \qquad (7\text{-}33)$$

The residue condition is therefore satisfied with the equals sign. It may also be shown that it is likewise satisfied at infinity. This completes the Darlington proof. To appreciate the notation so far consider the synthesis of the following p.r. impedance function as a two-port reactance network terminated in a 1-Ω resistor.

$$Z(s) = \frac{2s^3 + 2s^2 + 2s + 1}{2s^2 + 2s + 1} \qquad (7\text{-}34)$$

Writing this equation in the form described by Eq. (7-2) gives

$$Z(s) = \frac{(2s^2 + 1) + (2s^3 + 2s)}{(2s^2 + 1) + (2s)}$$

Thus

$$m_1 = 2s^2 + 1$$

$$n_1 = 2s^3 + 2s$$

$$m_2 = 2s^2 + 1$$

$$n_2 = 2s$$

The open-circuit parameters are now determined by making use of Eqs (7-12a) and (7-12b):

$$Z_{11}(s) = \frac{2s^2 + 1}{2s}$$

$$Z_{22}(s) = \frac{2s^2 + 1}{2s}$$

$$Z_{21}(s) = \frac{1}{2s}$$

$Z_{21}(s)$ is in this instance a perfect square and it is therefore unnecessary to augment $Z(s)$. Furthermore, since all open-circuit parameters are odd reactance functions, the first term for this class of network is easily satisfied. It therefore only remains to satisfy the residue condition.

Forming the residues at the origin yields

$$k_{11}^0 = \tfrac{1}{2}$$

$$k_{22}^0 = \tfrac{1}{2}$$

$$k_{21}^0 = \tfrac{1}{2}$$

The residue condition is therefore met at the origin.

Evaluating the residues at infinity gives

$$k_{11}^\infty = 1$$

$$k_{22}^\infty = 1$$

$$k_{21}^\infty = 0$$

The residue condition is also satisfied at infinity. Since the impedance matrix associated with $Z(s)$ is that of a reactance network, $Z(s)$ is realizable as a reactance network terminated in a 1-Ω resistor. It therefore only remains to develop an appropriate equivalent circuit for $Z(s)$.

SYNTHESIS OF TWO-PORT *LCR* NETWORKS WITH ATTENUATION POLES AT INFINITY

If all the attenuation poles of a network are at infinity (or at the origin) it may be synthesized as a reactance ladder network with series inductors and shunt capacitors terminated in a 1-Ω resistor. A schematic diagram for such a network is illustrated in Fig. 7-3. In this type of circuit the attenuation poles coincide with the poles of the driving point impedance. It may thus be synthesized in the first Cauer form in Chapter 3 which involves the removal of poles exclusively at infinity.

The attenuation poles may be evaluated by examining the open-circuit parameters of $Z(s)$ or by forming $S_{21}(s)$ in Table 1-1 in Chapter 1 in terms of $m_{1,2}(s)$ and $n_{1,2}(s)$:

$$S_{21}(s) = \frac{2Z_{21}(s)}{(Z_{11} + 1)(Z_{22} + 1) - Z_{21}^2(s)} \tag{7-35}$$

It is observed that the zeros of $S_{21}(s)$ are those of $Z_{21}(s)$, plus those associated with the private poles of $Z_{11}(s)$ and $Z_{22}(s)$.

Writing the open-circuit parameters in terms of $m_{1,2}(s)$ and $n_{1,2}(s)$ gives

$$S_{21}(s) = \frac{2n_2(s)\sqrt{m_1(s)m_2(s) - n_1(s)n_2(s)}}{m_1(s)n_2(s) + m_2(s)n_2(s) + n_1(s)n_2(s) + n_2^2(s)} \tag{7-36}$$

If $S_{21}(s)$ is evaluated for the open-circuit parameters defined by $Z(s)$ in Eq. (7-34) the result is

$$S_{21}(s) = \frac{1}{s^3 + 2s^2 + 2s + 1} \tag{7-37}$$

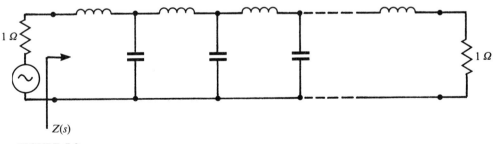

$Z(s)$

FIGURE 7-3
Two-port reactance network with transmission zeros at infinity.

Since all the attenuation poles are at infinity, $Z(s)$ can be synthesized in the simple ladder network in Fig. 7-4.

Forming a Cauer continued fraction expansion of $Z(s)$ in Eq. (7-34) gives

$$
\begin{array}{r}
2s^2 + 2s + 1 \overline{)\,2s^3 + 2s^2 + 2s + 1}\ (s \\
\underline{2s^3 + 2s^2 \quad\quad\ \ + s} \\
s + 1 \overline{)\,2s^2 + 2s + 1}\ (2s \\
\underline{2s^2 + 2s} \\
1\overline{)\,s + 1}\ (s \\
\underline{s} \\
1\overline{)\,1}\ (1 \\
\underline{1} \\
0
\end{array}
$$

The elements in Fig. 7-4 are therefore specified by

$$L_1 = 1\,\text{H}$$

$$C_2 = 2\,\text{F}$$

$$L_3 = 1\,\text{H}$$

$$R = 1\,\Omega$$

The open-circuit parameters of the simple T network in Fig. 7-4 are readily expressed (using the material in Chapter 4) as

$$Z_{11}(s) = Z_{22}(s) = \frac{2s^2 + 1}{2s}$$

$$Z_{21}(s) = \frac{1}{s}$$

in agreement with the original specification.

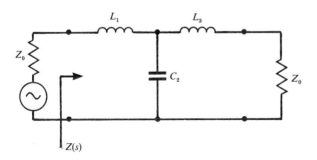

FIGURE 7-4
Two-port Cauer network for

$$Z(s) = \frac{2s^3 + 2s^2 + 2s + 1}{2s^2 + 2s + 1}$$

DARLINGTON SYNTHESIS PROCEDURE

In the approximation problem the specification is usually stated in terms of an amplitude squared function rather than as a simple amplitude function. The first task in the synthesis problem is therefore to obtain the input driving immittance of the network from a magnitude squared transfer function. For the lossless two-port network in Fig. 7-1 between 1 Ω terminations, S_{11} and S_{21} are related by the unitary condition in Chapter 1 by

$$|S_{11}(j\omega)|^2 = 1 - |S_{21}(j\omega)|^2 \qquad (7\text{-}38)$$

where ω is the normal frequency variable.

If $|S_{11}(j\omega)|$ is known it is possible to determine an appropriate value for $S_{11}(s)$ by a technique known as analytic continuation. The square of the reflection function is

$$|S_{11}(j\omega)|^2 = S_{11}(j\omega)S_{11}^*(j\omega) \qquad (7\text{-}39)$$

For a rational function with real coefficients, the conjugate of a function is equal to the function of the conjugate variable:

$$S_{11}^*(j\omega) = S_{11}(j\omega)^* \qquad (7\text{-}40)$$

Since the variable is imaginary, the conjugate of the variable is equal to the negative of the variable:

$$S_{11}(j\omega)^* = S_{11}(-j\omega) \qquad (7\text{-}41)$$

Thus

$$|S_{11}(j\omega)|^2 = S_{11}(j\omega)S_{11}(-j\omega) \qquad (7\text{-}42)$$

In terms of the variable s the preceding equation becomes

$$|S_{11}(j\omega)|^2 = S_{11}(s)S_{11}(-s)|_{s=j\omega} \qquad (7\text{-}43)$$

Combining Eqs (7-38) and (7-43) gives

$$S_{11}(s)S_{11}(-s)|_{s=j\omega} = 1 - |S_{21}(j\omega)|^2 \qquad (7\text{-}44)$$

The remaining problem is to separate $S_{11}(s)S_{11}(-s)$ into its constituents. This is done by dividing the poles and zeros between the LHP and RHP.

Once $S_{11}(s)$ is known the input immittance of the network is given by the following standard relation:

$$Y(s) = \frac{1 - S_{11}(s)}{1 + S_{11}(s)} \qquad (7\text{-}45)$$

The final step involves construction of $Y(s)$ as a reactance ladder network. Taking $S_{21}(s)$ in Eq. (7-37) as an example and making use of the unitary condition in Eq. (7-44) leads to

$$S_{11}(s)S_{11}(-s) = \frac{s^6}{(s^3 + 2s^2 + 2s + 1)(-s^3 + 2s^2 - 2s + 1)}$$

and
$$S_{11}(s)\frac{\pm s^3}{s^3 + 2s^2 + 2s + 1}$$

Using the negative sign gives the result in Eq. (7-34):

$$Z(s) = \frac{2s^3 + 2s^2 + 2s + 1}{2s^2 + 2s + 1}$$

Taking the positive sign leads to

$$Y(s) = \frac{2s^3 + 2s^2 + 2s + 1}{2s^2 + 2s + 1}$$

The approximation problem for $S_{21}(s)$ is dealt with in Chapter 8.

PROBLEMS

7-1 The impedance function of a two-port LCR network is

$$Z(s) = \frac{s^3 + 2s^2 + 2s + 2}{s^3 + 2s^2 + 2s}$$

Synthesize this impedance function as a two-port reactance network terminated in a 1-Ω resistor in terms of its open-circuit parameters. State its transmission zeros.

7-2 Obtain $S_{21}(s)$ for the impedance function of Prob. 7-1 and check that its zeros of transmission agree with those obtained in Prob. 7-1. Obtain a Cauer ladder expansion of $Z(s)$ by the removal of poles at the origin.

7-3 Show that an admittance network is always realizable as a two-port susceptance network terminated by a 1-Ω resistor [by writing $Y(s)$ in terms of its short-circuit parameters].

7-4 The admittance of a two-port LCR network is

$$Y(s) = \frac{2s^3 + 2s^2 + 2s + 1}{2s^2 + 2s + 1}$$

Obtain its short-circuit parameters and determine its transmission zeros.

7-5 Realize the admittance function in Prob. 7-4 as a two-port susceptance network terminated in a 1-Ω resistor.

7-6 Obtain the open-circuit parameters of the following driving point impedance by augmenting it according to (7-26):

$$Z(s) = \frac{2s + 2}{s + 2}$$

7-7 Find $H(s)$ and $H(-s)$ for

$$|H(j\omega)|^2 = \frac{4 + \omega^2}{1 + \omega^6}$$

7-8 Verify Eq. (7-33) using Eq. (7-30).

BIBLIOGRAPHY

Darlington, S., 'Synthesis of reactance 4-poles which produce prescribed insertion loss characteristics', *J. Math. Phys.*, Vol. 18, pp. 257–353, 1939.

CHAPTER
8

THE AMPLITUDE
APPROXIMATION

INTRODUCTION

The ideal low-pass transfer function is characterized by a magnitude function that is a constant in the passband and zero in the stopband. The corresponding phase is linear in the passband and the phase delay is a constant there. This ideal amplitude response is illustrated in Fig. 8-1. Since such a low-pass network cannot be represented by a quotient of finite-degree rational polynomials, it is necessary to seek some approximation to it. This may be done by having the amplitude, phase and delay stay within prescribed tolerances. The four class solutions to the approximation problem are the Butterworth (maximally flat), Chebyshev (equal ripple in the passband), inverse Chebyshev (equal ripple in the stopband) and elliptic (equal ripple in the pass- and stopbands). The frequencies at which the attenuation is zero are known as the attenuation zeros of the network; those at which it is infinite are known as the attenuation poles of the network. In a degree n network there are n transmission zeros in the passband and n attenuation poles in the stopband. This chapter deals in detail with the first of these three approximation problems and summarizes the fourth problem. It is of note that the former problems are all special cases of the latter one. In the Butterworth case the attenuation zeros lie at the origin and the attenuation poles all reside at infinite frequency. In the Chebyshev approximation problem the frequencies of the attenuation zeros lie at finite frequencies and those of the attenuation poles are

$S_{21}(\omega^2)$

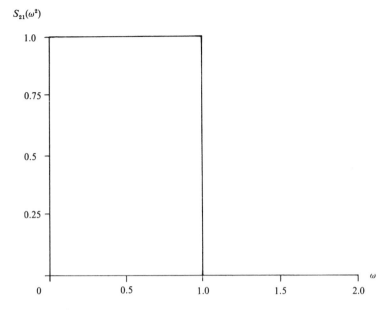

FIGURE 8-1
Schematic diagram of ideal low-pass amplitude transfer function.

all at infinity. The frequencies of the attenuation zeros of the inverse Chebyshev problem reside at the origin and those of its attenuation poles reside at finite frequencies in the stopband. In the elliptic approximation problem the frequencies of the attention zeros lie at finite frequencies in the passband and those of the attenuation poles are at finite frequencies in the stopband. The first two problems are also sometimes referred to as all-pole networks. The approximation problem tackled here is best dealt with in terms of a characteristic function and this is the notation adopted in this chapter. The task of finding an approximation to the ideal phase shift characteristics is dealt with in Chapter 9. Transfer functions are bounded real as discussed in Chapter 5.

THE APPROXIMATION PROBLEM

The approximation problem consists of deducing a bounded real amplitude squared transfer function which approximates the required filter specification. Scrutiny of this problem indicates that this quantity may be expressed as the ratio of two even order polynomials in ω.

$$S_{21}(j\omega)S_{21}^*(j\omega) = \frac{f(\omega^2)}{g(\omega^2)} \tag{8-1}$$

This statement may be understood by writing

$$F(s) = \text{Ev } F(s) + \text{Od } F(s)$$
$$F^*(s) = \text{Ev } F(s) - \text{Od } F(s)$$

and constructing

$$F(s)F^*(s) = [\text{Ev } F(s)]^2 - [\text{Od } F(s)]^2$$

Since the product of two even or two odd polynomials is an even polynomial, $F(s)F^*(s)$ is an even polynomial as asserted.

It is also of note, using the unitary condition, that the amplitude squared of the reflection coefficient of the network is

$$S_{11}(j\omega)S_{11}^*(j\omega) = \frac{h(\omega^2)}{g(\omega^2)} \qquad (8\text{-}2)$$

and that the polynomials $f(\omega^2)$, $g(\omega^2)$ and $h(\omega^2)$ are related by the Feldtkeller relationship (introduced in Chapter 1):

$$g(\omega^2) = f(\omega^2) + h(\omega^2) \qquad (8\text{-}3)$$

$g(s)$ is a strictly Hurwitz polynomial and $f(s)$ is an arbitrary one as may be understood by making use of the bilinear transformation between $S_{11}(s)$ and $Y_{11}(s)$ below.

$$S_{11}(s) = \frac{1 - Y_{11}(s)}{1 + Y_{11}(s)} \qquad (8\text{-}4)$$

It may be separately demonstrated that $h(s)$ is a pure odd or even polynomial.

In order for the function in Eq. (8-1) to represent a low-pass function it should not depart appreciably from unity in the passband and should be approximately zero in the stopband. The required relationship between the numerator and denominator polynomials may be understood by making use of the Feldtkeller condition. The result is

$$|S_{21}(j\omega)|^2 = \frac{f(\omega^2)}{f(\omega^2) + h(\omega^2)} \qquad (8\text{-}5a)$$

In modern filter theory this equation is usually written in terms of the ratio of the two polynomials $f(\omega^2)$ and $h(\omega^2)$ as

$$|S_{21}(j\omega)|^2 = \frac{1}{1 + K(\omega^2)} \qquad (8\text{-}5b)$$

where

$$K(\omega^2) = \frac{h(\omega^2)}{f(\omega^2)} \qquad (8\text{-}5c)$$

When $K(\omega^2)$ is zero, the attenuation is zero. This implies that its zeros coincide with the attenuation zeros of the network. Similarly, when it is infinite, the attenuation is infinite. This indicates that the poles of $K(\omega^2)$ correspond to the attenuation poles of the network. If $f(\omega^2)$ equals unity then $K(\omega^2)$ is a polynomial and all the attenuation poles lie at infinity. If $f(\omega^2)$ is a polynomial then some of the attenuation poles can be arbitrarily located on the $j\omega$ axis. $K(\omega^2)$ is known as the characteristic function of the transfer function. The approximation problem consists of finding suitable forms for this quantity. The four classic solutions are the Butterworth, Chebyshev, inverse Chebyshev and elliptic approximation problems. The ability of locating some attenuation poles at finite frequencies and others at infinity is also of some interest.

THE BUTTERWORTH
(OR MAXIMALLY FLAT)
LOW-PASS FILTER APPROXIMATION

The characteristic equation for the so-called Butterworth approximation is defined by

$$h(\omega^2) = (\omega^2)^n \qquad (8\text{-}6a)$$

$$f(\omega^2) = 1 \qquad (8\text{-}6b)$$

and $$K(\omega^2) = (\omega^2)^n \qquad (8\text{-}6c)$$

The characteristic equation associated with this approximation has all its n attenuation zeros at the origin and all its n attenuation poles at infinity. This situation is readily verified by investigating $K(\omega^2)$ or forming $|S_{21}(j\omega)|^2$ below:

$$|S_{21}(j\omega)|^2 = \frac{1}{1 + (\omega^2)^n} \qquad (8\text{-}7)$$

This result may be derived by expanding $K(\omega^2)$ in polynomial form:

$$K(\omega^2) = a_0 + a_2\omega^2 + a_4\omega^4 + a_{2n-2}\omega^{2n-2} + a_{2n}\omega^{2n} \qquad (8\text{-}8)$$

The coefficients of this polynomial may now be adjusted once the passband specification is spelled out:

$$K(\omega^2) = 0 \qquad \text{Remove This} \qquad \text{at } \omega = 0 \qquad (8\text{-}9a)$$

$$K(\omega^2) = 1 \qquad \text{at } \omega = 1 \qquad (8\text{-}9b)$$

$$\frac{\partial S_{21}(\omega^2)^2}{\partial \omega^2} = \frac{\partial^2 S_{21}(\omega^2)}{\partial(\omega^2)^2} = \frac{\partial^4 S_{21}(\omega^2)}{\partial(\omega^2)^4} = 0 \qquad \text{at } \omega = 0 \qquad (8\text{-}9c)$$

The first boundary condition indicates that

$$a_0 = 0 \qquad (8\text{-}10a)$$

and the second permits a_{2n} to be written in terms of $a_2, a_4, \ldots, a_{2n-2}$:

$$a_{2n} = 1 - a_2 - a_4 - \cdots - a_{2n-2} \qquad (8\text{-}10b)$$

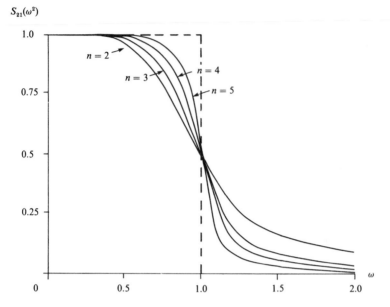

$S_{21}(\omega^2)$

FIGURE 8-2
Maximally flat low-pass amplitude transfer function approximation.

The third boundary condition minimizes the error at the origin and is known as a maximally flat approximation. Forming these derivatives at the origin one at a time using the binomial theorem readily indicates that

$$a_2 = a_4 = \cdots = a_{2n-2} = 0 \tag{8-10c}$$

and leads to the required form for $K(\omega^2)$ in (8-6c).
Other properties of this amplitude squared transfer function are that, at $\omega = 1$,

$$|S_{21}(j\omega)|^2 = \tfrac{1}{2} \tag{8-11a}$$

for all n and that, for $\omega \gg 1$, the response of the network is approximately described by

$$|S_{21}(j\omega)|^2 \approx \frac{1}{\omega^{2n}} \tag{8-11b}$$

Figure 8-2 depicts $|S_{21}(j\omega)|$ for parametric values of n. The order of the filter is uniquely fixed by the attenuation α_1 at a frequency ω_1 in the stopband.

THE CHEBYSHEV (OR EQUAL-RIPPLED) LOW-PASS FILTER APPROXIMATION

The maximally flat approximation to the ideal low-pass filter is best at $\omega = 0$ but deteriorates as ω approaches the cutoff frequency $\omega = 1$. A more interesting

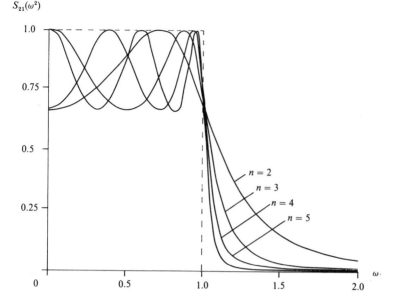

FIGURE 8-3
Equal-ripple low-pass amplitude approximation.

approximation is one that is equal rippled in the passband and is monotonically increasing in the stopband in the manner indicated in Fig. 8-3. The characteristic equation associated with this boundary condition is

$$h(\omega^2) = \varepsilon^2 T_n^2(\omega) \tag{8-12a}$$

$$f(\omega^2) = 1 \tag{8-12b}$$

$$K(\omega^2) = \varepsilon^2 T_n^2(\omega) \tag{8-12c}$$

and the corresponding amplitude squared transfer function is

$$|S_{21}(j\omega)|^2 = \frac{1}{1 + \varepsilon_2 T_n^2(\omega)} \tag{8-13}$$

$T_n(\omega)$ is a Chebyshev function of the first kind of order n defined by

$$T_n(\omega) = \cos(n \cos^{-1} \omega) \qquad \omega \leq 1 \tag{8-14}$$

$$T_n(\omega) = \cosh(n \cosh^{-1} \omega) \qquad \omega \geq 1 \tag{8-15}$$

In the interval $|\omega| < 1$ this function oscillates between ± 1 and has $(n-1)$ equal maxima or minima in the interval. At $\omega = \pm 1$ it assumes the values of ± 1. Outside this region it exceeds unity in value and approaches $\pm \infty$. Diagrams of five Chebyshev identities in the interval $|\omega| < 1$ are indicated in Fig. 8-4. The Chebyshev polynomials may be expanded as a power series in ω by having recourse to the following recurrence formula:

$$T_{n+1}(\omega) + T_{n-1}(\omega) = 2\omega T_n(\omega) \tag{8-16}$$

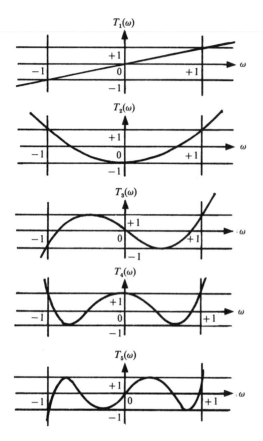

FIGURE 8-4
Chebyshev functions of the first kind.

The first two Chebyshev functions expressed as polynomials in ω may be deduced from Eq. (8-14) as

$$T_0(\omega) = 1 \qquad (8\text{-}17)$$

$$T_1(\omega) = \omega \qquad (8\text{-}18)$$

Higher-order polynomials are determined by combining the preceding three identities:

$$T_2(\omega) = 2\omega^2 - 1 \qquad (8\text{-}19)$$

$$T_3(\omega) = 4\omega^3 - 3\omega \qquad (8\text{-}20)$$

$$T_4(\omega) = 8\omega^4 - 8\omega^2 + 1 \qquad (8\text{-}21)$$

The characteristic function of order n in Eqs (8-12) has therefore n attenuation zeros at finite frequencies in the passband and a similar number of attenuation poles in the stopband at infinite frequency.

The amplitude squared transfer function at $\omega = 1$ is

$$|S_{21}(j\omega)|^2 = \frac{1}{1+\varepsilon^2} \qquad (8\text{-}22a)$$

independent of n.

At $\omega = 0$ it reduces to

$$|S_{21}(0)|^2 = 1 \qquad (8\text{-}22b)$$

or

$$|S_{21}(0)|^2 = \frac{1}{1+\varepsilon^2} \qquad (8\text{-}22c)$$

depending upon whether n is even or odd.

For $\omega \gg 1$ the scattering parameter becomes

$$|S_{21}(j\omega)| \approx \frac{1}{\varepsilon T_n(\omega)} \qquad (8\text{-}22d)$$

The order of the filter is therefore fixed by the attenuation in the stopband once ε is set by that in the passband.

INVERSE CHEBYSHEV FILTERS

The Butterworth and Chebyshev magnitude squared transfer functions dealt with so far are all poles ones. A magnitude squared transfer function that has attenuation poles at finite frequencies and has a flat band-pass response is the inverse Chebyshev one. This transfer function is rarely used in practice in that a standard Chebyshev prototype having the same order will satisfy the same specification. It is outlined here to illustrate the existence of prototype transfer functions with attenuation poles at finite frequencies.

The inverse Chebyshev transfer function is derived by subtracting the standard Chebyshev magnitude square transfer function in Eq. (8-13) from unity and replacing ω by $1/\omega$. The result is

$$|S_{21}(j\omega)|^2 = \frac{1}{1 + 1/[\varepsilon^2 T_n^2(1/\omega)]} \qquad (8\text{-}23)$$

The steps in mapping Eq. (8-13) into (8-23) are illustrated in Fig. 8-5a, b and c.

The characteristic equation has in this instance all of its zeros at the origin and all of its poles in the stopband:

$$K(\omega^2) = \frac{1}{\varepsilon^2 T_n^2(1/\omega)} \qquad (8\text{-}24)$$

The corresponding transfer function is equal to unity at $\omega = 0$:

$$|S_{21}(0)|^2 = 1 \qquad (8\text{-}25a)$$

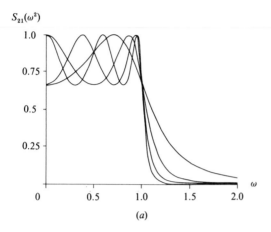

(a)

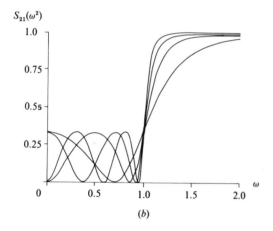

(b)

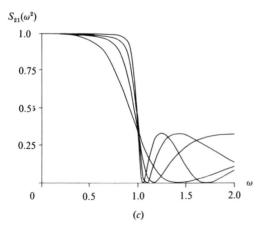

(c)

FIGURE 8-5
(a) Amplitude squared Chebyshev scattering parameter $|S_{21}(j\omega)|^2$.
(b) Amplitude squared scattering parameter obtained by forming $1 - |S_{21}(j\omega)|^2$. (c) Inverse Chebyshev amplitude squared scattering parameter obtained by replacing ω by $1/\omega$ in Fig. 8-5b.

and is equal to

$$|S_{21}(j1)|^2 = \frac{\varepsilon^2}{1+\varepsilon^2} \qquad (8\text{-}25b)$$

at $\omega = 1$.

In the stopband it ripples between $\varepsilon^2/(1+\varepsilon^2)$ and zero and is equal to either $\varepsilon^2/(1+\varepsilon^2)$ or zero at infinite frequency.

ELLIPTIC LOW-PASS APPROXIMATION

Filters that display equal ripples in the pass- and stopbands may be realized by employing elliptic rational functions. These filters provide an additional degree of freedom to those dealt with so far. The characteristic function in the case of the elliptic prototype is defined by

$$K(\omega^2) = \varepsilon^2 F_n^2(\omega) \qquad (8\text{-}26)$$

$F_n(\omega)$ is a rational function in ω which oscillates between ± 1 for $|\omega| \leq 1$ and for which $|F_n(\omega)|$ is $\geq 1/k_1$ for $|\omega| \geq k$.

It may be shown that the rational function that displays such properties is the elliptic one given by

$$F_n(\omega) = \text{sn}\left[\frac{nK_1}{K}\,\text{sn}^{-1}(\omega, k), k_1\right] \qquad \text{for } n \text{ odd} \qquad (8\text{-}27)$$

and

$$F_n(\omega) = \text{sn}\left[K_1 + \frac{nK_1}{K}\,\text{sn}^{-1}(\omega, k), k_1\right] \qquad \text{for } n \text{ even} \qquad (8\text{-}28)$$

where

$$K_1 = K(k_1) \qquad (8\text{-}29)$$

$$K = K(k) \qquad (8\text{-}30)$$

are the complete elliptic integrals of moduli k_1 and k respectively, each of which is bounded between 0 and 1.

The Jacobian elliptic sine function of modulus k is $\text{sn}(u, k)$ and the inverse elliptic function is $\text{sn}^{-1}(u, k)$. Figure 8-6 indicates the solution in the case of $n = 3$.

The three quantities n, k and k_1 are not completely independent and are related by

$$\frac{K_1'}{K'} = \frac{nK_1}{K} \qquad (8\text{-}31)$$

where

$$K_1' = K(k_1') \qquad (8\text{-}32)$$

$$K' = K(k') \qquad (8\text{-}33)$$

$F_3(\omega)$

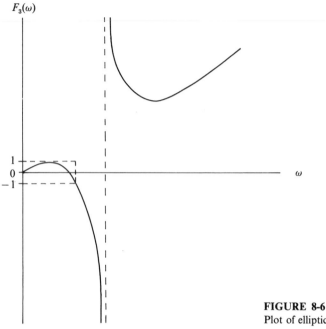

FIGURE 8-6
Plot of elliptic rational function for $n = 3$.

and

$$k_1' = (1 - k_1^2)^{1/2} \tag{8-34}$$

$$k' = (1 - k^2)^{1/2} \tag{8-35}$$

are the complementary moduli of k_1 and k respectively.

Two properties of the elliptic amplitude squared transfer function are that

$$|S_{21}(j1)|^2 = \frac{1}{1 + \varepsilon^2} \tag{8-36a}$$

and

$$|S_{21}(j/k_1)|^2 = \frac{1}{1 + \varepsilon^2/k_1^2} \tag{8-36b}$$

independent of n. Therefore ε is specified by the ripple level in the passband and k_1 is fixed by both the ripple level in the stopband and the lower band edge frequency k_1 of the stopband. Finally, the order n is set from a knowledge of k and k_1 by Eq. (8-33).

Furthermore, at the origin

$$|S_{21}(0)|^2 = 1 \tag{8-36c}$$

or

$$|S_{21}(0)|^2 = \frac{1}{1 + \varepsilon^2} \tag{8-36d}$$

depending on whether n is odd or even.

$S_{21}(\omega^2)$

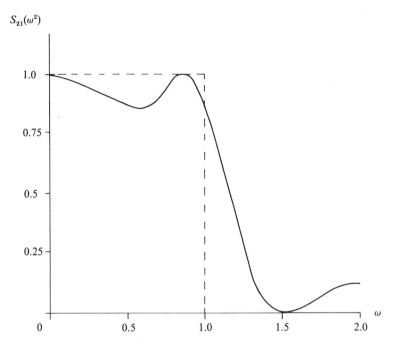

FIGURE 8-7
Elliptic amplitude approximation for $n = 3$.

The attenuation response for n odd has zero attenuation at $\omega = 0$ and infinite attenuation at $\omega = \infty$. Such a network can always be realized as a ladder network without mutual inductances or ideal transformers. This is, however, not in general the case for n even. Figure 8-7 indicates the frequency response for $n = 3$.

The Butterworth (or maximally flat), Chebyshev (equal ripple in passband) and inverse Chebyshev (equal ripple in stopband) are in fact all special cases of the elliptic prototype. This may be understood by examination of the low-pass amplitude response in Fig. 8-7. When ω_s and α_{min} both approach infinity, then the stopband loss is monotonic, whereas the passband is still equiripple. This situation corresponds to the Chebyshev characteristic. If, instead, ω_p and α_{max} approach zero (while ω_s and α_{min} are unchanged), then an inverse Chebyshev characteristic is realized. When α_{max} and ω_p approach zero while α_{min} and ω_s approach infinity, the elliptic response reduces to the Butterworth one.

APPROXIMATION PROBLEM USING CHARACTERISTIC FUNCTION TECHNIQUE

One advantage of working with the characteristic function is that its critical frequencies lie on the $j\omega$ axis whereas those of the transfer parameter reside on the left half of the s plane. Scrutiny of some approximation problems indicates that it is plausible to distribute the zeros of the characteristic function in the passband and the poles in the stopband. This remark therefore provides a simple technique for constructing some non-standard solutions. As an example consider the construction of a filter prototype for which $|S_{21}(s)|$ has all its attenuation zeros at the origin (with a quasi maximally flat passband) and has a single attenuation pole at finite frequencies (ω_1). One possible solution is

$$h(\omega^2) = b(\omega^2)^n$$

$$f(\omega^2) = (\omega^2 - \omega_1^2)^2$$

for which

$$K(\omega^2) = \frac{b(\omega^2)^n}{(\omega^2 - \omega_1^2)^2}$$

The lowest order n for which $K(\omega^2)$ has a simple attenuation pole at infinite frequency is $n = 3$. This characteristic equation has therefore a pair of attenuation poles at finite frequency, a single attenuation pole at infinite frequency and three attenuation zeros at the origin.

The coefficient b is an arbitrary constant that may be adjusted to fix the attenuation at the cutoff frequency (1 rad/s) to 3 dB for purposes of comparison

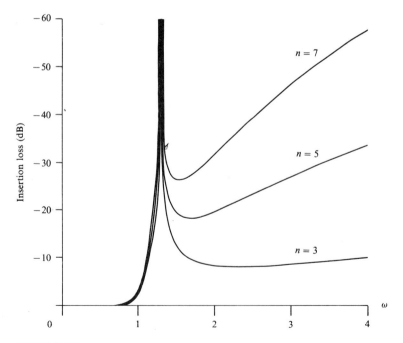

FIGURE 8-8
Frequency response of low-pass filter with a single attenuation pole at $\omega = 1.3\omega_c$ for $n = 5, 7, 9$.

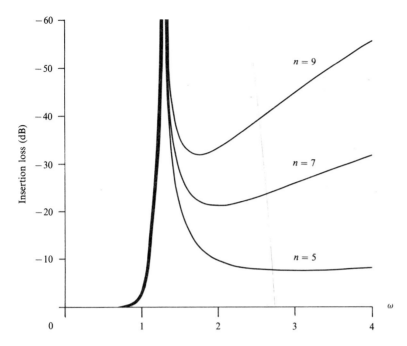

FIGURE 8-9
Frequency response of low-pass filter with a pair of attenuation poles at $\omega = 1.3\omega_c$ for $n = 3, 5, 7$.

with the classic Butterworth problem, i.e.

$$\frac{1}{1 + b(\omega^2)^n/(\omega^2 - \omega_1^2)^2} = \tfrac{1}{2}$$

Taking ω_1 as 1.3 (say) gives

$$b = 0.476$$

Figure 8-8 indicates the response of this type of filter.

Figure 8-9 depicts a similar result but with a pair of attenuation poles at $\omega = 1.3\omega_c$. Here

$$K(\omega^2) = \frac{b(\omega^2)^n}{(\omega^2 - 1.3^2)^4}$$

Figure 8-10 illustrates the case of a quasi maximally flat passband with attenuation poles at $\omega = 1.3\omega_c$ and $\omega = 2.1\omega_c$. The characteristic function is given in this instance by

$$K(\omega^2) = \frac{b(\omega^2)^n}{(\omega^2 - 1.3^2)^2(\omega^2 - 2.1^2)^2}$$

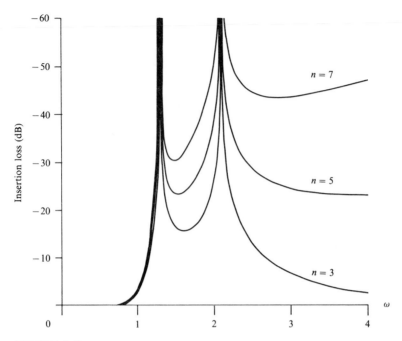

FIGURE 8-10
Frequency response of low-pass filter with attenuation poles at $\omega = 1.3\omega_c$ and $2.1\omega_c$ for $n = 3, 5, 7$.

POLE PLACEMENT TECHNIQUE

If

$$n \cosh^{-1}(\omega)$$

in the definition of $T_n(\omega)$ is replaced by

$$n \log_e(\omega + \sqrt{\omega^2 - 1})^n$$

then $T_n(\omega)$ may be written as

$$2T_n(\omega) = (\omega + \sqrt{\omega^2 - 1})^n + (\omega - \sqrt{\omega^2 - 1})^n$$

Forming the rational part (rat) of this quantity indicates that one possible identity for $T_n(\omega)$ is

$$T_n(\omega) = \text{rat}(\omega + \sqrt{\omega^2 - 1})^n$$

as is readily verified by forming $T_1(\omega)$, $T_2(\omega)$, etc.
    The rational part of the function

$$(\omega + \sqrt{\omega^2 - 1})^n$$

can therefore be considered as placing n poles at infinity in a low-pass characteristic function.

It may be separately shown that the rational part of

$$\frac{(2a_i^2 - 1)\omega^2 - a_i^2 + 2a_i\omega\sqrt{a_i^2 - 1}\sqrt{\omega^2 - 1}}{a_i^2 - \omega^2}$$

places a pair of complex conjugate poles at $\omega^2 = -a_i^2$.

In obtaining this factor note is made of the fact that this type of function has the property that its amplitude is unity in the passband.

The characteristic function associated with n poles at infinity frequency and m pairs of conjugate poles at $\omega^2 = -a_i^2$ is then obtained by taking the rational part of the product of these types of factors. Taking the case of one pole at infinite frequency and a pair of conjugate poles at $\omega^2 = -4$ gives

$$\frac{K(\omega)}{\varepsilon} = \text{rat}\left[\left(\frac{7\omega^3 - 4 + 4\sqrt{3}\,\omega\sqrt{\omega^2 - 1}}{4 - \omega^2}\right)(\omega + \sqrt{\omega^2 - 1})\right]$$

Taking the rational part of this quantity gives

$$K(\omega) = \frac{\varepsilon[7\omega^4 - 4\omega + 4\sqrt{3}\,\omega(\omega^2 - 1)]}{4 - \omega^2}$$

The required amplitude squared characteristic function is therefore given by

$$K(\omega^2) = \frac{\varepsilon^2[7\omega^4 - 4\omega + 4\sqrt{3}\,\omega(\omega^2 - 1)]^2}{\omega^4 - 8\omega^2 + 16}$$

It may be separately demonstrated that the low-pass rational characteristic function which has a single pair of attenuation poles at $\omega^2 = -a^2$ in the stopband and which has an equiripple one in the passband is

$$K(\omega^2) = \varepsilon^2\left[\frac{(a + \sqrt{a^2 - 1})^2\omega T_{n-1}(\omega) - 2a^2 T_{n-2}(\omega) + (a - \sqrt{a^2 - 1})^2\omega T_{n-3}(\omega)}{2(a^2 - \omega^2)}\right]^2$$

Taking $n = 3$ and $a^2 = -4$ and noting that

$$T_{n-1}(x) = T_2(x) = 2x^2 - 1$$

$$T_{n-2}(x) = T_1(x) = x$$

$$T_{n-3}(x) = T_0(x) = 1$$

readily gives the result deduced by taking the rational part of the appropriate factors.

ORDER OF BUTTERWORTH RESPONSE

The Butterworth response is specified by a single parameter n which is fixed by the attenuation α_1 and beyond in the stopband. A typical filter characteristic is

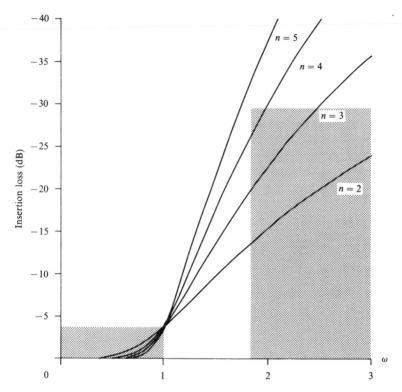

FIGURE 8-11
Maximally flat low-pass insertion loss approximation.

depicted in Fig. 8-11. The attenuation α in decibels at ω is defined from Eq. (8-7) as

$$\alpha(\text{dB}) \geq 10 \log_{10}[(1 + \omega^2)^n] \tag{8-37}$$

The order of the filter is therefore uniquely specified by the required attenuation α_1 at ω_1 by

$$n = \frac{\log_{10}(A - 1)}{2 \log_{10} \omega_1} \tag{8-38}$$

where

$$A = \text{antilog}_{10}\left(\frac{\alpha_1}{10}\right) \tag{8-39}$$

By way of an example it is desired to design a filter having a Butterworth response with an attenuation of at least 40 dB at a frequency 1.5 times the cutoff frequencies and beyond.

Making use of the preceding equations gives A as 10 000 and n as 11.35. The order of the required Butterworth filter is therefore 12.

ORDER OF CHEBYSHEV RESPONSE

The Chebyshev response is specified by two parameters, the ripple level ε in the passband and the attenuation α_1 at a frequency ω_1 and beyond in the stopband (Fig. 8-12). Both the insertion and attenuation losses are usually specified in decibels in the manner indicated in Fig. 8-10. The insertion loss α_0 at $\omega = 1$ is

$$\alpha_0(\text{dB}) = 10 \log_{10}(1 + \varepsilon^2) \tag{8-40}$$

independent of n, as is evident from the recurrence formula in Eq. (8-16). The attenuation α_1 at ω_1 in the stopband is given by

$$\alpha_1(\text{dB}) \geqq 10 \log_{10}[1 + \varepsilon^2 T_n^2(\omega_1)] \tag{8-41}$$

The order n is determined by the preceding equation with the help of Eq. (8-15) or the recurrence identity in Eq. (8-16):

$$n = \frac{\cosh^{-1}(\sqrt{A - 1}/\varepsilon)}{\cosh^{-1} \omega_1} \tag{8-42}$$

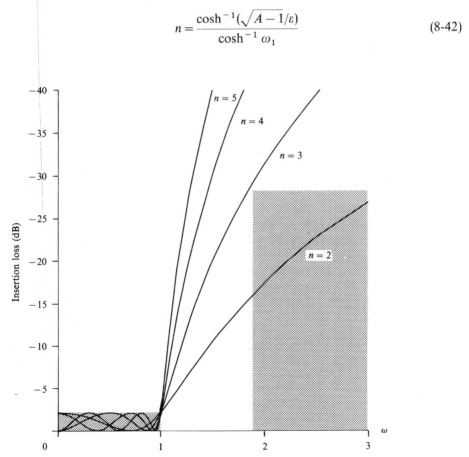

FIGURE 8-12
Equal-ripple low-pass insertion loss approximation.

where

$$A = \mathrm{antilog}_{10}\left(\frac{\alpha_1}{10}\right) \tag{8-43}$$

The specification of this type of filter is sometimes expressed in terms of the VSWR of the network instead of its ripple level. The relationship between the two is, in this instance, described by

$$\varepsilon = \frac{\mathrm{VSWR} - 1}{2\sqrt{\mathrm{VSWR}}} \tag{8-44}$$

where

$$\mathrm{VSWR} = \frac{R_{\mathrm{L}}}{R_0} = r_{n+1} \tag{8-45}$$

The relationship in Eq. (8-44) may be derived by making use of the unitary condition

$$S_{11}(s)S_{11}^*(s) + S_{21}(s)S_{21}^*(s) = 1 \tag{8-46}$$

and noting that

$$|S_{11}(s)| = \frac{\mathrm{VSWR} - 1}{\mathrm{VSWR} + 1} \tag{8-47}$$

Combining these relationships gives the required result in Eq. (8-44).

By way of an example it is required to design a low-pass filter with a 1-dB ripple and an attenuation of 40 dB at 1.5 times the cutoff frequency. Making use of Eq. (8-40) indicates that

$$1 = 10 \log_{10}(1 + \varepsilon^2)$$

and

$$\varepsilon = 0.509$$

Making use of Eqs (8-42) and (8-43) gives A as 10 000 and n as 6.22. Taking the nearest integer indicates that the order n of the filter may be taken as 7. It is apparent, in this example, that the Chebyshev filter requires only about half the number of elements than are needed in the case of the Butterworth filter with the same band-stop specification.

As a second example consider the design of a low-pass filter with a cutoff frequency of 4150 MHz, a passband VSWR of 1.25 and an attenuation of better than 25 dB at 4500 MHz and beyond. The ripple level for this filter is calculated from Eq. (8-44) as

$$\varepsilon = \frac{\mathrm{VSWR} - 1}{2\sqrt{\mathrm{VSWR}}} = 0.111$$

The value of A to be used in Eq. (8-42) is given by Eq. (8-43) as

$$A = \mathrm{antilog}_{10}\left(\frac{25}{10}\right) = 316.2$$

The normalized frequency ω_1 is

$$\omega_1 = \frac{4500}{4150} = 1.084$$

The minimum value of n is now calculated using Eq. (8-37) as

$$n = \frac{\cosh^{-1}(\sqrt{316.2 - 1})/0.111}{\cosh^{-1} 1.084} = 14.2$$

Since n must be an integer, a value of $n = 15$ is adequate.

ORDER OF ELLIPTIC FILTER

The order of the elliptic response may be deduced from a specification of the attenuations in the pass- and stopbands and the lower and upper frequencies of the transitional band. This calculation will now be illustrated in the case of a filter with a ripple level of 0.50 dB in the passband and a minimum attenuation of 18 dB in the stopband. The upper frequency of the transitional band is 1.4 times that of the cutoff frequency. The required frequency response is indicated in Fig. 8-13.

The calculation starts by evaluating the ripple level ε by having recourse to Eq. (8-36a). The result is

$$\varepsilon = 0.349\ 31$$

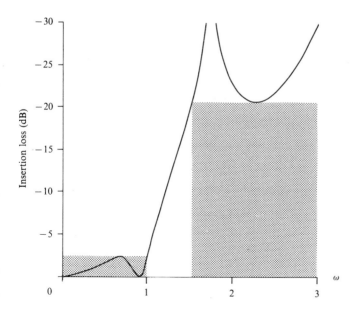

FIGURE 8-13
Elliptic low-pass insertion loss approximation.

The modulus k_1 is then calculated from Eq. (8-36b) as

$$k_1 = 0.044\,32$$

If the cutoff frequency of the low-pass prototype is taken as 1 rad/s then the modulus k is given by

$$k = \frac{1}{1.4} = 0.714\,29$$

The complementary modulis k_1' and k' are then determined using Eqs (8-34) and (8-35) as

$$k_1' = 0.999\,02$$

$$k' = 0.699\,85$$

The complete elliptic integrals K, K', K_1 and K_1' are next evaluated by having recourse to standard tables as

$$K = K(k) = 1.862\,82$$

$$K' = K(k') = 1.845\,53$$

$$K_1 = K(k_1) = 1.571\,57$$

$$K_1' = K(k_1) = 4.494\,61$$

n is therefore given by Eq. (8-31) as

$$n = 2.886\,74$$

The order of the required filter is therefore $n = 3$.

PROBLEMS

8-1 Show that the Butterworth high-pass transfer function obtained by replacing ω and $1/\omega$ in Eq. (8-7) is maximally flat at $\omega \to \infty$.

8-2 Show that the poles of the Butterworth transfer function lie on a circle.

8-3 Determine $T_6(\omega)$, $T_7(\omega)$ and $T_8(\omega)$.

8-4 Verify that $\cos(n \cos^{-1} x) = \cosh(n \cosh^{-1} x)$.

8-5 Show that $T_n(-\omega) = (-1)^n T_n(\omega)$. Thus $T_n^2(\omega)$ is an even function.

8-6 Verify the recurrent formula in Eq. (8-16) by putting $\omega = \cos \theta$ in Eq. (8-14) and forming $T_{n+1}(\omega) = \cos(n+1)\theta$ and $T_{n-1}(\omega) = \cos(n-1)\theta$.

8-7 Obtain a graph of frequency versus attenuation in decibels for a Butterworth transfer function by plotting $= 10 \log_{10} 1 + (\omega^2)^n$ versus the normalized frequency variable for $n = 1, 2, 3, 4$ and 5.

8-8 Repeat Prob. 8-7 for Chebyshev and inverse Chebyshev amplitude squared transfer functions.

8-9 Check that the 3-dB point for the inverse Chebyshev amplitude squared transfer function is

$$\omega_c = \frac{1}{\cosh(1/n \cosh^{-1} 1/\varepsilon)}$$

8-10 Using the unitary condition obtain $S_{11}(s)S_{11}(-s)$ for the inverse Chebyshev amplitude squared transfer function.

8-11 Sketch the transfer function in Fig. 8-6 with $C = 80$ and 120 instead of $C = 100$.

8-12 Calculate the order n of an elliptic low-pass filter having an insertion loss of 0.25 dB in the passband and an attenuation of 20 dB in the stopband. The upper frequency of the transitional band is 1.1 times the cutoff frequency (1 rad/s). Repeat with the factor 1.1 replaced by 1.2.

8-13 Verify Eq. (8-22).

BIBLIOGRAPHY

Belevitch, V., 'Chebyshev filters and amplifier networks', *Wireless Engr*, Vol. 29, pp. 106–110, April 1952.

Butterworth, S., 'On the theory of filter amplifiers', *Wireless Engr*, Vol. 7, pp. 536–541, 1930.

Chebyshev, P. L., *Théorie des Mécanismes connus sous le Nom de Parallélogrammes*, Vol. 1, Oeuvres, St Petersburg, 1899.

Green, E., 'Synthesis of ladder networks to give Butterworth or Chebyshev response in the pass-band', IEE Monograph, No. 88, January 1954.

Orchard, H. J., 'Formulae for ladder filters', *Wireless Engr*, Vol. 30, pp. 3–5, January 1953.

Orchard, H. J., 'Computations of elliptic functions of rational fractions of a quarter period', *IRE Trans. on Circuit Theory*, Vol. CT-5, pp. 352–355, 1958.

Piloty, H., 'Kanonische Kettenschaltungen fur Reaktanzvierpole mit vorgeschriebenen Betriebseigenschaften', *Telegr. Fernspr., Funk- und Fernsehtechnik*, Vol. 29, pp. 249–258, 279–290, 320–325, 1940.

Ragan, G. L., *Microwave Transmission Lines*, McGraw-Hill, New York, 1948.

Takahasi, H., 'On the ladder type filter networks with Chebyshev response', *J. Inst. Electrl Commun. Engrs Japan*, Vol. 34, pp. 65–74, February 1951.

Weinberg, L., 'Explicit formulas for Chebyshev and Butterworth ladder networks', *J. Appl. Phys.*, Vol. 28, pp. 1155–1160, October 1957.

Weinberg, L. and Slepian, P., 'Takahasi's results on Chebyshev and Butterworth ladder networks', *IRE Trans. on Circuit Theory*, Vol. CT-7, pp. 88–101, June 1960.

CHAPTER
9

AMPLITUDE, PHASE AND DELAY

INTRODUCTION

If a two-port network has a constant gain and linear phase (constant group delay) the relation between the various components of the demodulated output signal of an FM signal is the same as that of the original components. Any deviation from these ideal characteristics causes the demodulated signal to contain components not present in the original signal. Thus, both the phase and amplitude characteristics of a filter are important. The phase characteristic is often represented by the group velocity given by the derivative of the phase with respect to the frequency.

Figure 9-1 depicts the amplitude, phase and group delay of an ideal low-pass two-port network. In practice, the use of any of the amplitude squared transfer functions defined in Chapter 8 leads to some deviation in the phase of the network from its ideal characteristic. Just as it is not possible to synthesize a network with an ideal amplitude response, it is also not possible to construct one with an ideal phase characteristic. However, some measure of phase equalization is possible by terminating the network by an all-pass network (dealt with in this chapter) or by approximating the phase of the network and accepting the ensuing amplitude characteristic. The problem of constructing a transfer function with controllable phase and amplitude characteristics is outside the scope of this text.

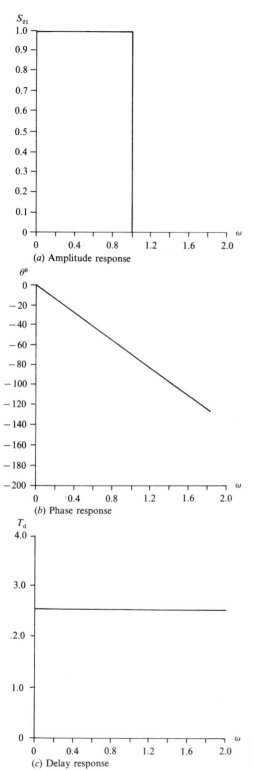

(a) Amplitude response

(b) Phase response

(c) Delay response

FIGURE 9-1
Responses of ideal low-pass filter.

AMPLITUDE, PHASE AND GROUP DELAY OF BUTTERWORTH LOW-PASS NETWORK

The amplitude, phase and group delay of a two-port network may be defined in terms of $S_{21}(s)$ by replacing s by $j\omega$:

$$S_{21}(s)|_{s=j\omega} = |S_{21}(j\omega)|\,\underline{/\beta} \tag{9-1}$$

where

$$|S_{21}(j\omega)|^2 = \frac{m_1^2(s) - n_1^2(s)}{m_2^2(s) - n_2^2(s)}\bigg|_{s=j\omega} \tag{9-2}$$

and

$$\beta = \tan^{-1}\frac{n_1(s)}{jm_1(s)}\bigg|_{s=j\omega} - \tan^{-1}\frac{n_2(s)}{jm_2(s)}\bigg|_{s=j\omega} \tag{9-3}$$

$n_{1,2}(s)$ and $m_{1,2}(s)$ are the odd and even parts of $S_{21}(s)$. The group delay is the derivative of β with respect to ω:

$$T_d = -\frac{d\beta}{d\omega} \tag{9-4}$$

For an $n = 1$ Butterworth network, $S_{21}(s)$ is given from Chapter 8 by

$$S_{21}(s) = \frac{1}{1+s} \tag{9-5}$$

The amplitude phase and group delay for this network are given by Eqs (9-1) to (9-4) by

$$|S_{21}(j\omega)|^2 = \frac{1}{1+\omega^2} \tag{9-6}$$

$$\beta = -\tan^{-1}\omega \tag{9-7}$$

$$T_d = -\frac{d\beta}{d\omega} = \frac{1}{1+\omega^2} \tag{9-8}$$

The relations for $n = 2$ are

$$S_{21}(s) = \frac{1}{s^2 + \sqrt{2}s + 1} \tag{9-9}$$

$$|S_{21}(j\omega)|^2 = \frac{1}{1+\omega^4} \tag{9-10}$$

$$\beta = -\tan^{-1}\frac{\sqrt{2}\,\omega}{1-\omega^2} \tag{9-11}$$

$$T_d = -\frac{d\beta}{d\omega} = \frac{\sqrt{2} + \sqrt{2}\,\omega^2}{1+\omega^4} \tag{9-12}$$

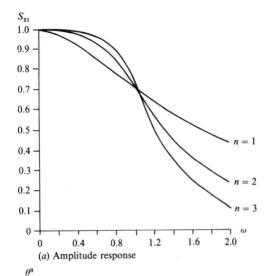

(a) Amplitude response

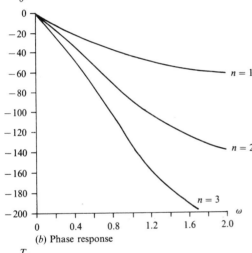

(b) Phase response

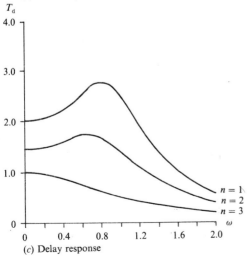

(c) Delay response

FIGURE 9-2

Responses of Butterworth low-pass filter for $n = 1$, 2 and 3.

For $n = 3$ the result is

$$S_{21}(s) = \frac{1}{s^3 + 2s^2 + 2s + 1} \tag{9-13}$$

$$|S_{21}(j\omega)|^2 = \frac{1}{1 + \omega^6} \tag{9-14}$$

$$\beta = -\tan^{-1} \frac{\omega^3 - 2\omega}{2\omega^2 - 1} \tag{9-15}$$

$$T_d = -\frac{d\beta}{d\omega} = \frac{2\omega^4 + \omega^2 + 2}{1 + \omega^6} \tag{9-16}$$

Figure 9-2a to c depicts the amplitude, phase and group delay for $n = 1$, 2 and 3.

MINIMUM AND NON-MINIMUM PHASE NETWORKS

Minimum and non-minimum networks may be defined by starting with an amplitude squared transfer function and invoking analytic continuation:

$$S_{21}(s)S_{21}(-s) = \frac{P(s)P(-s)}{Q(s)Q(-s)} \tag{9-17}$$

Since $S_{21}(s)$ is bounded real its poles must lie on the LHP but its zeros may lie on either the LHP or the RHP. The two possible solutions for $S_{21}(s)$ are therefore

$$S_{21}(s) = \frac{P(s)}{Q(s)} \tag{9-18}$$

or

$$S'_{21}(s) = \frac{P(-s)}{Q(s)} \tag{9-19}$$

The first solution realizes a minimum phase network, whereas the second one leads to a non-minimum phase one. This statement may be demonstrated by rewriting $S'_{21}(s)$ as

$$S'_{21}(s) = \frac{P(-s)}{Q(s)} \frac{P(s)}{P(s)} \tag{9-20}$$

or

$$S'_{21}(s) = S_{21}(s) \frac{P(-s)}{P(s)} \tag{9-21}$$

Thus, a non-minimum phase network can always be written as a minimum one within an all-pass network. This is a general result.

The definition of minimum and non-minimum phase networks may be understood by recognizing that an all-pass network can only add phase shift to the output wave of the network.

All-pass networks find application as phase equalizers, as will be seen later in this chapter.

ALL-PASS NETWORKS

The all-pass network in (9-21) is defined by

$$S_{21}(s) = \frac{P(-s)}{P(s)} \tag{9-22}$$

Since $S_{21}(s)$ is bounded real as discussed in Chapter 2, $P(s)$ is a strictly Hurwitz polynomial given in factored form by Eqs (2-10) and (2-11) in Chapter 2:

$$P(s) = \prod_{k=0}^{n} (s + \sigma_k) \prod_{k=0}^{m} (s + \sigma_k)^2 + \omega_k^2 \tag{9·23}$$

where n represents the number of real poles or zeros and m represents the number of non-real poles or zeros.

Expressing (9-22) in terms of (9-23) leads to

$$S_{21}(s) = \prod_{k=0}^{n} \frac{(-s + \sigma_k)}{(s + \sigma_k)} \prod_{k=0}^{m} \frac{[(-s + \sigma_k)^2 + \omega_k^2]}{[(s + \sigma_k)^2 + \omega_k^2]} \tag{9-24}$$

$S_{21}(s)$ may be represented by a cascade connection of all-pass networks. Terms like

$$\frac{-s + \sigma_k}{s + \sigma_k} \tag{9-25}$$

are known as all-pass C networks.

The zeros and poles of an all-pass C network lie on the real axis and are given by

$$z_k = -\sigma_k \tag{9-26}$$

$$p_k = \sigma_k \tag{9-27}$$

The pole zero diagram of this network is depicted in Fig. 9-3.

The amplitude, phase and delay of an all-pass C network are

$$|S_{21}(j\omega)| = 1 \tag{9-28}$$

$$\beta_k = \tfrac{1}{2} \tan^{-1} \frac{\omega}{\alpha_k} \tag{9-29}$$

$$T_k = \frac{2\sigma_k}{\sigma_k^2 + \omega^2} \tag{9-30}$$

Figure 9-4 indicates the delay T_k for the C-type all-pass network versus ω for parametric values of σ_k. The physical realization of this type of network is described in the next section.

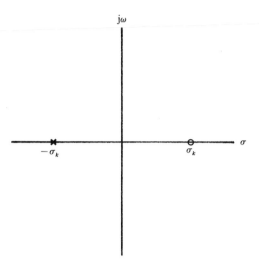

FIGURE 9-3
Pole zero diagram of all-pass C network.

Terms like

$$\frac{(-s + \sigma_k)^2 + \omega_k^2}{(s + \sigma_k)^2 + \omega_k^2} \tag{9-31}$$

are known as all-pass D networks.

The zeros and poles of the all-pass D network are

$$z_k = \sigma_k \pm j\omega_k \tag{9-32}$$

$$p_k = -\sigma_k \pm j\omega_k \tag{9-33}$$

The pole zero diagram for this type of network is depicted in Fig. 9-5.

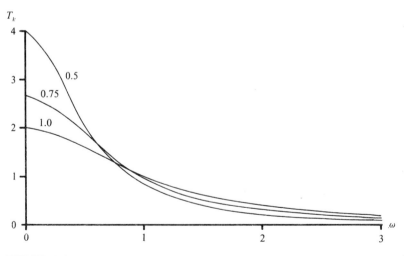

FIGURE 9-4
Delay response of all-pass C network.

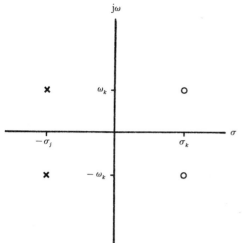

FIGURE 9-5
Pole zero diagram of all-pass D network.

The amplitude, phase and delay of this circuit are

$$|S_{21}(j\omega)| = 1 \tag{9-34}$$

$$\beta_k = 2 \tan^{-1} \frac{2\sigma_k\omega_k}{\sigma_k^2 + \omega_k^2 - \omega^2} \tag{9-35}$$

$$T_k = \frac{4\sigma_k(\omega_k^2 + \omega^2)}{(\omega_k^2 - \omega^2)^2 + 4\sigma_k^2\omega^2} \tag{9-36}$$

Figure 9-6 depicts T_k versus frequency for parametric values of σ_k and $\omega_k = 1$.

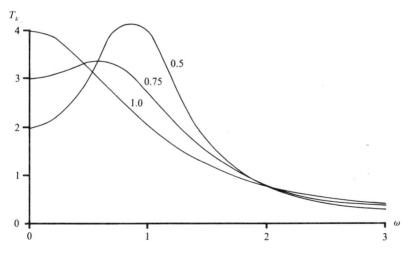

FIGURE 9-6
Delay response of all-pass D network.

PHASE EQUALIZATION

A linear phase filter may be constructed by using one network to meet the amplitude requirement of the specification and a second all-pass network to equalize the phase shift of the overall network. The underlining principle of phase equalization may be understood by writing the overall phase function as

$$\beta(\omega) = \theta(\omega) + \phi(\omega) \tag{9-37}$$

where $\beta(\omega)$ is the overall phase shift of the two networks, $\theta(\omega)$ is the phase shift of the network used to meet the amplitude characteristic and $\phi(\omega)$ is the phase shift of the all-pass network.

In the vicinity of ω equal to zero the phase angles may be expanded as

$$\beta(\omega) = \beta_1\omega + \beta_3\omega^3 + \beta_5\omega^5 + \cdots \tag{9-38}$$

$$\theta(\omega) = \theta_1\omega + \theta_3\omega^3 + \theta_5\omega^5 + \cdots \tag{9-39}$$

$$\phi(\omega) = \phi_1\omega + \phi_3\omega^3 + \phi_5\phi^5 + \cdots \tag{9-40}$$

all phase functions being odd functions of ω.

Combining the previous equations leads to

$$\beta(\omega) = (\theta_1 + \phi_1)\omega + (\theta_3 + \phi_3)\omega^3 + (\theta_5 + \phi_5)\omega^5 + \cdots \tag{9-41}$$

To obtain a linear phase at $\omega = 0$ all the coefficients in the above equations except the first one must be set to zero. For a simple all-pass D network there are two variables σ_k and ω_k which can be adjusted to meet this requirement:

$$\theta_3 + \phi_3 = 0 \tag{9-42}$$

$$\theta_5 + \phi_5 = 0 \tag{9-43}$$

ALL-PASS LATTICE NETWORKS

It will now be shown that the constant resistance network described in Chapter 6 is an all-pass network. This may be demonstrated by starting with the identity in Table 1-1:

$$S_{21}(s) = \frac{2R_0 Z_{21}(s)}{[Z_{11}(s) + R_0]^2 - Z_{21}^2(s)} \tag{9-44}$$

Writing $Z_{11}(s)$ and $Z_{21}(s)$ in terms of their eigenvalues in Eqs (6-7) and (6-9) in Chapter 6 leads to

$$S_{21}(s) = \frac{2R_0\{[Z_1(s) - Z_2(s)]/2\}}{\{[Z_1(s) + Z_2(s)]/2 + R_0\}^2 - \{[Z_1(s) - Z_2(s)]/2\}^2} \tag{9-45}$$

The constant resistance condition is defined by Eq. (6-44) in Chapter 6:

$$R_0^2 = Z_1(s)Z_2(s) \tag{9-46}$$

Combining the two preceding equations gives the required result:

$$S_{21}(s) = \frac{Z_1(s) - 1}{Z_1(s) + 1} \tag{9-47}$$

If $Z_1(s)$ is a reactance function

$$Z_1(s) = \frac{m(s)}{n(s)} \quad \text{or} \quad \frac{n(s)}{m(s)} \tag{9-48}$$

$$Z_2(s) = \frac{R_0^2}{Z_1(s)} \tag{9-49}$$

$S_{21}(s)$ is therefore put in the following form as a preamble to the synthesis of all-pass lattice networks:

$$S_{21}(s) = \frac{m(s) - n(s)}{m(s) + n(s)} \quad \text{or} \quad \frac{n(s) - m(s)}{n(s) + m(s)} \tag{9-50}$$

Thus

$$|S_{21}(j\omega)| = 1 \tag{9-51}$$

$$\beta = \tfrac{1}{2} \tan^{-1} \frac{n(s)}{jm(s)} \tag{9-52}$$

$$T_d = -\frac{d\beta}{d\omega} \tag{9-53}$$

A similar result is obtained by starting with $S_{21}(s)$ in terms of the short-circuit parameters.

SYNTHESIS OF ALL-PASS LATTICE NETWORKS

The most simple all-pass networks is the C type defined by Eq. (9-25):

$$S_{21}(s) = \frac{-s + \sigma}{s + \sigma} \tag{9-54}$$

It may be realized as a symmetrical network by writing $S_{21}(s)$ in the form of Eq. (9-47):

$$S_{21}(s) = \frac{\sigma/s - 1}{\sigma/s + 1} \tag{9-55}$$

Thus

$$Z_1(s) = \frac{\sigma}{s} = \frac{1}{sC} \tag{9-56}$$

where

$$C = \frac{1}{\sigma} \quad \text{F} \tag{9-57}$$

Making use of Eq. (9-49) yields

$$Z_2(s) = \frac{R_0^2}{Z_1(s)} = s\sigma R_0^2 = sL \tag{9-58}$$

where

$$L = \sigma R_0^2 \quad \text{H} \tag{9-59}$$

The equivalent all-pass C-lattice network is depicted in Fig. 9-7. $Z_1(s)$ and $Z_2(s)$ conform with the notation in Fig. 6-5 in Chapter 6.

The all-pass D network defined by (9-31) may also be realized as a lattice network. The transfer parameter for this network is given by

$$S_{21}(s) = \frac{s^2 - as + b}{s^2 + as + b} \tag{9-60}$$

where a and b are defined by (9-31) as

$$a = 2\sigma \tag{9-61}$$

$$b = \sigma^2 + \omega^2 \tag{9-62}$$

$S_{21}(s)$ is now put in the standard form of Eq. (9-47).

For a reactance network Eq. (9-48) applies:

$$m(s) = s^2 + b \tag{9-63}$$

$$n(s) = as \tag{9-64}$$

$m(s)$ and $n(s)$ are the even and odd parts of $P(s)$ and $Q(s)$.

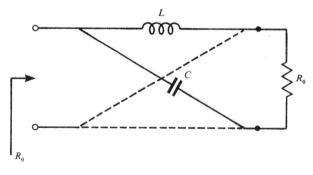

FIGURE 9-7
Lattice network for all-pass C network.

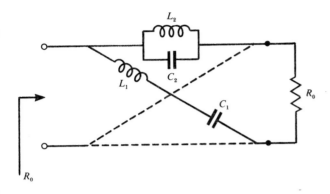

FIGURE 9-8
Lattice network for all-pass D network.

$Z_1(s)$ and $Z_2(s)$ are therefore determined by

$$Z_1(s) = \frac{s}{a} + \frac{b}{as} \tag{9-65}$$

$$Z_2(s) = \frac{R_0^2}{Z_1(s)} = \frac{1}{s/aR_0^2 + b/aR_0^2 s} \tag{9-66}$$

The all-pass D-lattice network is illustrated in Fig. 9-8 with

$$L_1 = a \tag{9-67}$$

$$C_1 = \frac{a}{b} \tag{9-68}$$

$$L_2 = \frac{aR_0^2}{b} \tag{9-69}$$

$$C_2 = \frac{1}{aR_0^2} \tag{9-70}$$

When the roots of an all-pass D network are real it may be synthesized by a cascade arrangement of two all-pass C networks. To illustrate this point consider the following all-pass D network.

$$S_{21}(s) = \frac{s^2 - 2s + 1}{s^2 + 2s + 1} \tag{9-71}$$

The roots of this transfer function are real. It may therefore be factored as

$$S_{21}(s) = \frac{(-s+1)^2}{(s+1)^2} \tag{9-72}$$

Suitable all-pass networks for $S_{21}(s)$ are shown in Fig. 9-9a and b.

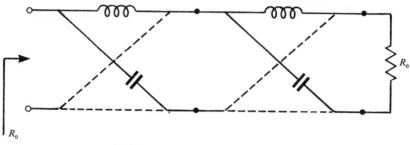

(a) All-pass C network

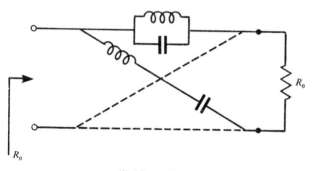

(b) All-pass D network

FIGURE 9-9
Realization of

$$S_{21}(s) = \frac{s^2 - 2s + 1}{s^2 + 2s + 1}$$

As a second example of the cascade synthesis of an all-pass network in terms of C and D networks consider the following all-pass function

$$S_{21}(s) = \frac{-s^3 + 2s^2 - 2s + 1}{s^3 + 2s^2 + 2s + 1} \tag{9-73}$$

This function may be factored as follows

$$S_{21}(s) = \frac{(-s+1)\,(s^2 - s + 1)}{(s+1)\,(s^2 + s + 1)} \tag{9-74}$$

The two-lattice realizations for this function are left as an exercise for the reader.

MAXIMALLY FLAT DELAY

Since an ideal delay (or phase) characteristic is not possible some kind of approximation to the required response is desirable. One approximation that leads to a maximally flat delay characteristic at the origin will now be described.

The ideal phase characteristic of a network is

$$S_{21}(s) = 1 \, e^{-s} \tag{9-75}$$

As a preamble to a Maclaurin expansion of this function it is useful to write the preceding equation as

$$S_{21}(s) = \frac{1}{\cosh s + \sinh s} \tag{9-76}$$

Since the denominator polynomial of $S_{21}(s)$ must be Hurwitz it may be written as

$$S_{21}(s) = \frac{a}{m(s) + n(s)} \tag{9-77}$$

where a is a constant chosen such that Eq. (9-77) has unit amplitude at $s = 0$ and $m(s)$ and $n(s)$ are obtained by forming a continued fraction expansion of the series forms of $\cosh s$ and $\sinh s$ and truncating the ensuing infinite expansion. The coefficients of $m(s)$ and $n(s)$ depend on the number of terms that are retained. It will be seen that this choice of $m(s)$ and $n(s)$ leads to a maximally flat delay transfer characteristic. This method is due to Orchard.

The series expansions of $\cosh s$ and $\sinh s$ are

$$\cosh s = 1 + \frac{s^2}{2!} + \frac{s^4}{4!} + \frac{s^6}{6!} + \cdots \tag{9-78}$$

$$\sinh s = s + \frac{s^3}{3!} + \frac{s^5}{5!} + \frac{s^7}{7!} + \cdots \tag{9-79}$$

The two preceding equations may now be used to construct $m(s)/n(s)$:

$$\frac{m(s)}{n(s)} = \frac{\cosh s}{\sinh s} = \frac{1}{s} + \cfrac{1}{\cfrac{3}{s} + \cfrac{1}{\cfrac{5}{s} + \cfrac{1}{\cfrac{7}{s} + \cdots}}} \tag{9-80}$$

The approximation employed is that of truncating this series with the first $(2n - 1)$ term. For example, for $n = 3$ the truncated continued fraction is

$$\frac{m(s)}{n(s)} = \frac{1}{s} + \cfrac{1}{\cfrac{3}{s} + \cfrac{1}{\cfrac{5}{s}}} = \frac{6s^2 + 15}{s^3 + 15s} \tag{9-81}$$

Thus

$$m(s) = 6s^2 + 15 \tag{9-82}$$

$$n(s) = s^3 + 15s \tag{9-83}$$

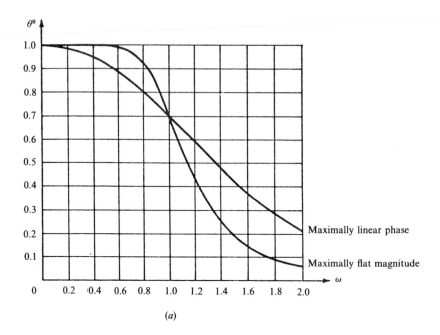

(a)

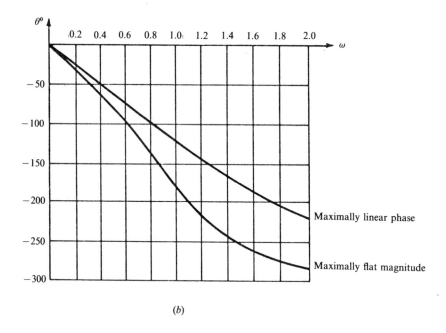

(b)

FIGURE 9-10
(a) Maximally flat phase characteristic for $n = 4$. (b) Amplitude characteristic for maximally flat phase network with $n = 4$.

TABLE 9-1
Coefficients of $B_n(s) = b_0 + b_1 s + \cdots + b_n s^n$

Order n	b_0	b_1	b_2	b_3	b_4	b_5	b_6	b_7
0	1							
1	1	1						
2	3	3	1					
3	15	15	6	1				
4	105	105	45	10	1			
5	945	945	420	105	15	1		
6	10 395	10 395	4 725	1 260	210	21	1	
7	135 135	135 135	62 370	17 325	3 150	378	28	1

and
$$S_{21}(s) = \frac{a}{s^3 + 6s^2 + 15s + 15} \tag{9-84}$$

where a is now found by ensuring that $S_{21}(s)$ is unity at $s = 0$. The final result is

$$S_{21}(s) = \frac{1}{s^3/15 + 6s^2/15 + s + 1} \tag{9-85}$$

To show that $S_{21}(s)$ has a maximally flat delay characteristic it is necessary to evaluate the phase and delay associated with the preceding function. This is left as an exercise for the student.

The denominator polynomials for $n = 1$ and 2 are

$$Q_1(s) = m(s) + n(s) = s + 1 \tag{9-86}$$

$$Q_2(s) = m(s) + n(s) = s^2 + 3s + 3 \tag{9-87}$$

A recurrence formula for $Q_n(s)$ is

$$Q_n(s) = (2n - 1)Q_{n-1}(s) + s^2 Q_{n-2}(s) \tag{9-88}$$

It may be shown that Q_n is related to a class of Bessel functions. Filters constructed using the above polynomials are sometimes referred to as Bessel ones. Table 9-1 depicts the first seven polynomials associated with $Q_n(s)$.

Figure 9-10a and b compares the amplitude and phase characteristics of $n = 4$ Butterworth and maximally flat delay networks.

PROBLEMS

9-1 Show that

$$S_{21}(s) = \frac{(1-s)^n}{(1+s)^n}$$

can be synthesized as a cascade arrangement of all-pass C networks.

9-2 Obtain the coefficients of the denominator polynomial of a maximally flat delay two-port network for $n = 3$, 4 and 5 by making use of Eqs (9-86) and (9-87). Check that the results agree with the polynomials given in Table 9-1.

9-3 Obtain an $n = 2$ low-pass network having a maximally flat delay.

9-4 Repeat Prob. 9-3 for $n = 3$.

9-5 Verify that the Bessel polynomials for $n = 1, 2$ and 3 lead to maximally flat delay characteristics.

9-6 Show that

$$\beta(s) = \tfrac{1}{2} \ln \frac{S_{21}(s)}{S_{21}(-s)}$$

9-7 Verify that Eqs (9-7), (9-11) and (9-15) satisfy (9-3).

9-8 Check that Eqs (9-8), (9-12) and (9-16) are consistent with (9-4).

9-9 Show that the time delays for Bessel filters of order 1 and 2 are

$$T_1 = \frac{1}{1 + \omega^2}$$

$$T_2 = \frac{9 + 3\omega^2}{9 + 3\omega^2 + \omega^4}$$

BIBLIOGRAPHY

Orchard, H. J., 'The roots of maximally flat delay polynomials', *IEEE Trans on Circuit Theory*, Vol. CT-12, pp. 452–454, September 1965.

CHAPTER
10

SYNTHESIS OF ALL-POLE LOW-PASS ELECTRICAL FILTERS

INTRODUCTION

Electrical filters are one of the most important class of circuits used in electrical engineering. Depending on system requirement they may have low-pass, band-pass, band-stop or high-pass frequency characteristics. Modern filter theory, due to Darlington, relies on synthesis, whereby an amplitude squared transfer function specification is realized as a two-port reactance network terminated in a 1-Ω resistance.

The first problem in network synthesis is to construct an amplitude squared transfer function to meet the specification. This problem has been tackled in Chapter 8. The second part of network synthesis consists of finding methods whereby the corresponding network can be deduced. This chapter is mainly concerned with the latter task for filters with Butterworth or Chebyshev specifications. The conventional approach to filter theory is to develop a low-pass prototype ladder network normalized to a 1-Ω termination and a cutoff frequency of 1 rad/s. Frequency and impedance transformations are then used to derive high-pass, band-pass and band-stop filters. This approach avoids the need to set down a multiplicity of tabulated results for the many different specifications met in practice.

Whereas the attenuation poles (transmission zeros) of Butterworth and Chebyshev low-pass amplitude squared transfer functions all lie at infinity (all-pole functons), this need not be the case in general. Inverse Chebyshev and quasi elliptic filters amplitude squared transfer functions with attenuation poles at finite frequencies are two possible examples. In the case of all-pole networks a first Cauer form extraction of the poles at infinity is all that is necessary to realize the required network. This may be understood by noting that the extraction of such poles does not impair the realizability of the remaining part of the function; nor does it affect the remaining value of its real part, since the real part contributed by such poles is identically zero. It can be shown, on the other hand, that the removal of complex poles may impair the realizability of the remainder function. The realization of networks with finite attenuation poles is separately treated in Chapter 11.

DARLINGTON SYNTHESIS PROCEDURE

In the approximation problem the specification is usually stated in terms of an amplitude squared transfer function rather than as a simple amplitude function. The first task in the synthesis problem is therefore to deduce the input driving immittance of the network from a knowledge of its magnitude squared transfer specification. For the lossless two-port network between 1-Ω terminations in Fig. 10-1 S_{11} and S_{21} are related by the unitary condition in Chapter 1 by

$$S_{11}(s)S_{11}(-s) = 1 - S_{21}(s)S_{21}(-s) \qquad (10\text{-}1)$$

The main problem is to separate $S_{11}(s)S_{21}(-s)$ into its constituents. This is done by dividing the poles and zeros between the LHP and RHP. The input immittance of the network is then given by the following standard bilinear relationship:

$$Y(s) = \frac{1 - S_{11}(s)}{1 + S_{11}(s)} \qquad (10\text{-}2)$$

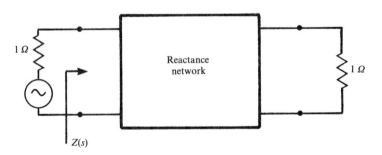

FIGURE 10-1
Doubly terminated filter circuit.

The final step involves construction of $Y(s)$ as a reactance ladder network terminated in a 1-Ω resistor in such a way as to realize the attenuation poles of the specification.

According to the discussion in Chapter 1, $S_{11}(s)$ is bounded real, the denominator polynomial of $S_{11}(s)$ is therefore Hurwitz, but the numerator one need not be Hurwitz. Thus, the poles of $S_{11}(s)$ are on the LHP and the zeros of $S_{11}(s)$ may have any complex plane location. The poles are therefore taken as the LHP poles of $S_{11}(s)S_{11}(-s)$, whereas the zeros are chosen in conjugate pairs from either the LHP or RHP zeros of $S_{11}(s)S_{11}(-s)$. For a minimum phase network the zeros are located on the LHP.

SYNTHESIS OF LADDER NETWORKS WITH BUTTERWORTH AMPLITUDE CHARACTERISTICS

Since the ideal low-pass characteristic is not possible some type of approximation to it is necessary. One magnitude squared transfer function due to Butterworth has been discussed in Chapter 8:

$$|S_{21}(j\omega)|^2 = \frac{1}{1 + (\omega^2)^n} \qquad (10\text{-}3)$$

This function has a 3-dB point at $\omega = 1$ for all n, and its amplitude response falls off at a rate of $6n$ decibels per octave. It also has the property that its first $2n - 1$ derivatives are equal to zero at the origin $\omega = 0$. This amplitude response is reproduced from Chapter 8 in Fig. 10-2 (expressed in decibels).

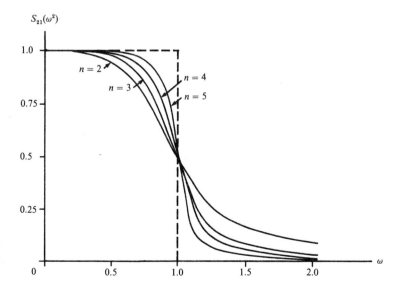

FIGURE 10-2
Low-pass Butterworth transfer characteristic.

Writing $S_{11}(s)$ in terms of $|S_{21}(j\omega)|^2$ by replacing $j\omega$ by s and having recourse to the unitary condition yields

$$S_{11}(s)S_{11}(-s) = \frac{(-s^2)^n}{1 + (-s^2)^n} \tag{10-4}$$

The poles of Eq. (10-4) are given by the roots of the denominator polynomial

$$1 + (-s^2)^n = 0 \tag{10-5a}$$

The roots of this type of polynomial may be deduced by having recourse to a root-finding subroutine. In this instance, however, the $2n$ roots of the preceding equation may be determined by writing Eq. (10-5a) as

$$(-1)^n p_k^{2n} = -1 \qquad \text{for } k = 1, 2, \ldots, 2n \tag{10-5b}$$

and resorting to the following identities:

$$-1 = \exp[j(2k - 1)\pi] \qquad \text{for } k = 1, 2, \ldots, 2n \tag{10-5c}$$

$$(-1)^n = \exp[j(-n\pi)] \tag{10-5d}$$

The required result is

$$p_k = \exp\left[j\left(\frac{2k + n - 1}{2n}\right)\pi\right] \qquad \text{for } k = 1, 2, \ldots, 2n \tag{10-5e}$$

The roots of this polynomial reside on a unit circle in the s plane and have symmetry with respect to both the real and imaginary axis. For n odd, a pair of roots lie on the real axis, but no roots lie on the imaginary axis for both n even or odd. The real and imaginary parts of Eq. (10-5e) are

$$\sigma_k = \cos\left(\frac{2k + n - 1}{2n}\pi\right) \tag{10-6a}$$

$$\omega_k = \sin\left(\frac{2k + n - 1}{2n}\pi\right) \tag{10-6b}$$

with $k = 1, 2, 3, 2n$.

The zeros of Eq. (10-4) are the roots of the numerator polynomial

$$(-s^2)^n = 0 \tag{10-7a}$$

and reside at the origin. Two possibilities are

$$\pm s^n = 0 \tag{10-7b}$$

As an example of the construction of $S_{11}(s)$ from an amplitude squared function consider the development of an $n = 3$ one with a Butterworth characteristic.

The poles of the amplitude squared reflection coefficient are the roots of Eq. (10-5a) with $n = 3$:

$$1 + (-s^2)^3 = 0$$

The roots of this equation may be deduced from Eq. (10-5e) or Eqs (10-6a,b). The result is

$$p_1 = \exp\left(j\,\frac{2\pi}{3}\right) = \cos 120 + j \sin 120 = -0.5 + j\,\frac{\sqrt{3}}{2}$$

$$p_2 = \exp(j\pi) = \cos 180 + j \sin 180 = -1$$

$$p_3 = \exp\left(j\,\frac{4\pi}{3}\right) = \cos 240 + j \sin 240 = -0.5 - j\,\frac{\sqrt{3}}{2}$$

$$p_4 = \exp\left(j\,\frac{5\pi}{3}\right) = \cos 300 + j \sin 300 = 0.5 - j\,\frac{\sqrt{3}}{2}$$

$$p_5 = 1$$

$$p_6 = \exp\left(j\,\frac{\pi}{3}\right) = \cos 60 + j \sin 60 = 0.5 + j\,\frac{\sqrt{3}}{2}$$

The above poles are displayed on a unit circle in Fig. 10-3. Since $S_{11}(s)$ is bounded real, its denominator polynomial is strictly Hurwitz. It is therefore constructed in terms of p_1, p_2 and p_3 and either choice of polynomials in Eq. (10-7b):

$$S_{11}(s) = \frac{\pm s^3}{(s+1)(s+0.5+j\sqrt{3}/2)(s+0.5-j\sqrt{3}/2)}$$

or

$$S_{11}(s) = \frac{\pm s^3}{1 + 2s + 2s^2 + s^3}$$

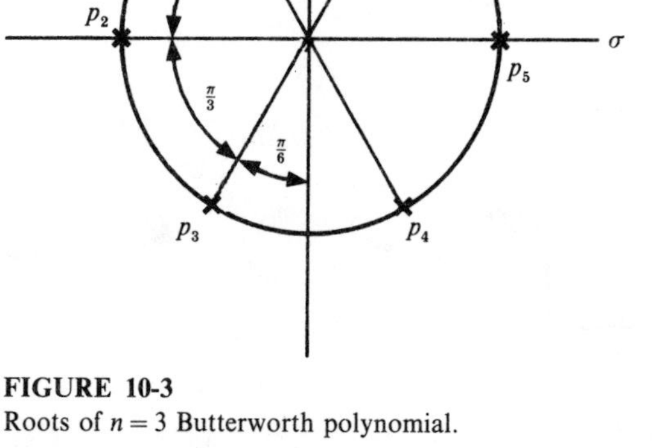

FIGURE 10-3
Roots of $n = 3$ Butterworth polynomial.

Once $S_{11}(s)$ has been deduced the input admittance of the network can be formed and the network can be synthesized by having recourse to Darlington's method. If the negative sign is used for $S_{11}(s)$ the result is

$$\frac{Y_{in}(s)}{1} = \frac{2s^3 + 2s^2 + 2s + 1}{2s^2 + 2s + 1}$$

A canonical realization for $Y_{in}(s)$ may now be developed by performing a Cauer-type ladder expansion of the admittance function that realizes the attenuation poles of $S_{21}(s)$:

$$
\begin{array}{r}
2s^2 + 2s + 1 \overline{)\ 2s^3 + 2s^2 + 2s + 1}\ (s \to y \\
\underline{2s^3 + 2s^2 + s} \\
s + 1 \overline{)\ 2s^2 + 2s + 1}\ (2s \to y \\
\underline{2s^2 + 2s} \\
1 \overline{)\ s + 1}\ (s \to y \\
\underline{s} \\
1 \overline{)\ 1}\ (1 \to R \\
\underline{1}
\end{array}
$$

The low-pass filter is thus synthesized as a two-port reactance network terminated in a 1-Ω resistor in the manner indicated in Fig. 10-4.

The general form for $S_{11}(s)$ is

$$S_{11}(s) = \frac{\pm s^n}{a_n s^n + a_{n-1} s^{n-1} + \cdots + a_1 s + a_0}$$

Table 10-1 depicts the coefficients of the denominator polynomial of $S_{11}(s)$.

Taking the positive sign for $S_{11}(s)$ leads to

$$\frac{Y_{in}(s)}{1} = \frac{2s^2 + 2s + 1}{2s^3 + 2s^2 + 2s + 1}$$

Since $Y_{in}(s)$ has a zero at infinity it is necessary to form $Z_{in}(s)$ to extract a pole there:

$$\frac{Z_{in}(s)}{1} = \frac{2s^3 + 2s^2 + 2s + 1}{2s^2 + 2s + 1}$$

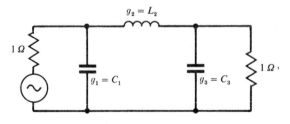

FIGURE 10-4
Butterworth $n = 3$ low-pass prototype for $S_{11}(s)$ realization.

TABLE 10-1
Coefficients of Butterworth polynomials
$$a(s) = a_0 + a_1 s + a_2 s^2 + \cdots + a_{n-1} s^{n-1} + a_n s^n$$

n	a_0	a_1	a_2	a_3	a_4	a_5	a_6	a_7	a_8	a_9	a_{10}
1	1	1	1								
2	1	1.414 21	1								
3	1	2.000 00	2.000 00	1							
4	1	2.613 13	3.414 21	2.613 13	1						
5	1	3.236 07	5.236 07	5.236 07	3.236 07	1					
6	1	3.863 79	7.461 62	9.141 62	7.464 10	3.863 70	1				
7	1	4.493 96	10.097 84	14.591 79	14.591 79	10.097 84	4.493 96	1			
8	1	5.125 83	13.137 07	21.846 15	25.688 36	21.846 15	13.137 07	5.125 83	1		
9	1	5.758 77	16.581 72	31.163 44	41.986 39	41.986 39	31.163 44	16.581 72	5.758 77	1	
10	1	6.392 45	20.431 73	42.802 06	64.882 40	74.233 43	64.882 40	42.802 06	20.431 73	6.392 45	1

Source: Matthaei, G. L., Young, L. and Jones, E. M. T., *Microwave Filters, Impedance-matching Networks and Coupling Structures*, Artech House, Norwood, MA, 1980..

A ladder expansion of $Z_{in}(s)$ in a first Cauer form which realizes the attenuation poles of the specification is

$$
2s^2 + 2s + 1 \;) \; 2s^3 + 2s^2 + 2s + 1 \; (\; s \\
\underline{2s^3 + 2s^2 + s} \\
s + 1 \;) \; 2s^2 + 2s + 1 \; (\; 2s \\
\underline{2s^2 + 2s} \\
1 \;) \; s + 1 \; (s \\
\underline{s} \\
1 \;) \; 1 \; (\; 1 \\
\underline{1} \\
0
$$

The ladder network associated with this solution is illustrated in Fig. 10-5. The transmission coefficients of the networks in Figs 10-4 and 10-5 differ by an all-pass network in the manner discussed in Eq. (9-21) in Chapter 9. The synthesis

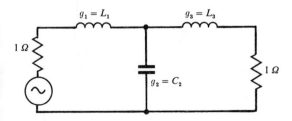

FIGURE 10-5
Butterworth $n = 3$ low-pass prototype for $-S_{11}(s)$ realization.

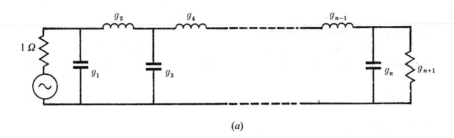

(a)

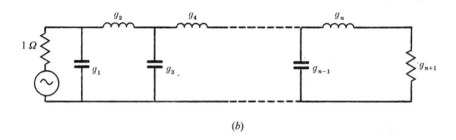

(b)

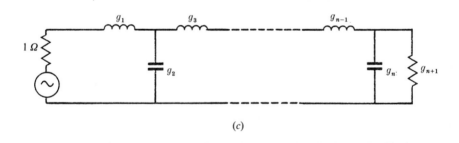

(c)

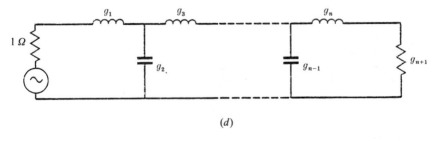

(d)

FIGURE 10-6
Canonical realization of normalized low-pass network of (a) $S_{11}(s)$ for n even and (b) n odd and (c) $-S_{11}(s)$ for n even and (d) n odd.

TABLE 10-2
Element values of Butterworth filters for $g_0 = g_{n+1} = 1\,\Omega$ and $\omega_0 = 1\,\text{rad/s}$

Value of n	g_1	g_2	g_3	g_4	g_5	g_6	g_7	g_8	g_9	g_{10}
1	2.0000									
2	1.4142	1.4142								
3	1.0000	2.0000	1.0000							
4	0.7654	1.8478	1.8478	0.7654						
5	0.6180	1.6180	2.0000	1.6180	0.6180					
6	0.5176	1.4142	1.9319	1.9319	1.4142	0.5176				
7	0.4450	1.2470	1.8019	2.0000	1.8019	1.2470	0.4450			
8	0.3902	1.1111	1.6629	1.9616	1.9616	1.1111	0.3902			
9	0.3473	1.0000	1.5321	1.8794	2.0000	1.8794	1.5321	1.0000	0.3473	
10	0.3129	0.9080	1.4142	1.7820	1.9754	1.9754	1.7820	1.4142	0.9080	0.3129

Source: as Table 10-1.

of filter networks with different values of n proceeds in a similar fashion but one recurrence formula for the elements of the filter is

$$g_k = 2 \sin\left(\frac{2k-1}{2n}\right) \qquad \text{for } k = 1, 2, \ldots, n \qquad (10\text{-}8)$$

For $n = 3$ this equation gives $g_1 = 1$, $g_2 = 2$, $g_3 = 1$ in agreement with the result derived above.

Figure 10-6 depicts the canonical realizations associated with $\pm S_{11}(s)$ for n even and odd and Table 10-2 gives the elements of this type of network for $n = 1$ to 10.

SYNTHESIS OF LADDER NETWORKS WITH CHEBYSHEV CHARACTERISTICS

The amplitude squared transfer function in the case of the Chebyshev approximation is defined in Chapter 8 by

$$|S_{21}(j\omega)|^2 = \frac{1}{1 + \varepsilon^2 T_n^2(\omega)} \qquad (10\text{-}9)$$

This amplitude response (expressed in decibels) is reproduced from Chapter 8 in Fig. 10-7. Invoking the unitary condition and analytic continuation gives

$$S_{11}(s)S_{11}(-s) = \frac{\varepsilon^2 T_n^2(s/j)}{1 + \varepsilon^2 T_n^2(s/j)} \qquad (10\text{-}10)$$

$S_{21}(\omega^2)$

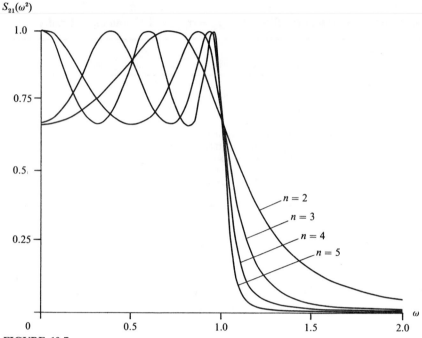

FIGURE 10-7
Low-pass Chebyshev transfer characteristic for $n = 3$ and 4.

It is now necessary to separate $S_{11}(s)$ and $S_{11}(-s)$ by finding the factors of the preceding equation:

$$1 + \varepsilon^2 T_n^2(s/\mathrm{j}) = 0 \qquad (10\text{-}11a)$$

$$\varepsilon^2 T_n^2 = 0 \qquad (10\text{-}11b)$$

The detailed solution to these characteristic equations will not be developed but the roots of the denominator and numerator polynomials may be given by

$$P_k = -\sin\left(\frac{2k-1}{2n}\pi\right)\sinh\left(\frac{1}{n}\sinh^{-1}\frac{1}{\varepsilon}\right)$$

$$+ \mathrm{j}\cos\left(\frac{2k-1}{2n}\pi\right)\cosh\left(\frac{1}{n}\cosh^{-1}\frac{1}{\varepsilon}\right) \quad \text{for } k = 1, 2, 3, \ldots, 2n \qquad (10\text{-}12a)$$

$$z_k = \mathrm{j}\cos\left(\frac{2k-1}{2n}\pi\right) \qquad k = 1, 2, 3, \ldots, 2n \qquad (10\text{-}12b)$$

respectively.

By way of an example of the development of a network with a Chebyshev characteristic, consider the construction of one of degree 3 with a ripple level of 1 dB.

The required poles are given by the roots of Eq. (10-11a) or (10-12a) as

$$p_1 = -\sin\frac{\pi}{6}\sinh\left(\frac{1}{3}\sinh^{-1}\frac{1}{\varepsilon}\right) + j\cos\frac{\pi}{6}\cosh\left(\frac{1}{3}\sinh^{-1}\frac{1}{\varepsilon}\right)$$

$$p_2 = -\sin\frac{3\pi}{6}\sinh\left(\frac{1}{3}\sinh^{-1}\frac{1}{\varepsilon}\right) + j\cos\frac{3\pi}{6}\cosh\left(\frac{1}{3}\sinh^{-1}\frac{1}{\varepsilon}\right)$$

$$p_3 = -\sin\frac{5\pi}{6}\sinh\left(\frac{1}{3}\sinh^{-1}\frac{1}{\varepsilon}\right) + j\cos\frac{5\pi}{6}\cosh\left(\frac{1}{3}\sinh^{-1}\frac{1}{\varepsilon}\right)$$

$$p_4 = -\sin\frac{7\pi}{6}\sinh\left(\frac{1}{3}\sinh^{-1}\frac{1}{\varepsilon}\right) + j\cos\frac{7\pi}{6}\cosh\left(\frac{1}{3}\sinh^{-1}\frac{1}{\varepsilon}\right)$$

$$p_5 = -\sin\frac{9\pi}{6}\sinh\left(\frac{1}{3}\sinh^{-1}\frac{1}{\varepsilon}\right) + j\cos\frac{9\pi}{6}\cosh\left(\frac{1}{3}\sinh^{-1}\frac{1}{\varepsilon}\right)$$

$$p_6 = -\sin\frac{11\pi}{6}\sinh\left(\frac{1}{3}\sinh^{-1}\frac{1}{\varepsilon}\right) + j\cos\frac{11\pi}{6}\cosh\left(\frac{1}{3}\sinh^{-1}\frac{1}{\varepsilon}\right)$$

The LHP poles are determined from the above set by p_1, p_2 and p_3:

$$p_1 = \frac{1}{2}\sinh\left(\frac{1}{3}\sinh^{-1}\frac{1}{\varepsilon}\right) + j\frac{\sqrt{3}}{2}\cosh\left(\frac{1}{3}\sinh^{-1}\frac{1}{\varepsilon}\right)$$

$$p_2 = -\sinh\left(\frac{1}{3}\sinh^{-1}\frac{1}{\varepsilon}\right)$$

$$p_3 = \frac{1}{2}\sinh\left(\frac{1}{3}\sinh^{-1}\frac{1}{\varepsilon}\right) - j\frac{\sqrt{3}}{2}\cosh\left(\frac{1}{3}\sinh^{-1}\frac{1}{\varepsilon}\right)$$

The roots of this polynomial lie on an ellipse in the manner indicated in Fig. 10-8.

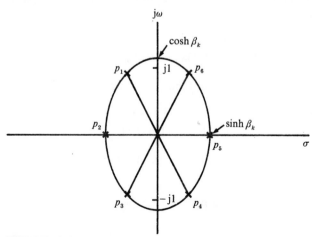

FIGURE 10-8
Roots of $n = 3$ Chebyshev polynomial.

For a 1-dB ripple level ε is fixed as

$$\varepsilon = 0.509$$

and p_1, p_2 and p_3 are given by

$$p_1 = -0.494\,17$$

$$p_2 = -0.247\,08 + j0.965\,99$$

$$p_3 = -0.247\,08 - j0.965\,99$$

The denominator polynomial of $S_{11}(s)$ is therefore described by

$$s^3 + 0.988\,34s^2 + 1.238\,41s + 0.491\,31$$

Table 10-3 summarizes the coefficients of this polynomial for $n = 1$ to 10 for ripple levels of $\frac{1}{2}$, 1 and 2 dB. The numerator polynomial $P(s)$ may be directly constructed from either the LHP or RHP roots of Eq. (10-11b).

$$\pm T_3(s/j) = 0$$

The roots of the preceding equation may be evaluated by Eq. (10-12b) or by having recourse to the recurrence form for $T_n(\omega)$ in Chapter 8. Making use of the latter relationship with $\varepsilon = 0.509$ leads to

$$P(s) = \pm 0.509(4s^3 + 3s)$$

The required form for $S_{11}(s)$ is therefore

$$S_{11}(s) = \frac{P(s)}{Q(s)} = \frac{\pm 0.509(4s^3 + 3s)\alpha}{s^3 + 0.988\,34s^2 + 1.2384s + 0.491\,31}$$

The multiplication constant α is introduced to ensure that $S_{11}(s)$ is bounded real (unity) at $s = j\infty$. The required result is

$$S_{11}(s) = \frac{\pm(s^3 + 0.75s)}{s^3 + 0.988\,34s^2 + 1.2384s + 0.491\,31}$$

Taking the negative sign yields the input admittance of the network as

$$\frac{Y_{in}(s)}{1} = \frac{2s^3 + 0.988\,34s^2 + 1.988\,41s + 0.491\,31}{0.988\,34s^2 + 0.488\,41s + 0.491\,31}$$

Forming a Cauer ladder network by removal of poles at infinity which realizes the attenuation poles of $S_{21}(s)$ gives

$$C_1 = 2.024 \text{ F}$$

$$L_2 = 0.994 \text{ H}$$

$$C_3 = 2.024 \text{ F}$$

TABLE 10-3
Coefficients of the Chebyshev polynomials
$$b(s) = b_0 + b_1 s + b_2 s^2 + \cdots + b_{n-1} s^{n-1} + b_n a_n$$

$\frac{1}{2}$-dB ripple ($\varepsilon = 0.349\,31$)

n	b_0	b_1	b_2	b_3	b_4	b_5	b_6	b_7	b_8	b_9	b_{10}
1	2.862 78	1									
2	1.516 20	1.425 63	1								
3	0.715 69	1.534 90	1.252 91	1							
4	0.379 05	1.025 46	1.716 87	1.197 39	1						
5	0.178 92	0.752 52	1.309 58	1.937 37	1.172 49	1					
6	0.094 76	0.432 37	1.171 86	1.589 76	2.171 85	1.159 18	1				
7	0.044 73	0.282 07	0.755 65	1.647 90	1.869 41	2.412 65	1.151 22	1			
8	0.023 69	0.152 54	0.573 56	1.148 59	2.184 02	2.149 22	2.656 75	1.146 08	1		
9	0.011 18	0.094 12	0.340 82	0.983 62	1.611 39	2.781 50	2.429 33	2.902 73	1.142 57	1	
10	0.005 92	0.049 29	0.237 27	0.626 97	1.527 43	2.144 24	3.440 93	2.709 74	3.149 88	1.140 07	1

1-dB ripple ($\varepsilon = 0.508\,85$)

n	b_0	b_1	b_2	b_3	b_4	b_5	b_6	b_7	b_8	b_9	b_{10}
1	1.965 23	1									
2	1.102 51	1.097 73	1								
3	0.491 31	1.238 41	0.988 34	1							
4	0.275 63	0.742 62	1.453 93	0.952 81	1						
5	0.122 83	0.580 53	0.974 40	1.688 82	0.936 82	1					
6	0.068 91	0.307 68	0.939 35	1.202 14	1.930 83	0.928 25	1				
7	0.030 71	0.213 67	0.548 62	1.357 54	1.428 79	2.176 08	0.923 12	1			
8	0.017 23	0.107 35	0.447 83	0.846 82	1.836 90	1.655 16	2.423 03	0.919 81	1		
9	0.007 68	0.070 61	0.244 19	0.786 31	1.201 61	2.378 12	1.881 48	2.670 95	0.917 55	1	
10	0.004 31	0.034 50	0.182 45	0.455 39	1.244 49	1.612 99	2.981 51	2.107 85	2.919 47	0.915 93	1

2-dB ripple ($\varepsilon = 0.764\,78$)

n	b_0	b_1	b_2	b_3	b_4	b_5	b_6	b_7	b_8	b_9	b_{10}
1	1.307 56	1									
2	0.823 06	0.803 82	1								
3	0.326 89	1.022 19	0.737 82	1							
4	0.205 77	0.516 80	1.256 48	0.716 22	1						
5	0.081 72	0.459 35	0.693 48	1.499 54	0.706 46	1					
6	0.051 44	0.210 27	0.771 46	0.867 02	1.745 86	0.701 23	1				
7	0.020 42	0.166 09	1.144 44	1.039 22	1.993 53	0.697 89	1				
8	0.012 86	0.072 94	0.358 70	1.598 22	1.579 58	1.211 71	2.242 25	0.696 07	1		
9	0.005 11	0.054 38	0.168 45	0.644 47	0.856 87	2.076 75	1.383 75	2.491 29	0.694 68	1	
10	0.003 22	0.023 34	0.144 01	0.317 76	1.038 91	1.158 53	2.636 25	1.555 74	2.740 60	0.693 69	1

Source: as Table 10-1.

Taking the positive sign for $S_{11}(s)$ gives

$$L_1 = 2.024 \text{ H}$$

$$C_2 = 0.994 \text{ F}$$

$$L^3 = 2.024 \text{ H}$$

Figures 10-9 and 10-10 depict the ladder networks for the above two solutions. These two networks again differ by an all-pass network as discussed in connection with Figs 10-4 and 10-5. For different values of n, g_k may be expressed by

$$g_1 = \frac{2\sin(\pi/2n)}{\sinh[(1/n)\sinh^{-1}(1/\varepsilon)]} \tag{10-13a}$$

and

$$g_k g_{k+1} = \frac{\left(\dfrac{2k-1}{2n}\pi\right)\sin\left(\dfrac{2k+1}{2n}\pi\right)}{\sinh^2\left(\dfrac{1}{n}\sinh^{-1}\dfrac{1}{\varepsilon}\right) + \sin^2\left(\dfrac{k\pi}{n}\right)} \tag{10-13b}$$

with $k = 1, 2, \ldots, (n-1)$. The filter is symmetrical, for n odd:

$$g_1 = g_n, \, g_2 = g_{n-1}, \ldots, g_k = g_{n-k+1}, \ldots \tag{10-13c}$$

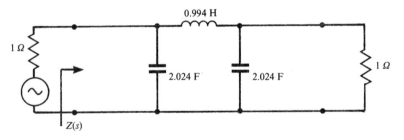

FIGURE 10-9
Ladder network for $n = 3$ Chebyshev filter with 1-dB ripple for $S_{11}(s)$ realization.

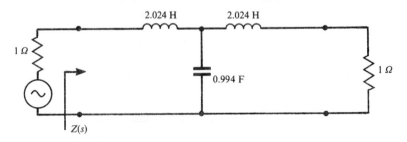

FIGURE 10-10
Ladder network for $n = 3$ Chebyshev filter with 1-dB ripple for $-S_{11}(s)$ realization.

Table 10-4 displays the g values for different values of ε for n between 1 and 10. If n is odd,

$$|S_{21}(0)| = 1 \qquad (10\text{-}14a)$$

and the terminating resistance g_{n+1} in Fig. 10-6 is

$$g_{n+1} = 1 \qquad (10\text{-}14b)$$

If n is even, $|S_{21}(0)|$ is

$$|S_{21}(0)|^2 = \frac{1}{1 + \varepsilon^2} \qquad (10\text{-}15a)$$

TABLE 10-4
Element values of Chebyshev filters having $g_0 = 1$ and $\omega_0 = 1$ rad/s

0.01-dB ripple

Value of n	g_1	g_2	g_3	g_4	g_5	g_6	g_7	g_8	g_9	g_{10}	g_{11}
1	0.0960	1.0000									
2	0.4488	0.4077	1.1007								
3	0.6291	0.9702	0.6291	1.0000							
4	0.7128	1.2003	1.3212	0.6476	1.1007						
5	0.7563	1.3049	1.5773	1.3049	0.7563	1.0000					
6	0.7813	1.3600	1.6896	1.5350	1.4970	0.7098	1.1007				
7	0.7969	1.3924	1.7481	1.6331	1.7481	1.3924	0.7969	1.0000			
8	0.8072	1.4130	1.7824	1.6833	1.8529	1.6193	1.5554	0.7333	1.1007		
9	0.8144	1.4270	1.8043	1.7125	1.9057	1.7125	1.8043	1.4270	0.8144	1.0000	
10	0.8196	1.4369	1.8192	1.7311	1.9362	1.7590	1.9055	1.6527	1.5817	0.7446	1.1007

0.1-dB ripple

Value of n	g_1	g_2	g_3	g_4	g_5	g_6	g_7	g_8	g_9	g_{10}	g_{11}
1	0.3052	1.0000									
2	0.8430	0.6220	1.3554								
3	1.0315	1.1474	1.0315	1.0000							
4	1.1088	1.3061	1.7703	0.8180	1.3554						
5	1.1468	1.3712	1.9750	1.3712	1.1468	1.0000					
6	1.1681	1.4039	2.0562	1.5170	1.9029	0.8618	1.3554				
7	1.1811	1.4228	2.0966	1.5733	2.0966	1.4228	1.1811	1.0000			
8	1.1897	1.4346	2.1199	1.6010	2.1699	1.5640	1.9444	0.8778	1.3554		
9	1.1956	1.4425	2.1345	1.6167	2.2053	1.6167	2.1345	1.4425	1.1956	1.0000	
10	1.1999	1.4481	2.1444	1.6265	2.2253	1.6418	2.2046	1.5821	1.9628	0.8853	1.3554

0.2-dB ripple

Value of n	g_1	g_2	g_3	g_4	g_5	g_6	g_7	g_8	g_9	g_{10}	g_{11}
1	0.4342	1.0000									
2	1.0378	0.6745	1.5386								
3	1.2275	1.1525	1.2275	1.0000							
4	1.3028	1.2844	1.9761	0.8468	1.5386						
5	1.3394	1.3370	2.1660	1.3370	1.3394	1.0000					
6	1.3598	1.3632	2.2394	1.4555	2.0974	0.8838	1.5386				
7	1.3722	1.3781	2.2756	1.5001	2.2756	1.3781	1.3722	1.0000			
8	1.3804	1.3875	2.2963	1.5217	2.3413	1.4925	2.1349	0.8972	1.5386		
9	1.3860	1.3938	2.3093	1.5340	2.3728	1.5340	2.3093	1.3938	1.3860	1.0000	
10	1.3901	1.3983	2.3181	1.5417	2.3904	1.5536	2.3720	1.5066	2.1514	0.9034	1.5386

TABLE 10-4 (*continued*)

0.5-dB ripple

Value of n	g_1	g_2	g_3	g_4	g_5	g_6	g_7	g_8	g_9	g_{10}	g_{11}
1	0.6986	1.0000									
2	1.4029	0.7071	1.9841								
3	1.5963	1.0967	1.5963	1.0000							
4	1.6703	1.1926	2.3661	0.8419	1.9841						
5	1.7058	1.2296	2.5408	1.2296	1.7058	1.0000					
6	1.7254	1.2479	2.6064	1.3137	2.4758	0.8696	1.9841				
7	1.7372	1.2583	2.6381	1.3444	2.6381	1.2583	1.7372	1.0000			
8	1.7451	1.2647	2.6564	1.3590	2.6964	1.3389	2.5093	0.8796	1.9841		
9	1.7504	1.2690	2.6678	1.3673	2.7239	1.3673	2.6678	1.2690	1.7504	1.0000	
10	1.7543	1.2721	2.6754	1.3725	2.7392	1.3806	2.7231	1.3485	2.5239	0.8842	1.9841

1.0-dB ripple

1	1.0177	1.0000									
2	1.8219	0.6850	2.6599								
3	2.0236	0.9941	2.0236	1.0000							
4	2.0991	1.0644	2.8311	0.7892	2.6599						
5	2.1349	1.0911	3.0009	1.0911	2.1349	1.0000					
6	2.1546	1.1041	3.0634	1.1518	2.9367	0.8101	2.6599				
7	2.1664	1.1116	3.0934	1.1736	3.0934	1.1116	2.1664	1.0000			
8	2.1744	1.1161	3.1107	1.1839	3.1488	1.1696	2.9685	0.8175	2.6599		
9	2.1797	1.1192	3.1215	1.1897	3.1747	1.1897	3.1215	1.1192	2.1797	1.0000	
10	2.1836	1.1213	3.1286	1.1933	3.1890	1.1990	3.1738	1.1763	2.9824	0.8210	2.6599

2.0-dB ripple

1	1.5296	1.0000									
2	2.4881	0.6075	4.0957								
3	2.7107	0.8327	2.7107	1.0000							
4	2.7925	0.8806	3.6063	0.6819	4.0957						
5	2.8310	0.8985	3.7827	0.8985	2.8310	1.0000					
6	2.8521	0.9071	3.8467	0.9393	3.7151	0.6964	4.0957				
7	2.8655	0.9119	3.8780	0.9535	3.8780	0.9119	2.8655	1.0000			
8	2.8733	0.9151	3.8948	0.9605	3.9335	0.9510	3.7477	0.7016	4.0957		
9	2.8790	0.9171	3.9056	0.9643	3.9598	0.9643	3.9056	0.9171	2.8790	1.0000	
10	2.8831	0.9186	3.9128	0.9667	3.9743	0.9704	3.9589	0.9554	3.7619	0.7040	4.0957

3.0-dB ripple

1	1.9953	1.0000									
2	3.1013	0.5339	5.8095								
3	3.3487	0.7117	3.3487	1.0000							
4	3.4389	0.7483	4.3471	0.5920	5.8095						
5	3.4817	0.7618	4.5381	0.7618	3.4817	1.0000					
6	3.5045	0.7685	4.6061	0.7929	4.4641	0.6033	5.8095				
7	3.5182	0.7723	4.6386	0.8039	4.6386	0.7723	3.5182	1.0000			
8	3.5277	0.7745	4.6575	0.8089	4.6990	0.8018	4.4990	0.6073	5.8095		
9	3.5340	0.7760	4.6692	0.8118	4.7272	0.8118	4.6692	0.7760	3.5340	1.0000	
10	3.5384	0.7771	4.6768	0.8136	4.7425	0.8164	4.7260	0.8051	4.5142	0.6091	5.8095

Source: as Table 10-1.

and the terminating resistance g_{n+1} in Fig. 10-6 is

$$g_{n+1} = (\varepsilon + \sqrt{1 + \varepsilon^2})^2 \qquad (10\text{-}15b)$$

or $$g_{n+1} = \frac{1}{(\varepsilon + \sqrt{1 + \varepsilon^2})^2} \qquad (10\text{-}15c)$$

depending on whether $S_{11}(s)$ is positive or negative. For a 1-dB ripple, ε is 0.509 and $g_{n+1} = 2.6599$ in agreement with Table 10-4. The two conditions in Eqs (10-15b, c) are deduced by using the relationship for $|S_{11}(0)|$ below:

$$|S_{11}(0)| = \frac{1 - g_{n+1}}{1 + g_{n+1}} \qquad (10\text{-}16)$$

to form, with the aid of the unitary condition, a second relationship for $|S_{21}(0)|$ in terms of g_{n+1}:

$$|S_{21}(0)|^2 = \frac{4g_{n+1}}{(1 + g_{n+1})^2} \qquad (10\text{-}17)$$

The required results in Eqs (10-15b, c) are then deduced by equating Eqs (10-15a) and (10-17).

SYNTHESIS OF LADDER NETWORKS WITH BESSEL AMPLITUDE CHARACTERISTICS

The synthesis of the Bessel amplitude squared transfer function (maximally flat phase filter) as a two-port reactance network terminated in a 1-Ω resistor follows directly the method employed in the cases of the Butterworth and Chebyshev realizations. Taking $n = 2$ as an example, and making use of Table 9-1 in Chapter 9 gives

$$S_{21}(s)S_{21}(-s) = \frac{\alpha^2}{(s^2 + 3s + 3)(s^2 - 3s + 3)}$$

$$S_{11}(s) = \frac{\pm s(s + \sqrt{3})}{s^2 + 3s + 3}$$

Employing the unitary condition with $\alpha^2 = 9$ to ensure that $S_{21}(s)$ and $S_{11}(s)$ are bounded real leads to

$$S_{11}(s)S_{11}(-s) = \frac{s^2(s^2 - 3)}{(s^2 + 3s + 3)(s^2 - 3s + 3)}$$

Forming $S_{11}(s)$ from the LHP poles and zeros of $S_{11}(s)S_{11}(-s)$ yields

$$S_{11}(s) = \frac{\pm s(s + \sqrt{3})}{s^2 + 3s + 3}$$

TABLE 10-5

Element values of maximally flat group-delay filters with $g_0 = g_{n+1} = 1\,\Omega$ and $\omega_0 = 1\,\text{rad/s}$

n	g_1	g_2	g_3	g_4	g_5	g_6	g_7
1	2.000						
2	1.577	0.423					
3	1.255	0.553	0.192				
4	1.060	0.512	0.318	0.110			
5	0.930	0.458	0.331	0.209	0.072		
6	0.838	0.412	0.316	0.236	0.418	0.051	
7	0.768	0.347	0.294	0.238	0.178	0.110	0.038

Source: Orchard, H. J., 'Formulas for ladder filters', *Wireless Engr*, Vol. 30, pp. 257–353, September 1939.

Using the negative sign for $S_{11}(s)$ in constructing $Y(s)$ gives

$$Y(s) = \frac{2s^2 + 4.732s + 3}{1.268s + 3}$$

A Cauer form expansion of $Y(s)$ is

$$
\begin{array}{r}
1.268s + 3 \,\overline{)\, 2s^2 + 4.732s + 3} \,(1.577s \\
\underline{2s^2 + 4.732s} \\
3 \,\overline{)\, 1.268s + 3} \,(0.423s \\
\underline{1.268s} \\
3 \,\overline{)\, 3} \,(1 \\
\underline{3} \\
0
\end{array}
$$

Thus $g_1 = 1.577$, $g_2 = 0.423$ and $g_3 = 1$, in agreement with Table 10-5.

PROBLEMS

10-1 Obtain Butterworth low-pass prototypes for $n = 2$ and 4.

10-2 Construct Chebyshev low-pass prototypes for $n = 2$ and 4 for $\varepsilon = 0.10$.

10-3 Design a Butterworth low-pass prototype with a cutoff frequency of 1 rad/s and an attenuation of better than 25 dB at 1.1 rad/s.

10-4 Repeat Prob. 10-3 for an attenuation of better than 25 dB at 1.2 rad/s.

10-5 A Chebyshev is required with a cutoff frequency of 1 rad/s, a VSWR of 1.10 and an attenuation of better than 30 dB at 1.3 rad/s. Find the order of the filter.

10-6 Repeat Prob. 10-5 for a VSWR of 1.05 and 1.15.

10-7 Design a Chebyshev filter with the specification in Prob. 10-5.

10-8 Check Eqs (10-15*b*, *c*).

10-9 Derive Eq. (10-17).

10-10 Using Eqs (10-13a, b) verify that Chebyshev filters are symmetrical provided n is odd.

10-11 Verify that Butterworth filters are symmetrical for n odd and even.

10-12 Verify that Eq. (10-8) is consistent with Table 10-1.

10-13 Verify that Eqs (10-13a, b) are consistent with Table 10-4 for ripple levels of $\varepsilon = \frac{1}{2}$, 1 and $1\frac{1}{2}$ dB.

10-14 By dividing the real parts of Eq. (10-12a) by $\cosh[(1/n)\sinh^{-1}(1/\varepsilon)]$ and the imaginary parts by $\sinh[(1/n)\sinh^{-1}(1/\varepsilon)]$ show that the poles of $S_{11}(s)$ lie on an ellipse.

BIBLIOGRAPHY

Matthaei, G. L., Young, L. and Jones, E. M. T. *Microwave Filters, Impedance-matching Networks and Coupling Structures*, Artech House, Norwood, MA, 1980.

Orchard, H. J., 'Formulas for ladder filters', *Wireless Engr*, Vol. 30, pp. 257–353, September, 1939.

CHAPTER
11

SYNTHESIS OF FILTER NETWORKS WITH ATTENUATION POLES AT FINITE FREQUENCIES

INTRODUCTION

By far the more interesting class of filter circuits is that for which the amplitude squared transfer function has attenuation poles at finite frequencies as well as at infinite frequency. The passband may have attenuation zeros either at the origin or at finite frequencies. Since the synthesis of one-port LCR functions is generally a more demanding task than that of one-port LC ones, the usual approach is to form one from the entries of its open- or short-circuited parameters. These latter quantities are of course one-port reactance functions and may be realized using the standard techniques already established in connection with this class of network. A knowledge of the first and second Foster and Cauer forms respectively and the zero shifting technique is sufficient for this purpose. This is usually done by forming relationships between the odd and even parts of the one-port LCR immittance and the open- and short-circuit parameters of the same two-port network, or equivalently, those of the corresponding reflection coefficient. Both approaches are described. In the case of odd degree symmetrical networks, the realization of an odd or even reactance function is another possibility. The first two procedures are illustrated by the realization of a transfer function with its attenuation zeros in the passband at the origin and a pair of attenuation poles at finite frequency and a single attenuation pole at infinite frequency in the stopband. It is also illustrated for the same degree network but with attenuation zeros at finite frequencies in the passband. The synthesis of the latter network poses no additional difficulties. The direct realization of an *LCR* immittance function in terms of a two-port reactance network terminated in a 1-Ω resistor will, however, always succeed by starting with a Brune preamble followed by a Brune cycle. However, this canonical solution is outside the remit of this work.

BADER IDENTITIES

Instead of realizing an *LCR* immittance directly, for which the synthesis procedure has incidentally not been established in this text, it is more usual to form one or the other of the immittance parameters of the network in such a way as to exhibit the required attenuation poles of the specification. This may be either done by having recourse to the Darlington relationships or by forming additional ones between the scattering parameters and the open- or short-circuit parameters. The former relationships have been given in Chapter 7. The latter ones will now be deduced.

The derivation starts by writing S_{11} and S_{21} as the ratios of two polynomials:

$$S_{11} = \frac{f}{g} \tag{11-1a}$$

$$S_{11} = \frac{h}{g} \tag{11-1b}$$

and noting that h is an arbitrary polynomial:

$$h = h_e + h_o \tag{11-2a}$$

g is a strictly Hurwitz polynomial:

$$g = g_e + g_o \tag{11-2b}$$

and f is an even or odd one:

$$f = f_e \text{ or } f_o \tag{11-2c}$$

The degree of the network is that of g; the degrees of the other polynomials cannot exceed that of g. The polynomials f, g and h are related by the Feldtkeller conditions in Chapter 1.

Making use of the relationships between the scattering and open-circuit variables in a reciprocal two-port network readily gives

$$\frac{1}{S_{21}} = \frac{g}{f} = \frac{(Z_{11} + 1)(Z_{22} + 1) - Z_{21}^2}{2Z_{21}} \tag{11-3}$$

$$\frac{S_{11}}{S_{22}} = \frac{h}{f} = \frac{(Z_{11} - 1)(Z_{22} + 1) - Z_{21}^2}{2Z_{21}} \tag{11-4}$$

If f is an even polynomial, the usual situation met in the specification of a low-pass prototype, the even and odd parts of these relatioships become

$$\frac{g_e}{f} = \frac{Z_{11} + Z_{22}}{2Z_{21}} \tag{11-5}$$

$$\frac{g_o}{f} = \frac{Z_{11}Z_{22} - Z_{21}^2 + 1}{2Z_{21}} \tag{11-6}$$

and

$$\frac{h_e}{f} = \frac{Z_{11} - Z_{22}}{2Z_{21}} \tag{11-7}$$

$$\frac{h_o}{f} = \frac{Z_{11}Z_{22} - Z_{21}^2 - 1}{2Z_{21}} \tag{11-8}$$

respectively provided, Z_{11}, Z_{22} and Z_{21} are odd functions.

Adding and subtracting Eqs (11-5) and (11-7) yields

$$\frac{Z_{11}}{Z_{21}} = \frac{g_e + h_e}{f} \tag{11-9}$$

$$\frac{Z_{22}}{Z_{21}} = \frac{g_e - h_e}{f} \tag{11-10}$$

and making use of Eqs (11-6) and (11-8) indicates that

$$Z_{21} = \frac{f}{g_o - h_o} \tag{11-11}$$

The required one-port open-circuited relationships are now obtained by eliminating Z_{21} in Eqs (11-9) and (11-10):

$$Z_{11} = \frac{g_e + h_e}{g_o - h_o} \tag{11-12}$$

$$Z_{22} = \frac{g_e - h_e}{g_o - h_o} \tag{11-13}$$

Likewise, it is easily demonstrated that the one-port short-circuit parameters are related to the odd and even parts of the numerator and denominator polynomials of $S_{11}(s)$ by

$$Y_{11} = \frac{g_e - h_e}{g_o + h_o} \tag{11-14}$$

$$Y_{22} = \frac{g_e + h_e}{g_o + h_o} \tag{11-15}$$

and that the transfer parameter Y_{21} is given by

$$Y_{21} = \frac{f}{g_o + h_o} \tag{11-16}$$

Taking the $n = 3$ low-pass Butterworth problem with the frequency response in Fig. 11-1 as an example for which the reflection coefficient is

$$S_{11}(s) = \frac{\pm s^3}{s^3 + 2s^2 + 2s + 1}$$

gives

$$h_e = 0$$
$$h_o = \pm s^3$$
$$g_e = 2s^2 + 1$$
$$g_o = s^3 + 2s$$

Scrutiny of $S_{21}(s)$ indicates that f is an even polynomial as asserted. The two

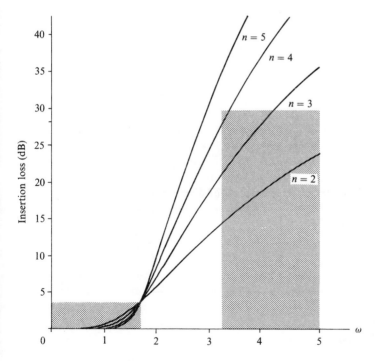

FIGURE 11-1
Low-pass $n = 3$ frequency response with attenuation poles at infinite frequency.

possible solutions are

$$Z_{11}(s) = \frac{2s^2 + 1}{2s}$$

$$Z_{11}(s) = \frac{2s^2 + 1}{2s^3 + 2s}$$

The odd reactance function whose degree is one less than that of the required network is now selected. This statement may be understood by recalling that the condition

$$Z_{11}(s) = \frac{V_1}{I_1}\bigg|_{I_2 = 0} \qquad (11\text{-}17)$$

suppresses the last element of the network. In a symmetrical network, however, the missing element coincides with the first one extracted from $Z_{11}(s)$. A first Cauer

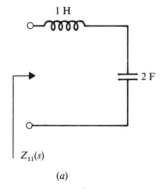

(a)

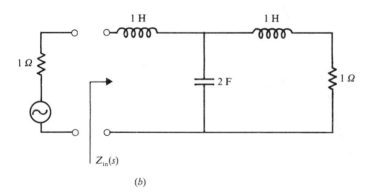

(b)

FIGURE 11-2
Realization of (a) $Z_{11}(s)$ reactance function and (b) $n = 3$ two-port LCR network.

form realizations of $Z_{11}(s)$ readily leads to

$$g_1 = 1 \text{ H}$$

$$g_2 = 2 \text{ H}$$

and the last element is specified by inspection as

$$g_3 = g_1 = 1 \text{ H}$$

in agreement with the classic result. Figure 11-2a illustrates the schematic diagram of $Z_{11}(s)$; the corresponding two-port LCR network is indicated in Fig. 11-2b.

If the first element is to be realized as a shunt one then it is desirable to work with the short-circuit parameters. The two possibilities are either

$$Y_{11}(s) = \frac{2s^2 + 1}{2s}$$

or

$$Y_{11}(s) = \frac{2s^2 + 1}{2s^3 + 2s}$$

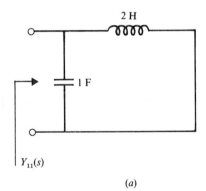

(a)

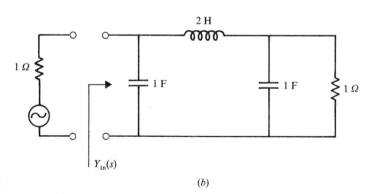

(b)

FIGURE 11-3
Realization of (a) $Y_{11}(s)$ susceptance function and (b) $n = 3$ two-port LCR network.

The first of these two relationships may be used to construct the desired result. Figure 11-3a depicts the schematic diagram of $Y_{11}(s)$ and Fig. 11-3b that of the required two-port *LCR* circuit.

SYNTHESIS OF TWO-PORT FILTERS WITH FINITE ATTENUATION POLES

The realization of filter prototypes with attenuation poles at finite frequencies will now be outlined. The amplitude response of this type of filter has been given some consideration in Chapter 9. The development of one filter with the scattering parameters below:

$$S_{21}(s)S_{21}(-s) = \frac{1}{1 + b(-s^2)^n/(s^2 + 1.69)^2}$$

$$S_{11}(s)S_{11}(-s) = \frac{b(-s^2)^n}{(s^2 + 1.69)^2 + b(-s^2)^n}$$

for $n = 3$ and $b = 0.4761$ will now be illustrated using both the Darlington and Bader approaches.

This amplitude transfer function has a pair of attenuation poles at $s^2 = -1.69$ and a single attenuation pole at infinity in the stopband, and is maximally flat in the passband with its attenuation zeros at the origin. Its frequency response is indicated in Fig. 11-4. The synthesis starts by deducing $S_{11}(s)$ from a knowledge of $S_{11}(s)S_{11}(-s)$.

Evaluating the denominator polynomial gives

$$s^6 - 2.1004s^4 - 7.0993s^2 - 5.9989 = 0$$

The required roots may be determined by having recourse to a standard root-finding subroutine:

$$s_1 = 0.29467 + j1.05575$$
$$s_2 = 0.29467 - j1.05575$$
$$s_3 = -0.29467 + j1.05575$$
$$s_4 = -0.29467 - j1.05575$$
$$s_5 = 2.03863$$
$$s_6 = -2.03863$$

Adopting the LHP roots in forming the denominator polynomial of $S_{11}(s)$ and noting that the numerator polynomial may be constructed from either $\pm s^3$ gives

$$S_{11}(s) = \frac{\pm s^3}{s^3 + 2.62796s^2 + 2.40288s + 2.44928}$$

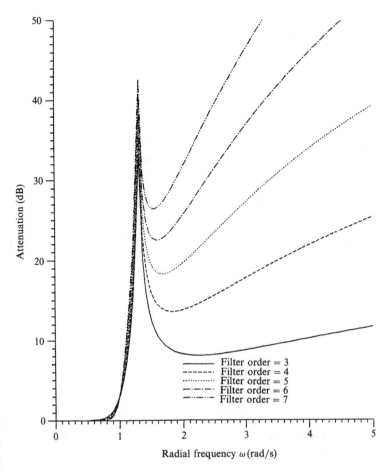

FIGURE 11-4
Low-pass $n = 3$ frequency response with attenuation poles at finite frequencies.

Writing $S_{11}(s)$ in the form of Eqs (11-2) gives the odd and even parts of h and g as

$$h_o = \pm s^3$$
$$h_e = 0$$
$$g_o = s^3 + 2.402\,88s$$
$$g_e = 2.627\,96s^2 + 2.449\,28$$

and the polynomial h is an even polynomial as asserted. The corresponding reactance function is

$$Z_{11}(s) = \frac{s^2 + 0.932\,00}{0.914\,35s}$$

or

$$Z_{11}(s) = \frac{2s^3 + 2.402\,88s}{2.627\,96s^2 + 2.449\,28}$$

FIGURE 11-5
Pole zero diagram of $Z_{11}(s)$.

The solution whose degree coincides with one less than that of the transfer function is now selected and is realized as a two-port network with the attenuation poles of $S_{21}(s)S_{21}(-s)$. In order to locate a pair of attenuation poles at $s^2 = -1.69$ it is necessary to place a pair of complex zeros there by partially extracting a pole at infinity. Inspection of the pole zero diagram in Fig. 11-5 indicates that this may be achieved by shifting the pair of complex zeros at $s^2 = -0.932\,00$ to $s^2 = -1.69$:

$$\frac{s^2 + 0.932\,00}{0.914\,35s} - ks\Big|_{s^2 = -1.69} = 0$$

The residue k is therefore given by

$$k = 0.4905$$

and the remainder reactance $Z_r(s)$ is

$$Z_r(s) = Z_{11}(s) - 0.4905s$$

Evaluating this quantity gives

$$Z_r(s) = \frac{s^2 + 1.69}{1.657\,80s}$$

$Z_r(s)$ has now a pair of complex conjugate zeros at $s^2 = -1.69$ as asserted. Forming the corresponding susceptance function

$$Y_r(s) = \frac{1}{Z_r(s)} = \frac{1.657\,80s}{s^2 + 1.69}$$

produces a pair of complex poles there which can readily be extracted. Its residue is

$$2k = 1.657\,80$$

The extraction of this midseries susceptance in the form of a series resonator in shunt with the network completes the synthesis of $Z_{11}(s)$.

Figure 11-6 depicts the topology of this reactance function. The realization of the network with the required transfer function is met by recalling that the condition in Eq. (11-17) suppresses the last element of the network. Making use of this observation and the fact that the network is symmetric gives the residue of the last element as

$$k = 0.4905$$

The network is thus symmetrical. Figure 11-7 depicts the required result. Figure 11-8 indicates the dual result.

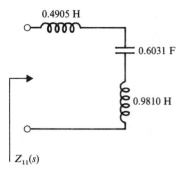

FIGURE 11-6
One-port realization of $Z_{11}(s) = (s^2 + 0.932\,00)/0.914\,35s$.

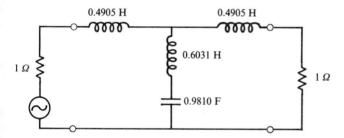

FIGURE 11-7
Low-pass $n = 3$ prototype using midseries resonator.

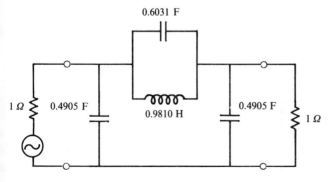

FIGURE 11-8
Low-pass $n = 3$ prototype using midshunt resonator.

   In the Darlington method the synthesis proceeds by making use of the bilinear transformation between reflection and impedance. Taking the positive sign gives $Z_{in}(s)$ as

$$Z_{in}(s) = \frac{2s^3 + 2.627\,96s^2 + 2.402\,88s + 2.449\,28}{2.627\,96s^2 + 2.402\,88s + 2.449\,28}$$

The pole zero diagram associated with this immittance function may be examined by factoring $Z_{in}(s)$:

$$Z_{in}(s) = \frac{(s + 1.774)(s + 0.0685 + j1.0179)(s + 0.0685 - j1.0179)}{(s + 0.4573 + j0.8503)(s + 0.4573 - j0.8503)}$$

Figure 11-9 illustrates the corresponding pole zero arrangement.
   Expressing $Z(s)$ in the Darlington form,

$$Z(s) = \frac{m_1(s) + n_1(s)}{m_2(s) + n_2(s)}$$

gives the two possibilities for the open-circuit parameter $Z_{11}(s)$ of a two-port reactance network terminated in a 1-Ω resistor as

$$Z_{11}(s) = \frac{m_1(s)}{n_2(s)}$$

$$Z_{11}(s) = \frac{n_1(s)}{m_2(s)}$$

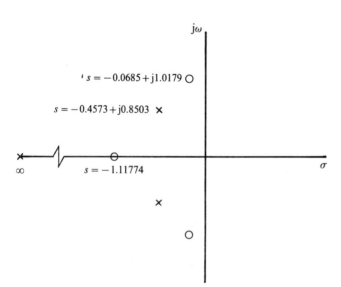

FIGURE 11-9
Pole zero diagram of $Z_{in}(s)$.

where

$$m_1(s) = 2.627\,96s^2 + 2.449\,28$$

$$n_1(s) = 2s^3 + 2.402\,88s$$

$$m_2(s) = 2.627\,96s^2 + 2.449\,28$$

$$n_2(s) = 2.402\,88s$$

The result is

$$Z_{11}(s) = \frac{s^2 + 0.932\,00}{0.914\,35s}$$

or

$$Z_{11}(s) = \frac{2.627\,96s^2 + 2.449\,28}{2s^3 + 2.449\,28s}$$

in keeping with the two relationships obtained using the Bader method.

SYNTHESIS OF $n = 3$
ELLIPTIC NETWORK

As another example of the Bader synthesis technique, consider the following reflection coefficient associated with a degree 3 elliptic filter with a pair of complex poles at $s^2 = -3.035\,56$:

$$S_{11}(s) = \frac{\pm(s^3 + 0.797\,310s)}{s^3 + 2.272\,671s^2 + 2.889\,021s + 3.007\,524}$$

The transfer characteristic of this network is indicated in Fig. 11-10. Here

$$h_e = 0$$

$$h_o = \pm(s^3 + 0.797\,310s)$$

$$g_e = 2.272\,67s^2 + 3.007\,52$$

$$g_o = s^3 + 2.889\,02s$$

and

$$Z_{11}(s) = \frac{2.272\,67s^2 + 3.007\,52}{2.095\,92s}$$

or

$$Z_{11}(s) = \frac{2.272\,67s^2 + 3.007\,52}{2s^2 + 3.686\,33s}$$

Taking the solution with the degree one less than that of $S_{11}(s)$ and partially extracting a pole at infinity in order to realize a pair of zeros at $s^2 = -3.035\,56$ gives

$$Z_{11}(s) - ks|_{s^2 = -3.035\,56} = 0$$

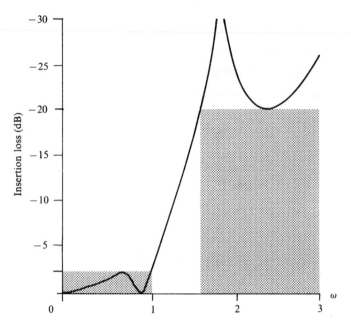

FIGURE 11-10
Low-pass $n = 3$ frequency response with attenuation poles at infinite frequency.

The required residue is

$$k = 0.612\,85$$

The remainder reactance is now described by

$$Z_r(s) = Z_{11}(s) - 0.612\,85s$$

Evaluating this quantity leads to

$$Z_r(s) = \frac{s^2 + 3.035\,56}{2.112\,84s}$$

The required complex conjugate poles may now be extracted by forming $Y_r(s)$ from a knowledge of $Z_r(s)$:

$$Y_r(s) = \frac{1}{Z_r(s)} = \frac{2.112\,84s}{s^2 + 3.035\,56}$$

The residue of the complex conjugate poles at $s^2 = -3.035\,56$ is given by

$$2k = 2.112\,84$$

Figure 11-11a depicts the required reactance network and Fig. 11-11b indicates that of the corresponding circuit.

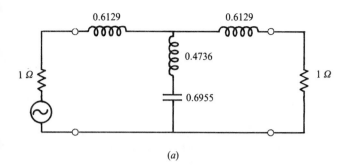

(a)

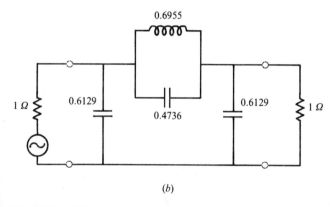

(b)

FIGURE 11-11
Realization of (a) $Z_{11}(s)$ associated with $n = 3$ elliptic filter and (b) $n = 3$ low-pass elliptic filter.

SYNTHESIS OF SYMMETRICAL CIRCUITS USING THE EIGENVALUE APPROACH

The eigenvalue approach, described by Rhodes, proceeds by noting the standard relationships between the open-circuited parameters and the impedance eigenvalues:

$$Z_{11} = \frac{\zeta_1 + \zeta_2}{2} \qquad (11\text{-}18a)$$

$$Z_{21} = \frac{\zeta_1 - \zeta_2}{2} \qquad (11\text{-}18b)$$

or

$$\zeta_1 = Z_{11} + Z_{21} \qquad (11\text{-}19a)$$

$$\zeta_2 = Z_{11} - Z_{21} \qquad (11\text{-}19b)$$

Making use of Eqs (11-11) and (11-12) gives

$$\zeta_1 = \frac{g_e + h_e + f_e}{g_o - h_o} \tag{11-20a}$$

$$\zeta_2 = \frac{g_e + h_e - f_e}{g_o - h_o} \tag{11-20b}$$

for the open- and short-circuited eigennetworks respectively. The open-circuited impedance eigenvalue (ζ_1) is applicable for n odd and the short-circuited eigenvalue (ζ_2) is used for n even.

Taking the classic Butterworth problem with the plus sign as an example gives

$$\zeta_1 = \frac{s^2 + 1}{s}$$

$$\zeta_2 = s$$

These impedance eigenvalues coincide with those associated with the eigennetworks obtained by bisecting the $n = 3$ network in Fig. 11-2a.

BIBLIOGRAPHY

Bader, W., 'Polynomvierpole vorgeschriebener Frequenzabhängigkeit', Arch. Elekrotech., Vol. 34, pp. 181–209, 1940.

Bader, W., 'Kopplungsfreie Kettenschaltungen', Telegr. Fernspr., Funk- und Fernsehtechnik, Vol. 31, pp. 177–189, 1942.

Meinguet, J., and Belevitch, V., 'On the realizability of ladder filters', IRE Trans. on Circuit Theory, Vol. CT-5, pp. 253, December 1958.

Saal, R., and Ulbrich, E., 'On the design of filters by synthesis', IRE Trans. on Circuit Theory, Vol. CT-5, pp. 284, December 1958.

12

FREQUENCY AND IMPEDANCE TRANSFORMATIONS

INTRODUCTION

High-pass, band-pass and band-stop filters can be derived from the low-pass prototype developed in the previous chapters by a technique known as frequency transformation. Using frequency transformations, the elements of the normalized low-pass prototype are changed into elements of the high-pass, band-pass and band-stop networks which includes denormalization of the cutoff frequency so that the filters need only be scaled for impedance.

The frequency transformations between the low-pass frequency variable and the other filter types may be introduced not only into the elements of the low-pass filter elements but also into any of the low-pass functions. This feature may be employed to deduce the scattering parameters of the high-pass, band-pass and band-stop filter prototypes in terms of the low-pass filter variables.

LOW-PASS TO HIGH-PASS
FREQUENCY TRANSFORMATION

The mapping between the low- and high-pass prototypes may be determined by recognizing that the branches of reactance networks are reactance functions. A suitable trial function for the immittances of a high-pass filter is

$$\frac{as^2 + b}{cs} g_r \qquad \text{for } r = 1, 2, \ldots, n \tag{12-1}$$

where g_r is the low-pass variable described in Chapter 10.

In order to obtain the coefficients a, b and c, use is made of the fact that the immittances of the low- and high-pass networks must be equal in order for them to exhibit similar transmission and reflection losses.

In the low- to high-pass transformation the origin $s = 0$ maps to $s = j\infty$ provided all immittances are zero:

$$s' g_r = \frac{as^2 + b}{cs} g_r = 0 \tag{12-2}$$

where s' is the low-pass variable and s is the high-pass one. Thus

$$a = 0 \tag{12-3}$$

Making use of this condition in (12-1) gives

$$\frac{b/c}{s} g_r \tag{12-4}$$

where b/c is now obtained by mapping the cutoff frequency $s' = +j$ of the low-pass prototype at the 3-dB frequency to that of the 3-dB point of the high-pass prototype at $s = \pm j\omega_0$:

$$\pm j g_r = \frac{b/c}{\pm j\omega_0} g_r \tag{12-5}$$

A suitable solution which leads to a p.r. function for the entries of the high-pass prototype is

$$\frac{b}{c} = \omega_0 \tag{12-6}$$

The mapping between the low-pass and high-pass prototypes is therefore achieved by substituting

$$\frac{\omega_0}{s} \qquad \text{for } s' \tag{12-7}$$

in each immittance function of the low-pass prototype network.

This transformation maps the segment $0 < |\omega'| < 1$ on the s' plane to $\omega < |\omega| < \infty$ on the s plane in the manner indicated in Fig. 12-1.

The relationship between the low- and high-pass networks is obtained by noting that immittance is invariable under frequency transformation.

For the inductance the impedance is

$$z = s'L \tag{12-8}$$

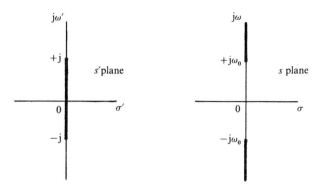

FIGURE 12-1
Low-pass to high-pass frequency transformation.

Introducing the frequency transformation in the above equation leads to

$$z = \left(\frac{\omega_0}{s}\right)L = \frac{1}{sC_h} \tag{12-9}$$

or the series inductance is transformed into a series capacitance:

$$C_h = \frac{1}{\omega_0 L} \tag{12-10}$$

For the shunt capacitance the admittance is

$$y = s'C \tag{12-11}$$

Using the frequency transformation gives

$$y = \left(\frac{\omega_0}{s}\right)C = \frac{1}{sL_h} \tag{12-12}$$

This indicates that the shunt capacitance is transformed into a shunt inductance:

$$L_h = \frac{1}{\omega_0 C} \tag{12-13}$$

Figure 12-2 illustrates the $n = 3$ high-pass network obtained in this way.

The transformation between the low-pass and high-pass variables in (12-7) may be used to relate not only the elements of the low-pass and high-pass prototypes but may also be used to relate any low-pass and high-pass functions. To demonstrate this point consider the synthesis of a high-pass filter based on the input admittance of an $n = 3$ Butterworth two-port low-pass network defined by

$$Y_{in}(s) = \frac{2s^3 + 2s^2 + 2s + 1}{2s^2 + 2s + 1}$$

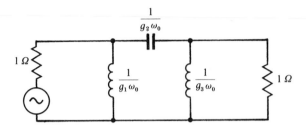

FIGURE 12-2
High-pass $n = 3$ network.

Replacing s' by s/ω_0 yields

$$Y_{in}(s) = \frac{2\omega_0^3 + 2\omega_0^2 s + 2\omega_0 s^2 + s^3}{2\omega_0^2 s + 2\omega_0 s^2 + s^3} \qquad (12\text{-}14)$$

Forming a Cauer-type ladder extension for $Y_{in}(s)$ gives

$$2\omega_0^2 s + 2\omega_0 s^2 + s^3)\ 2\omega_0^3 + 2\omega_0^2 s + 2\omega_0 s^2 + s^3\ (\frac{\omega_0}{s}$$
$$\underline{2\omega_0^3 + 2\omega_0^2 s + \omega_0 s^2}$$
$$\omega_0 s^2 + s^3)\ 2\omega_0^2 s + 2\omega_0 s^2 + s^3\ (\frac{2\omega_0}{s}$$
$$\underline{2\omega_0^2 s + 2\omega_0 s^2}$$
$$s^3)\ \omega_0 s^2 + s^3\ (\frac{\omega_0}{s}$$
$$\underline{\omega_0 s^2}$$
$$s^3)s^3$$
$$\underline{s^3}$$
$$0$$

Comparing this expansion with that of the low-pass prototype indicates that the C and L values of the low-pass network are mapped into the L and C values of the high-pass prototype according to Eqs (12-10) and (12-13).

LOW-PASS TO BAND-PASS TRANSFORMATION

The low-pass to band-pass transformation may also be derived by ensuring that the immittances of the low-pass and band-pass branches are equal at both their centre and band-edge frequencies. A suitable reactance function for the band-pass network is

$$\frac{as^2 + b}{cs}\,g_r \qquad \text{for } r = 1, 2, \ldots, n \qquad (12\text{-}15)$$

The coefficients a, b and c are deduced by applying the following three boundary conditions:

$$s' = 0 \quad \text{maps into } s = j\omega_0 \qquad (12\text{-}16)$$

$$s' = j1 \quad \text{maps into } s = j\omega_2 \qquad (12\text{-}17)$$

$$s' = -j1 \text{ maps into } s = j\omega_1 \qquad (12\text{-}18)$$

Making use of the first condition gives

$$\frac{-a\omega_0^2 + b}{jc\omega_0} = 0 \qquad (12\text{-}19)$$

or
$$b = a\omega_0^2 \qquad (12\text{-}20)$$

Thus (12-15) becomes

$$\frac{a(s^2 + \omega_0^2)}{cs} g_r \qquad \text{for } r = 1, 2, \ldots, n \qquad (12\text{-}21)$$

Utilizing the second and third boundary conditions gives the following two relationships:

$$j1 = \frac{a(-\omega_2^2 + \omega_0^2)}{jc\omega_2} \qquad (12\text{-}22)$$

$$-j1 = \frac{a(-\omega_1^2 + \omega_0^2)}{jc\omega_1} \qquad (12\text{-}23)$$

ω_0 is now determined by combining these two equations:

$$\frac{\omega_2^2 - \omega_0^2}{\omega_2} = \frac{-(\omega_1^2 - \omega_0^2)}{\omega_1} \qquad (12\text{-}24)$$

Evaluating the preceding equality indicates that

$$\omega_0^2 = \omega_1\omega_2 \qquad (12\text{-}25)$$

a/c may now be evaluated from either Eq. (12-22) or (12-23) as

$$\frac{c}{a} = \omega_2 - \omega_1 \qquad (12\text{-}26)$$

Introducing the last three equations into (12-21) gives the required result:

$$s' = \frac{\omega_0}{\text{BW}}\left(\frac{s}{\omega_0} + \frac{\omega_0}{s}\right) \qquad (12\text{-}27)$$

where
$$\text{BW} = \omega_2 - \omega_1 \qquad (12\text{-}28)$$

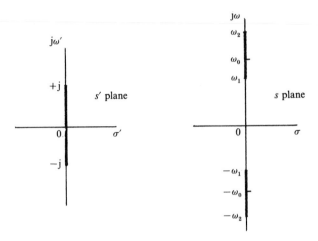

FIGURE 12-3
Low-pass to band-pass frequency transformation.

This transformation maps the segment $0 < |\omega'| < 1$ on the s' plane to the segment $|\omega_2| < |\omega| < |\omega_1|$ on the s plane in the manner shown in Fig. 12-3.

Using the fact that the impedance is invariant under frequency transformations the series inductance of the low-pass prototype maps into a series LC resonator for the band-pass filter:

$$z = s'L = \left[\frac{\omega_0}{\text{BW}} \left(\frac{s}{\omega_0} + \frac{\omega_0}{s} \right) \right] L \tag{12-29}$$

This equation may be written as

$$z = sL_{\text{s}} + \frac{1}{sC_{\text{s}}} \tag{12-30}$$

where

$$L_{\text{s}} = \frac{L}{\text{BW}} \tag{12-31}$$

$$C_{\text{s}} = \frac{\text{BW}}{\omega_0^2 L} \tag{12-32}$$

Applying the frequency transformation to the shunt capacitance of the low-pass prototype gives

$$y = s'C = \left[\frac{\omega_0}{\text{BW}} \left(\frac{s}{\omega_0} + \frac{\omega_0}{s} \right) \right] C \tag{12-33}$$

This equation may be written as

$$y = sC_{\text{p}} + \frac{1}{sL_{\text{p}}} \tag{12-34}$$

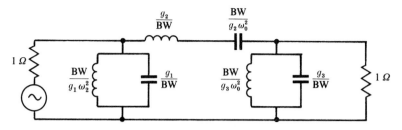

FIGURE 12-4
Band-pass $n = 3$ network.

where

$$C_p = \frac{C}{BW} \qquad (12\text{-}35a)$$

$$L_p = \frac{BW}{C\omega_0^2} \qquad (12\text{-}35b)$$

The shunt capacitance therefore maps into a shunt LC circuit. Figure 12-4 depicts the $n = 3$ band-pass circuit arrangement.

LOW-PASS TO BAND-STOP TRANSFORMATION

The mapping between the low-pass and band-stop filters proceeds in a similar fashion to that used to map the low-pass to the high-pass and band-pass ones. The boundary conditions are given in this instance by

$$s' = j\infty \quad \text{maps into} \quad s = j\omega_0$$
$$s' = j1 \quad \text{maps into} \quad s = j\omega_2$$
$$s' = -j1 \text{ maps into} \quad s = j\omega_1$$

and the trial function is taken as

$$s'g_r \to \frac{cs}{as^2 + b} g_r \qquad (12\text{-}36)$$

The result is

$$s' \to \frac{BW}{\omega_0(s/\omega_0 + \omega_0/s)} \qquad (12\text{-}37)$$

where the variables have the meaning stated earlier.

This transformation maps the segment $|\omega'| < 1$ to $|\omega_2| > |\omega| > |\omega_1|$ as indicated in Fig. 12-5.

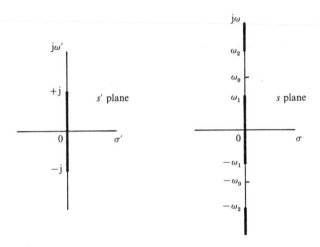

FIGURE 12-5
Low-pass to band-stop
frequency transformation.

Making use of the fact that immittance is invariant under frequency transformations shows that the series inductance maps into a shunt-tuned circuit with element values

$$C_p = \frac{1}{L\,BW} \tag{12-38}$$

$$L_p = \frac{L\,BW}{\omega_0^2} \tag{12-39}$$

whereas the shunt capacitance maps into a series-tuned circuit with element values

$$C_s = \frac{C\,BW}{\omega_0^2} \tag{12-40}$$

$$L_s = \frac{1}{C\,BW} \tag{12-41}$$

The circuit arrangement for the band-stop filter with $n = 3$ is shown in Fig. 12-6.

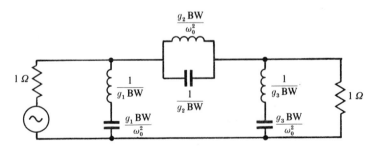

FIGURE 12-6
Band-stop $n = 3$ network.

FREQUENCY SCALING

Although the various frequency transformations introduced so far have incorporated frequency scaling of the resultant filter networks, the low-pass prototype is still scaled to 1 rad/s. To obtain an arbitrary cutoff frequency the following frequency transformation is used:

$$s \to \frac{s}{\omega_0} \tag{12-42}$$

where ω_0 is dimensionless.

Using the fact that immittance remains invariant under frequency transformation leads to

$$sL \to \left(\frac{s}{\omega_0}\right)L \tag{12-43}$$

$$L \to \frac{L}{\omega_0} \tag{12-44}$$

Similarly, for the capacitance:

$$sC \to \left(\frac{s}{\omega_0}\right)C \tag{12-45}$$

or

$$C \to \frac{C}{\omega_0} \tag{12-46}$$

Resistors obviously remain unaffected by frequency scaling.

IMPEDANCE SCALING

The immittances synthesized so far have all been normalized to a 1-Ω generator impedance so that the terminating elements of the corresponding filters have in each instance been realized as 1-Ω resistor. If it is necessary to operate a filter between different impedance levels (the usual case) then it is necessary to renormalize each element of the original filter in some appropriate manner. Impedance scaling may be understood by considering the $n = 3$ example in Chapter 10 but normalized to a 50-Ω (say) instead of to a 1-Ω generator impedance:

$$\frac{Z_{in}(s)}{50} = \frac{2s^3 + 2s^2 + 2s + 1}{2s^2 + 2s + 1} \tag{12-47}$$

Synthesizing this impedance function in a first Cauer form readily gives the following element values for the network:

$$L = 50 \text{ H}$$
$$C = 0.04 \text{ F}$$
$$L = 50 \text{ H}$$
$$R = 50 \text{ }\Omega$$

Comparing these element values with those obtained in Chapter 10 indicates that the denormalized inductors, capacitors and resistor are related to the normalized ones by the following substitutions:

$$L \to Z_0 L \qquad (12\text{-}48)$$

$$C \to \frac{C}{Z_0} \qquad (12\text{-}49)$$

$$1 \to Z_0 \qquad (12\text{-}50)$$

L, C and 1 are the normalized inductive, capacitive and resistive elements of the normalized prototype, $Z_0 L$, C/Z_0 and Z_0 are those of the denormalized one.

FILTER PROTOTYPES USING FREQUENCY AND IMPEDANCE TRANSFORMATIONS

In the first example of the use of frequency transformations consider the design of a high-pass $n = 3$ Butterworth filter terminated in a 50-Ω load with a cutoff frequency of 1000 rad/s. The low-pass prototype has $C_1 = 1$ F, $L_2 = 2$ H and $C_3 = 1$ F. This circuit is depicted in Fig. 12-7.

For this prototype the shunt capacitors are transformed into shunt inductances using Eq. (12-13):

$$L_1' = \frac{1}{\omega_0 C_1} = 10^{-3} \text{ H}$$

$$L_3' = \frac{1}{\omega_0 C_3} = 10^{-3} \text{ H}$$

The series inductor is transformed into a series capacitor with the aid of Eq. (12-10):

$$C_2' = \frac{1}{\omega_0 L_2} = 500 \times 10^{-6} \text{ F}$$

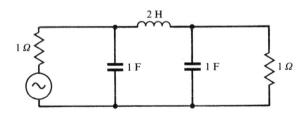

FIGURE 12-7
Low-pass $n = 3$ Butterworth prototype.

Impedance scaling the L' and C' values using Eqs (12-48) and (12-49) gives

$$L'_1 = R_0 L_1 = 50 \times 10^{-3} \text{ H}$$

$$L'_3 = R_0 L_3 = 50 \times 10^{-3} \text{ H}$$

$$C'_2 = \frac{C_2}{R_0} = 10 \times 10^{-6} \text{ F}$$

The high-pass filter obtained in this way is depicted in Fig. 12-8.

As a second example consider the design of a band-pass filter with an $n = 3$ Butterworth response, a centre frequency of 1000 rad/s, a bandwidth of 1000 rad/s and a 50-Ω termination.

The low-pass to band-pass transformation maps the series inductance into a series LC resonator. Using Eqs (12-35a, b) yields

$$L_{s2} = \frac{L_2}{\text{BW}} = 2 \times 10^{-3} \text{ H}$$

$$C_{s2} = \frac{\text{BW}}{\omega_0^2 L_2} = 0.5 \times 10^{-3} \text{ F}$$

The low-pass to band-pass transformation maps the shunt capacitors into parallel LC resonators according to Eqs (12-35a and b):

$$L_{p1} = \frac{\text{BW}}{C_1 \omega_0^2} = 1 \times 10^{-3} \text{ H}$$

$$C_{p1} = \frac{C_1}{\text{BW}} = 1 \times 10^{-3} \text{ F}$$

$$L_{p3} = \frac{\text{BW}}{C_3 \omega_0^2} = 1 \times 10^{-3} \text{ H}$$

$$C_{p3} = \frac{C_3}{\text{BW}} = 1 \times 10^{-3} \text{ F}$$

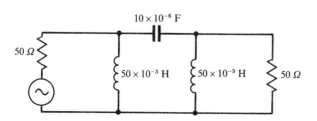

FIGURE 12-8
High-pass $n = 3$ Butterworth filter with cutoff frequency of 1000 rad/s and 50-Ω terminations.

Scaling the elements of the band-pass filter to cater for a 50-Ω termination instead of a 1-Ω one leads to the required result

$$L'_{p1} = R_0 L_{p1} = 50 \times 10^{-3} \text{ H}$$

$$C'_{p1} = \frac{C_{p1}}{R_0} = 20 \times 10^{-6} \text{ F}$$

$$L'_{s2} = R_0 L_{s2} = 100 \times 10^{-3} \text{ H}$$

$$C'_{s2} = \frac{C_{s1}}{R_0} = 0.01 \times 10^{-3} \text{ F}$$

$$L'_{p3} = R_0 L_{p3} = 50 \times 10^{-3} \text{ H}$$

$$C'_{p3} = \frac{C_{p3}}{R_0} = 20 \times 10^{-6} \text{ F}$$

This circuit is illustrated in Fig. 12-9.

As a final example, consider the design of an $n = 3$ Butterworth band-stop filter with a centre frequency of 10 000 rad/s and a bandwidth of 100 rad/s terminated in a 50-Ω load.

In this example the frequency transformations in Eqs (12-30) to (12-41) are combined with impedance scaling in Eqs (12-48) and (12-49).

The series inductance of the low-pass prototype is mapped into a shunt LC resonator defined by

$$L'_{p2} = \frac{L_2 \text{ BW } R_0}{\omega_0^2} = 100 \times 10^{-6} \text{ H}$$

$$C'_{p2} = \frac{1}{L_2 \text{ BW } R_0} = 100 \times 10^{-6} \text{ F}$$

The shunt capacitors of the low-pass prototype are mapped into series LC shunt networks:

$$L'_{s1} = \frac{R_0}{C_1 \text{ BW}} = 0.20 \times 10^{-3} \text{ H}$$

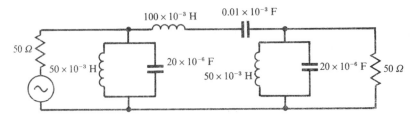

FIGURE 12-9
Band-pass $n = 3$ Butterworth filter with $\omega_0 = 1000$ rad/s, $\omega_2 - \omega_1 = 1000$ rad/s and $R_0 = 50 \, \Omega$.

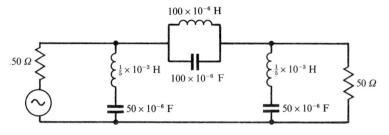

FIGURE 12-10
Band-stop $n = 3$ Butterworth filter with $\omega_0 = 1000$ rad/s, $\omega_2 - \omega_1 = 1000$ rad/s and $R_0 = 50\ \Omega$.

$$C'_{s1} = \frac{C_1\ \mathrm{BW}}{\omega_0^2 R_0} = 50 \times 10^{-6}\ \mathrm{F}$$

$$L'_{s3} = \frac{R_0}{C_3\ \mathrm{BW}} = 0.20 \times 10^{-3}\ \mathrm{H}$$

$$C'_{s3} = \frac{C_3\ \mathrm{BW}}{\omega_0^2 R_0} = 50 \times 10^{-6}\ \mathrm{F}$$

The equivalent circuit for this network is depicted in Fig. 12-10.

BAND-PASS, BAND-STOP AND HIGH-PASS TRANSFER FUNCTIONS WITH CHEBYSHEV CHARACTERISTICS

The characteristics of the four standard types of equal-ripple filter responses considered in this section are illustrated in Fig. 12-11a to d. The relationships between the low-pass amplitude squared transfer function and the other three amplitude squared functions may be determined by starting with that of the low-pass function discussed in Chapter 8:

$$S_{21}(s)S_{21}(-s) = \frac{1}{1 + \varepsilon^2 T_n^2(s/\mathrm{j})} \tag{12-51}$$

The frequency transformation between the low-pass and band-pass prototypes is given in Eq. (12-29). Substituting this relation into Eq. (12-51) leads to

$$S_{21}(s)S_{21}(-s) = \frac{1}{1 + \varepsilon^2 T_n^2[(\omega_0/\mathrm{j}\ \mathrm{BW})(s/\omega_0 + \omega_0/s)]} \tag{12-52}$$

Making use of analytical continuity by replacing s by $\mathrm{j}\omega$ yields the required result:

$$S_{21}(\mathrm{j}\omega)S_{21}(-\mathrm{j}\omega) = \frac{1}{1 + \varepsilon^2 T_n^2[(\omega_0/\mathrm{BW})(\omega/\omega_0 - \omega_0/\omega)]} \tag{12-53}$$

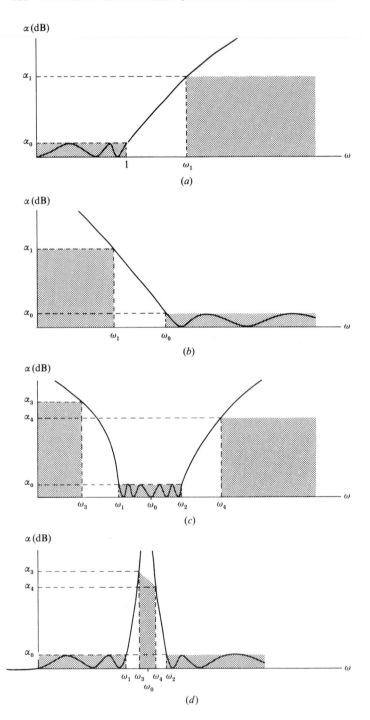

FIGURE 12-11

Attenuation response of (a) low-pass filter, (b) high-pass filter, (c) band-pass filter and (d) band-stop filter with equal-ripple characteristic.

where ω_0 and BW have the meaning in Eqs (12-26) and (12-28). The attenuation response of the band-pass transfer function expressed in decibels is

$$\alpha(\text{dB}) = 10 \log_{10}\left\{1 + \varepsilon^2 T_n^2\left[\frac{\omega_0}{\text{BW}}\left(\frac{\omega}{\omega_0} - \frac{\omega_0}{\omega}\right)\right]\right\} \tag{12-54}$$

At $\omega = \omega_1$ and ω_2,

$$\alpha_0(\text{dB}) = 10 \log_{10}(1 + \varepsilon^2)$$

for all n.

The order of the filter is therefore determined by the attenuation $\alpha_{3,4}$ in the stopband. This band-pass characteristic is illustrated in Fig. 12-11c.

It is readily demonstrated that the transfer functions for band-stop and high-pass filters are expressed in decibels by

$$\alpha(\text{dB}) = 10 \log_{10}\left\{1 + \varepsilon^2 T_n^2\left[\frac{\text{BW}}{\omega_0(\omega/\omega_0 - \omega_0/\omega)}\right]\right\} \tag{12-55}$$

and $$\alpha(\text{dB}) = 10 \log_{10}\left[1 + \varepsilon^2 T_n^2\left(\frac{\omega_0}{\omega}\right)\right] \tag{12-56}$$

These two transfer functions are illustrated in Fig. 12-11b and d.

ORDERS OF EQUAL-RIPPLE BAND-PASS, BAND-STOP AND HIGH-PASS FILTERS

The orders of band-pass, band-stop and high-pass filters with equal ripples are readily obtained by solving Eqs (12-54), (12-55) and (12-56) in terms of the specification α_1 at ω_1 or $\alpha_{3,4}$ at $\omega_{3,4}$. Starting with Eq. (12-54) yields

$$T_n\left[\frac{\omega_0}{\text{BW}}\left(\frac{\omega_{3,4}}{\omega_0} - \frac{\omega_0}{\omega_{3,4}}\right)\right] = \frac{\sqrt{A-1}}{\varepsilon} \tag{12-57}$$

where

$$A = \text{antilog}_{10}\left(\frac{\alpha_{3,4}}{10}\right) \tag{12-58}$$

Solving for $T_n(x)$ by making use of Eq. (10-51) in Chapter 10 leads to

$$n = \frac{\cosh^{-1}(\sqrt{A-1}/\varepsilon)}{\cosh^{-1}[(\omega_0/\text{BW})(\omega_{3,4}/\omega_0 - \omega_0/\omega_{3,4})]} \tag{12-59}$$

In the case of the band-stop filter defined by Eq. (12-55) the result is

$$n = \frac{\cosh^{-1}(\sqrt{A-1}/\varepsilon)}{\cosh^{-1}\{\text{BW}/[\omega_0(\omega_{3,4}/\omega_0 - \omega_0/\omega_{3,4})]\}} \tag{12-60}$$

with A defined by Eq. (12-58).

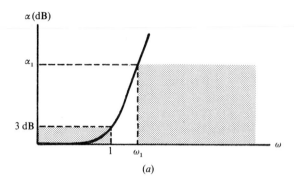

(a)

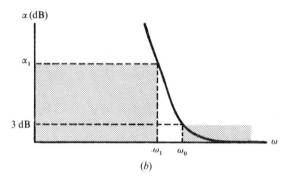

(b)

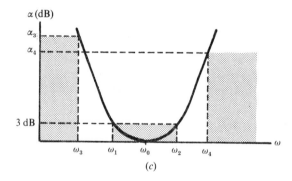

(c)

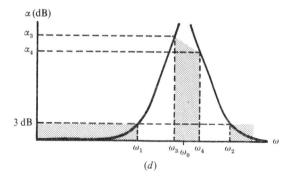

(d)

FIGURE 12-12
Attenuation of (a) low-pass filter,
(b) high-pass filter,
(c) band-pass filter and
(d) band-stop filter with
Butterworth characteristic.

The result for the high-pass filter deduced from Eq. (12-56) is

$$n = \frac{\cosh^{-1}(\sqrt{A - 1}/\varepsilon)}{\cosh^{-1}(\omega_0/\omega)} \qquad (12\text{-}61)$$

with A defined by

$$A = \text{antilog}_{10}\left(\frac{\alpha_1}{\omega_1}\right) \qquad (12\text{-}62)$$

The more severe specification at ω_3 or ω_4 determines the order n of the filters in Eqs (12-59) and (12-60). ε is in each instance determined by the pass-band requirement.

Consider now the design of an equal-ripple high-pass filter with a ripple level of 1 dB, a cutoff frequency of 30 MHz and an attenuation of at least 40 dB at 29 MHz.

The ripple coefficient ε is evaluated by making use of Eq. (12-56) at $\omega = \omega_0$:

$$\alpha_0(\text{dB}) = 10 \log_{10}(1 + \varepsilon^2)$$

The required result is

$$\varepsilon = 0.509$$

The factor A is now calculated using Eq. (12-62):

$$A = \text{antilog}_{10}\left(\frac{40}{10}\right) = 10\,000$$

Finally, n is determined from Eq. (12-61) as

$$n = \frac{\cosh^{-1} 196}{\cosh^{-1}(30/29)} = 22.96$$

The order of the filter is therefore

$$n = 23$$

The design of the filter proceeds by forming the low-pass prototype with $n = 23$, normalized to 1 rad/s, and thereafter invoking the appropriate frequency transformation between the low-pass and high-pass prototypes. The network is then impedance scaled as required.

Figure 12-12 summarizes the situation for filters with Butterworth-like specifications.

FILTER SPECIFICATION IN TERMS OF ITS REFLECTION COEFFICIENT

Sometimes a filter specification is defined in terms of its reflection coefficient instead of its transfer one. Since all the cases proceed in the same fashion only the maximally flat high-pass filter will be considered. The relation between the two descriptions is obtained from the unitary condition in Eq. (12-31):

$$S_{11}(s)S_{11}(-s) = \frac{(-1)^n(\omega_0/s)^{2n}}{1+(-1)^n(\omega_0/s)^{2n}} \tag{12-63}$$

Making use of analytic continuity by replacing s by $j\omega$ gives

$$|S_{11}(j\omega)|^2 = \frac{(\omega_0/\omega)^{2n}}{1+(\omega_0/\omega)^{2n}} \tag{12-64}$$

In decibels the preceding equation becomes

$$\alpha(\text{dB}) = 10\log_{10}\left[1+\left(\frac{\omega}{\omega_0}\right)^{2n}\right] \tag{12-65}$$

The order n of the filter is obtained from its specification α_1 at ω_1:

$$n = \frac{\log_{10}(A-1)}{\log_{10}(\omega_1/\omega_0)^2}$$

where

$$A = \text{antilog}_{10}\left(\frac{\alpha_1}{10}\right) \tag{12-67}$$

where α is known as the return loss of the filter.

DECIBEL-FOR-DECIBEL TRADE-OFF

If the reflection and transmission losses in the passband and stopband of a two-port filter network over the frequency intervals $0-\omega_1$ and $0-\omega_3$ are small, as is the case in practical circuits, it may be shown that the return loss in decibels in the passband may be traded decibel for decibel for the attenuation in decibels in the stopband. This property will now be demonstrated in the case of the Chebyshev transfer function approximation problem. The derivation starts by constructing the following quantity:

$$\frac{S_{11}(s)S_{11}(-s)}{S_{21}(s)S_{21}(-s)} = \varepsilon^2 T_n^2(s/j) \tag{12-68}$$

If ω_1 is chosen as the bandwidth corresponding to the maximum ripple level in the passband and ω_3 as the bandwidth corresponding to the minimum attenuation in the stopband, then

$$\left[\frac{S_{11}(s)S_{11}(-s)}{S_{21}(s)S_{21}(-s)}\right]_{s=j\omega_1} \times \left[\frac{S_{21}(s)S_{21}(-s)}{S_{11}(s)S_{11}(-s)}\right]_{s=j\omega_3} = \frac{T_n^2(s/j)_{s=j\omega_1}}{T_n^2(s/j)_{s=j\omega_3}}$$
$$= F(\omega_1,\omega_3), \text{ say} \tag{12-69}$$

Taking the logarithm on both sides yields

$$(R_1+A_3)-(A_1+R_3) = F \tag{12-70}$$

where

$$R_1 = -10 \log_{10}[S_{11}(s)S_{11}(-s)|_{s=j\omega_1}] \quad \text{dB} \quad (12\text{-}71)$$

$$A_1 = -10 \log_{10}[S_{21}(s)S_{21}(-s)|_{s=j\omega_1}] \quad \text{dB} \quad (12\text{-}72)$$

$$R_3 = -10 \log_{10}[S_{11}(s)S_{11}(-s)|_{s=j\omega_3}] \quad \text{dB} \quad (12\text{-}73)$$

$$A_3 = -10 \log_{10}[S_{21}(s)S_{21}(-s)|_{s=j\omega_3}] \quad \text{dB} \quad (12\text{-}74)$$

$$F = -10 \log_{10}[F(\omega_1, \omega_3)] \quad \text{dB}$$

R_1 and R_3 are the return losses in decibels at ω_1 and ω_3, A_1 and A_3 are the transmission losses in decibels at ω_1 and ω_3 and F is a dimensionless quantity in decibels known as the characteristic factor.

The attenuations $A_{1,3}$ in the passband and the reflection loss $R_{1,3}$ in the stopband are determined from the unitary condition by

$$A_1 = 10 \log_{10}[1 - 10^{-R_1/10}] \quad (12\text{-}75)$$

$$R_3 = 10 \log_{10}[1 - 10^{-A_3/10}] \quad (12\text{-}76)$$

For most applications, the return loss in the passband and the attenuation in the stopband are at least 10 dB or better. Thus

$$R_1 = 10 \text{ dB}$$
$$A_3 = 10 \text{ dB}$$
$$A_1 = 0.45 \text{ dB}$$
$$R_3 = 0.45 \text{ dB}$$

and $\qquad\qquad F = 19.1 \text{ dB}$

For higher values of return loss and attenuation, it is easy to recognize that

$$A_1 + R_3 \approx 0 \quad (12\text{-}77)$$
$$R_1 + A_3 \approx F \quad (12\text{-}78)$$

Thus, a trade-off exists, decibel for decibel, between the return loss in the passband and the transmission loss in the stopband at a pair of specified frequencies, ω_1 and ω_3. The choice of these frequencies is arbitrary as long as Eq. (12-77) is satisfied.

It is usual to choose ω_1 with respect to the equi-ripple peaks in the passband. Thus the ripple level determines R_1.

PROBLEMS

12-1 The input impedance of a low-pass filter is

$$Y(s) = \frac{2s^3 + 2s^2 + 2s + 1}{2s^2 + 2s + 1}$$

Find the input impedance of this network under band-pass, band-stop and high-pass transformations.

12-2 Find $S_{11}(s)$ for each $Y(s)$ in Prob. 12-1.

12-3 The amplitude squared transfer function for a Butterworth low-pass filter is

$$S_{21}(s)S_{21}(-s) = \frac{1}{1 + (-1)^n(s)^{2n}}$$

Obtain the corresponding high-pass, band-pass and band-stop transfer parameters.

12-4 Repeat Prob. 12-3 for a Chebyshev amplitude response

$$S_{21}(s)S_{21}(-s) = \frac{1}{1 + \varepsilon^2 T_n^2(s/j)}$$

12-5 Derive the frequency transformation between the low-pass and band-stop filters from first principles.

12-6 Obtain an $n = 3$ high-pass Butterworth filter with a cutoff radian frequency of 9000 rad/s terminated in 50-Ω loads.

12-7 Design an $n = 3$ band-stop filter with a Butterworth characteristic with a centre frequency of 10 000 rad/s, a bandwidth of 2000 rad/s and loads of 50 Ω.

12-8 A low-pass transfer function is given by

$$S_{21}(s) = \frac{1}{a_n s^n + a_{n-1} s^{n-1} + \cdots + 1}$$

Find the corresponding high-pass, band-pass and band-stop transfer functions.

12-9 Design a maximally flat band-pass filter for which $\omega_1 = 1.0$ MHz, $\omega_2 = 1.2$ MHz, $\alpha_3 \geqq 25$ dB at $\omega_3 = 0.80$ MHz, $\alpha_4 \geqq 30$ dB at $\omega_4 = 1.40$ MHz.

12-10 Determine the order of a maximally flat band-pass filter for which $\omega_1 = 5.9$ MHz, $\omega_2 = 6.4$ GHz, $\alpha_3 \geqq 30$ dB at $\omega_3 = 5.5$ GHz, $\alpha_4 \geqq 32$ dB at $\omega_4 = 6.7$ GHz.

12-11 Repeat Prob. 12-10 for an equal-ripple band-pass filter with a ripple level of 0.50 dB.

12-12 Find the order of a high-pass filter with a cutoff frequency $\omega_1 = 3$ MHz and an attenuation $\alpha_2 \geqq 45$ dB at $\omega_2 = 2.9$ MHz. Repeat with $\omega_2 = 2.8$ MHz.

12-13 Construct a high-pass filter with the specification in Prob. 12-12 to work between 50-Ω loads.

12-14 Obtain the order of maximally flat band-pass and band-stop filters in terms of a reflection loss specification.

12-15 Repeat Prob. 12-14 in the case of equal-ripple band-pass and band-stop filters.

CHAPTER
13

SYNTHESIS OF BROADBAND MATCHING NETWORKS

INTRODUCTION

In many engineering problems it is necessary to match, over some frequency interval, a complex load consisting of a resistance in shunt with a capacitor to a load resistance. This is a recurring problem in amplifier design. The theoretical limitation of this matching problem has been investigated by Bode. He showed that the ideal reflection coefficient for this type of network is a constant (rather than zero) over the passband and unity in the stopband. The optimum approximation is again a Chebyshev one.

The realization of an equalizer network differs in a major way from that of a filter network between equal real terminations in that the reflection coefficient in the passband is dependent upon the order of the network. Its realization starts by expressing the reflection coefficient (or transfer function) in the usual way as the ratio of two polynomials as a preamble to a Cauer continued fraction expansion of the ensuing immittance function. This must be done in such a way that the first element coincides with the susceptance of the load. Since the optimum reflection coefficient is not uniquely defined but is also dependent upon the load and the order of the equalizer, it is necessary to minimize it before the design is fully defined. This optimization, due to Fano, is included in the chapter as a preamble to design.

Whereas the matching problem is often posed in band-pass form it is sufficient to deal with the low-pass problem. After the low-pass network has been derived, a simple low-pass to band-pass transformation may be used to centre the network. An ideal transformer (or equivalent T or π networks) is used for matching the real parts of the generator and load impedances.

BODE'S PARALLEL RC LOAD

The matching problem considered in this chapter is that between a load consisting of a capacitance C in shunt with a resistor R and a generator of internal impedance Z_0 over a frequency interval $\omega_2-\omega_1$. This is a common problem in the design of negative resistance amplifiers and other circuits. The equalizer in Fig. 13-1 is assumed to be a reactance network.

Since the first element in the filter must be the capacitor C, the problem is formulated in terms of S_{22} rather than S_{11} (by recalling that for a lossless network):

$$|S_{22}(j\omega)|^2 = |S_{11}(j\omega)|^2 = 1 - |S_{21}(j\omega)|^2 \qquad (13-1)$$

The maximum gain bandwidth that can be achieved by this means has been defined by Bode by the following integral equation:

$$\int_0^\infty \log_e \left| \frac{1}{S_{22}(j\omega)} \right| d\omega \leq \frac{\pi}{RC} \qquad (13-2)$$

This integral equation suggests that the maximum bandwidth is obtained with this type of network by having $|S_{22}|_{max}$ a constant over the required frequency interval and unity elsewhere. Assuming this idealized situation the above integral becomes

$$(\omega_2 - \omega_1) \log_e \frac{1}{|S_{22}|_{max}} = \frac{\pi}{RC} \qquad (13-3)$$

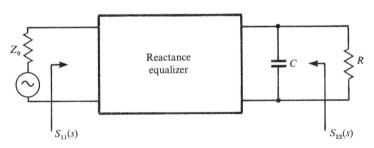

FIGURE 13-1
Schematic diagram of complex RC load and equalizer.

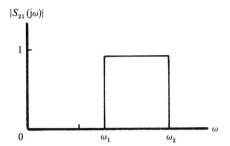

FIGURE 13-2
Idealized frequency response of band-pass equalizer network.

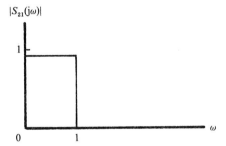

FIGURE 13-3
Idealized frequency response of low-pass equalizer network.

The idealized frequency response in Fig. 13-2 cannot be achieved in practice because it requires a matching network with an infinite number of matching elements. However, Eq. (13-3) is useful in that it serves to establish the result in the ideal situation.

Often, the matching problem is posed in band-pass form, but as bandwidth is invariant under frequency transformation, as discussed in Chapter 10, it is only necessary to consider the low-pass case. Figure 13-3 depicts the idealized low-pass response. After the low-pass matching network has been obtained, a simple low-pass to band-pass transformation (as described in Chapter 11) will centre the band as required.

THE CHEBYSHEV RESPONSE

One amplitude approximation for $|S_{21}(j\omega)|$ for which the transfer coefficient approximates a constant over the passband and is unity elsewhere is

$$|S_{21}(j\omega)|^2 = \frac{1}{1 + K^2 + \varepsilon^2 T_n^2(\omega)} \tag{13-4}$$

The transfer function in Fig. 13-4 must be realized in such a way that the first element coincides with the susceptance of the capacitance in Fig. 13-1 normalized to $1\ \Omega$ and a passband of 1 rad/s. This element is defined in terms of C and R and $\omega_2 - \omega_1$ as

$$g_1 = (\omega_2 - \omega_1)RC \tag{13-5}$$

The reflection coefficient associated with the transfer function in Eq. (13-4) is obtained from the unitary condition in Eq. (13-1) as

$$|S_{22}(j\omega)|^2 = \frac{K^2 + \varepsilon^2 T_n^2(\omega)}{1 + K^2 + \varepsilon^2 T_n^2(\omega)} \tag{13-6}$$

Using analytic continuation leads to

$$S_{22}(s)S_{22}(-s) = \frac{K^2 + \varepsilon^2 T_n^2(s/j)}{1 + K^2 + \varepsilon^2 T_n^2(s/j)} \tag{13-7}$$

The poles and zeros of the preceding function are

$$1 + K^2 + \varepsilon^2 T_n^2(s/j) = 0 \tag{13-8}$$

$$K^2 + \varepsilon^2 T_n^2(s/j) = 0 \tag{13-9}$$

The roots of these two equations are determined by

$$T_n(s/j) = \mp j\,\frac{1 + K^2}{\varepsilon^2} \tag{13-10}$$

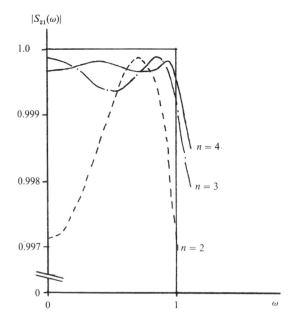

FIGURE 13-4
Chebyshev transfer characteristic for low-pass equalizer network.

$$T_n(s/j) = \mp j \frac{K}{\varepsilon} \qquad (13\text{-}11)$$

The preceding equations have the form of Eq. (10-11) in Chapter 10. Thus the poles and zeros of $S_{22}(s)$ are obtained by inspection from Eq. (7-34) in that chapter as

$$p_k = -\sin\left(\frac{2k-1}{2n}\pi\right)\sinh\left(\frac{1}{n}\sinh^{-1}\frac{\sqrt{1+K^2}}{\varepsilon}\right)$$
$$+ j\cos\left(\frac{2k-1}{2n}\pi\right)\cosh\left(\frac{1}{n}\sinh^{-1}\frac{\sqrt{1+K^2}}{\varepsilon}\right) \qquad (13\text{-}12)$$

$$z_k = -\sin\left(\frac{2k-1}{2n}\pi\right)\sinh\left(\frac{1}{n}\sinh^{-1}\frac{K}{\varepsilon}\right)$$
$$+ j\cos\left(\frac{2k-1}{2n}\pi\right)\sinh\left(\frac{1}{n}\sinh^{-1}\frac{K}{\varepsilon}\right) \qquad (13\text{-}13)$$

with $\qquad k = 1, 2, \ldots, 2n$

$S_{22}(s)$ may now be formed from $S_{22}(s)S_{22}(-s)$ by associating the LHP poles with $S_{22}(s)$ and the RHP poles with $S_{22}(-s)$. The zeros of $S_{22}(s)$ may be selected from either the LHP or RHP as discussed in Chapter 2. However, a minimum phase network is defined by using the LHP roots to construct $S_{22}(s)$. $S_{22}(s)$ therefore takes the following standard form:

$$S_{22}(s) = \frac{(s-z_1)(s-z_2)\cdots(s-z_n)}{(s-p_1)(s-p_2)\cdots(s-p_n)} \qquad (13\text{-}14)$$

where both the poles and zeros lie on the LHP.

Finally, the input admittance $Y(s)$ is obtained in terms of $S_{22}(s)$ as

$$Y(s) = \frac{1-S_{22}(s)}{1+S_{22}(s)} \qquad (13\text{-}15)$$

A Cauer-type continued fraction expansion of $Y(s)$ must now be found in such a way that the first element in the expansion corresponds to g_1 in Eq. (13-5). In addition, since $|S_{22}|_{\max}$ in Eq. (13-6) is not uniquely defined, it is necessary to minimize it over the frequency interval of the equalizer before $Y(s)$ can be synthesized. This optimization problem is dealt with in the next section.

ORDER n OF EQUALIZER

In order to proceed with the optimization problem it is useful to introduce the following substitutions:

$$\frac{1+K^2}{\varepsilon^2} = \sinh^2(na) \qquad (13\text{-}16)$$

$$\frac{K^2}{\varepsilon^2} = \sinh^2(nb) \qquad (13\text{-}17)$$

Thus, Eq. (13-12) and (13-13) become

$$p_k = -\sin\left(\frac{2k-1}{2n}\pi\right)\sinh(a) + j\cos\left(\frac{2k-1}{2n}\pi\right)\cosh(a) \qquad (13\text{-}18)$$

$$z_k = -\sin\left(\frac{2k-1}{2n}\pi\right)\sinh(b) + j\cos\left(\frac{2k-1}{2n}\pi\right)\cosh(b) \qquad (13\text{-}19)$$

with $\qquad k = 1, 2, \ldots, 2n$

The maximum and minimum values of $S_{22}(s)$ may also be expressed in terms of a and b from Eq. (13-6) as

$$\left|S_{22}\right|_{min} = \frac{\sinh(nb)}{\sinh(na)} \qquad (13\text{-}20)$$

$$\left|S_{22}\right|_{max} = \frac{\cosh(nb)}{\cosh(na)} \qquad (13\text{-}21)$$

In what follows it is also necessary to express g_1 in Eq. (13-5) in terms of Eqs (13-16) and (13-17). Closed formulas for the elements of the low-pass matching network, which avoid the necessity for expressing the input immittance of the network as the ratio of two polynomials and then breaking these down into continued fraction form, as stated below without proof. However, in order to familiarize the reader with the notation this will be derived here for $n = 1$.

The LHP poles and zeros of $S_{22}(s)$ for $n = 1$ are given from Eqs (13-18) and (13-19) by

$$p_1 = -\sinh(a) \qquad (13\text{-}22)$$

$$z_1 = -\sinh(b) \qquad (13\text{-}23)$$

Forming $S_{22}(s)$ in Eq. (13-14) leads to

$$S_{22}(s) = \mp\frac{s + \sinh(b)}{s + \sinh(a)} \qquad (13\text{-}24)$$

The minus sign is used in forming the preceding equation to accommodate the configuration of the circuit in Fig. 13-1.

Making use of Eq. (13-24) in (13-15) yields $Y(s)$ as

$$Y(s) = \frac{2s + \sinh(a) + \sinh(b)}{\sinh(a) - \sinh(b)} \qquad (13\text{-}25)$$

A continued fraction expansion of $Y(s)$ leads to the required result:

$$g_1 = \frac{2}{\sinh(a) - \sinh(b)} \qquad (13\text{-}26)$$

The terminating element is

$$g_{n+1} = \frac{\sinh(a) + \sinh(b)}{\sinh(a) - \sinh(b)} \qquad (13\text{-}27)$$

The derivation of the general result is beyond the scope of this text but may be expressed as

$$g_1 = \frac{2 \sin(\pi/2n)}{\sinh(a) - \sinh(b)} \tag{13-28}$$

$$g_r g_{r+1} = \frac{4 \sin\{[(2r-1)/2n]\pi\} \sin\{[(2r+1)/2n]\pi\}}{\sinh^2(a) + \sinh^2(b) + \sin^2[(r/n)\pi] - 2 \sinh(a) \sinh(b) \cos[(r/n)\pi]} \tag{13-29}$$

with

$$r = 1, 2, \ldots, (n-1$$

and

$$g_{n+1} = \frac{g_n}{g_1} \frac{\sinh(a) + \sinh(b)}{\sinh(a) - \sinh(b)} \tag{13-30}$$

These equations may be used for design once n is specified. Equations (13-26) and (13-27) are compatible with Eqs (13-28) to (13-30) for $n = 1$.

The low-pass prototype for n even and odd obtained from Eq. (13-28) to (13-30) are illustrated in Fig. 13-5a and b. It may also be shown that the terminating resistance or conductance defined by Eq. (13-30) is equal to the VSWR corresponding to Eq. (13-21) for n even and to Eq. (13-20) for n odd:

$$g_{n+1} = \frac{1 + |S_{22}|_{\text{max}}}{1 - |S_{22}|_{\text{max}}}, \qquad n \text{ even} \tag{13-31}$$

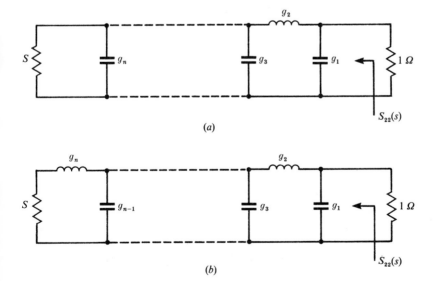

(a)

(b)

FIGURE 13-5
Ladder representation of equalizer with (a) n equal odd and (b) n equal even.

$$g_{n+1} = \frac{1 + |S_{22}|_{max}}{1 - |S_{22}|_{max}}, \quad n \text{ odd} \tag{13-32}$$

These two equations provide a good check on the numerical working of any example.

In the preceding notation, there are two variables, a and b, set by g_1 in Eq. (13-28) to be specified in addition to n before the network is completely specified, but only one quantity, g_1, is given. Thus, there are an infinite number of possible matching networks but only one is optimum. This optimum network may be determined by minimizing $|S_{22}|_{max}$ in Eq. (13-21) subject to g_1 in Eq. (13-28).

In what follows, the minimization condition is obtained by a method of undetermined multipliers. This commences by setting

$$b = \lambda a \tag{13-33}$$

The value λ of the undetermined multiplier is of no interest; it merely expresses the fact that b is soluble through a multiplication constant.

Differentiating Eq. (13-21) with respect to a (after replacing b by λa) leads to

$$\frac{d|S_{22}|_{max}}{da} = \frac{n\lambda \cosh(na)\sinh(n\lambda a) - \cosh(n\lambda a)\sinh(na)}{\cosh^2(na)} \tag{13-34}$$

Setting the preceding equation to zero yields the turning point as

$$\lambda = \frac{\tanh(na)}{\tanh(n\lambda a)} \tag{13-35}$$

Since g_1 in Eq. (13-28) is a constant it may also be differentiated to yield

$$\cosh(a) - \lambda \cosh(\lambda a) = 0 \tag{13-36}$$

This second equation expresses λ as

$$\lambda = \frac{\cosh(a)}{\cosh(\lambda a)} \tag{13-37}$$

Eliminating λ between Eqs (13-35) and (13-37) and replacing λa by b gives the required result as

$$\frac{\tanh(na)}{\cosh(a)} = \frac{\tanh(nb)}{\cosh(b)} \tag{13-38}$$

A simultaneous solution of Eqs (13-28) and (13-38) leads to values of a and b for parametric values of n. Sinh(a) and sinh(b) are obtained from a knowledge of g_1 from Fig. 13-6 and Eq. (13-28). $|S_{22}|_{max}$ may be evaluated using Eq. (13-21) or Fig. 13-7. The g values are next found by using Eqs (13-28) to (13-30) or by forming a Cauer form expansion of $Y(s)$ in Eq. (13-15). Finally, the network is impedance scaled and mapped into its band-pass configuration using the results in Chapter 12.

As an example of the use of the explicit formulas consider the design of an equalizer between a load consisting of a 20-Ω resistor in parallel with an 11-pF

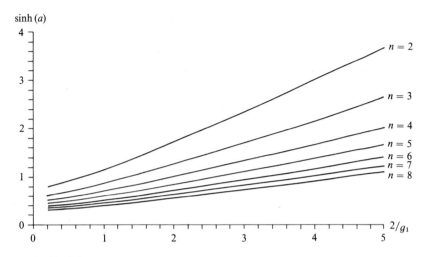

FIGURE 13-6
g_1 versus $\sinh(a)$ for equalizer with Chebyshev transfer characteristic.

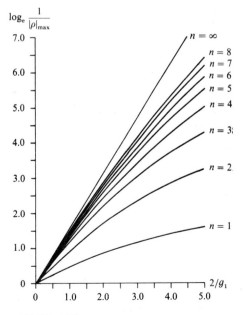

FIGURE 13-7
g_1 versus $|S_{22}|_{\max}$ for equalizer with Chebyshev transfer characteristic.

capacitor and a generator of internal impedance of 50 Ω. This network, shown in Fig. 13-8a, is to be matched between 525 and 1000 MHz with a VSWR of less than 1.10 (equivalent to a reflection coefficient of less than 0.05).

The first step in the design of the equalizer is to normalize the susceptance to 1 Ω and to a cutoff frequency of 1 rad/s. Thus g_1 becomes

$$g_1 = (\omega_2 - \omega_1)RC = 0.59$$

The scaled network is indicated in Fig. 13-8b.

The order of the equalizer is next determined by making use of Fig. 13-7. This illustration indicates that an $n = 3$ equalizer meets the VSWR of the specification. Referring to Fig. 13-7 gives

$$\log_e \frac{1}{|S_{22}|_{max}} = 3.33$$

The equivalent VSWR is

$$VSWR|_{max} = \frac{1 + |S_{22}|_{max}}{1 - |S_{22}|_{max}} = 1.074$$

a and b are now obtained from Fig. 13-6 and Eq. (13-28) as

$$\sinh(a) = 1.86, \qquad a = 1.3797$$

$$\sinh(b) = 0.165, \qquad b = 0.1642$$

The minimum reflection coefficient may be obtained from a knowledge of a and b by having recourse to Eq. (13-20):

$$|S_{22}|_{min} = \frac{\sinh(3b)}{\sinh(3a)} = 0.0164$$

The corresponding VSWR is

$$VSWR|_{min} = \frac{1 + |S_{22}|_{min}}{1 - |S_{22}|_{min}} = 1.033$$

The normalized low-pass elements of the equalizer are now obtained from Eqs (13-28) and (13-29) from a knowledge of n, a and b as

$$g_1 = 0.59$$

$$g_2 = 0.864$$

$$g_3 = 0.5098$$

FIGURE 13-8

Schematic diagram of (a) complex RC load and equalizer, (b) complex RC load normalized to 1 Ω and 1 rad/s, (c) low-pass equalizer, (d) low-pass equalizer after impedance scaling, (e) low-pass equalizer after impedance denormalization showing use of ideal transformer to cater for mismatch between real parts of generator and load impedances and (f) band-pass equalizer after frequency mapping between low-pass and band-pass networks. (R. Levy).

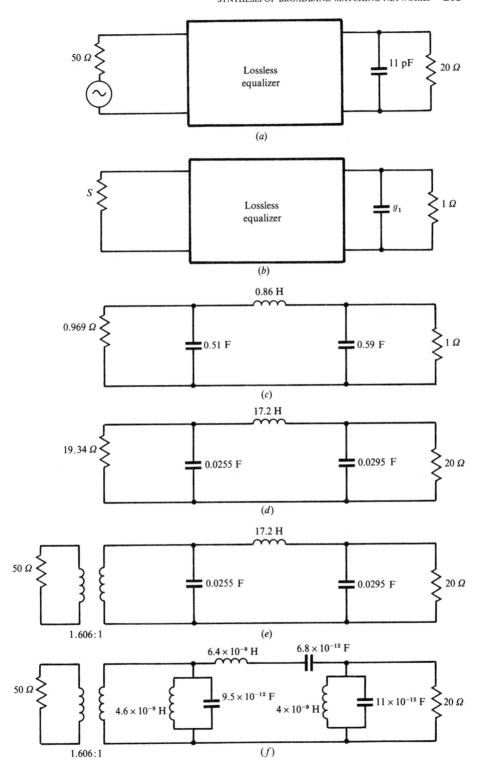

The terminating conductance is given from Eq. (13-30) as

$$g_{n+1} = 1.032$$

which is equivalent to a resistor of

$$r_{n+1} = 0.969$$

It is observed that g_1 is in agreement with the specification and that g_{n+1} is compatible with Eq. (13-32).

The network in Fig. 13-8c is now impedance scaled using the transformations in Chapter 12 as shown in Fig. 13-8d. An ideal transformer of turns ratio 1.606:1 is next introduced at the input terminals of the network to cater for the 50-Ω generator impedance in Fig. 13-8a. This is indicated in Fig. 13-8e. Finally, a low-pass to band-pass frequency transformation is used to frequency scale and centre the qualizer. The complete arrangement is illustrated in Fig. 13-8f. In agreement with the specification, the capacitance at the load is 11 pF.

For completeness, the elements of the low-pass equalizer prototype will now be derived without the aid of the closed-form formulas. This may be done by determining n, a and b, as before, and thereafter forming $S_{22}(s)$ in Eq. (13-14) in terms of the LHP poles and zeros defined by Eqs (13-18) and (13-19).

The LHP poles and zeros for $n = 3$ are given from Eqs (13-18) and (13-19) in terms of a and b by

$$p_1 = -\tfrac{1}{2}\sinh(a) + j\frac{\sqrt{3}}{2}\cosh(a)$$

$$p_2 = -\sinh(a)$$

$$p_3 = -\tfrac{1}{2}\sinh(a) - j\frac{\sqrt{3}}{s}\cosh(a)$$

$$z_1 = -\tfrac{1}{2}\sinh(b) + j\frac{\sqrt{3}}{2}\cosh(b)$$

$$z_2 = -\sinh(b)$$

$$z_3 = -\tfrac{1}{2}\sinh(b) - j\frac{\sqrt{3}}{2}\cosh(b)$$

The reflection coefficient $S_{22}(s)$ is therefore expressed by Eq. (13-14) as

$$S_{22}(s) =$$

$$\pm\frac{[s + \tfrac{1}{2}\sinh(b) - j\sqrt{3}/2\cosh(b)][s + \tfrac{1}{2}\sinh(b) + j\sqrt{3}/2\cosh(b)][s + \sinh(b)]}{[s + \tfrac{1}{2}\sinh(a) - j\sqrt{3}/2\cosh(a)][s + \tfrac{1}{2}\sinh(a) + j\sqrt{3}/2\cosh(a)][s + \sinh(a)]}$$

Substituting for a and b in the preceding equation yields

$$S_{22}(s) = \frac{s^3 + 0.33s^2 + 0.799s + 0.127}{s^3 + 3.72s^2 + 7.664s + 7.819}$$

Either this function or Eq. (13-6) may be used to display $|S_{22}(j\omega)|$ versus ω.

Taking the minus sign in forming $Y(s)$ leads to

$$Y(s) = \frac{2s^3 + 4.05s^2 + 8.46s + 7.95}{3.39s^2 + 6.87s + 7.69}$$

The ladder network may now be synthesized from the preceding equation by forming a continuous fraction expansion of $Y(s)$.

$$3.39s^2 + 6.87s + 7.69)2s^3 + 4.05s^2 + 8.46s + 7.95(0.59s$$
$$\underline{2s^3 + 4.05s^2 + 4.54s}$$
$$\qquad 3.92s + 7.95)3.39s^2 + 6.86s + 7.69(0.86s$$
$$\qquad \underline{3.39s^2 + 6.86s}$$
$$\qquad\qquad 7.69)3.92s + 7.95(0.510s$$
$$\qquad\qquad \underline{3.92s}$$
$$\qquad\qquad\qquad 7.95)7.69(0.967$$
$$\qquad\qquad\qquad \underline{7.69}$$
$$\qquad\qquad\qquad\qquad 0$$

It is observed that the g values generated in this way are in agreement, within rounding of errors, with those obtained using the closed formulas.

GAIN OF SIMPLE NEGATIVE RESISTANCE AMPLIFIER

The schematic diagram of a simple reflection type negative resistance amplifier using an ideal circulator is illustrated in Fig. 13-9. The gain of such an amplifier may be obtained by replacing R by $-R$ in the definition of the reflection coefficient of the one-port circuit terminating port 2 of the circulator.

The transducer gain of the circulator terminated in a positive resistance R is

$$|S_{22}(R)|^2 = \left|\frac{R - Z_0}{R + Z_0}\right|^2 \tag{13-39}$$

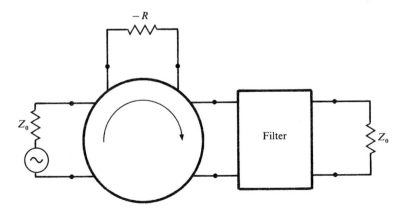

FIGURE 13-9
One-port negative resistance amplifier using three-port circulator.

If the positive real termination R is replaced by $-R$ the preceding equation becomes

$$|S_{22}(-R)|^2 = \left|\frac{R + Z_0}{R - Z_0}\right|^2 \qquad (13\text{-}40)$$

Thus, the power reflection coefficient of a negative resistance amplifier is defined as the reciprocal of that associated with a positive real termination.

$$S_{22}(-R) = \frac{1}{S_{22}(R)} \qquad (13\text{-}41)$$

Equation (13-41) relates the performance of the negative resistance reflection amplifier to that of the same network operating as an impedance matching network terminated in a positive resistance with the same value as that of the negative resistance amplifier. Thus, the problem of finding the maximum gain over a given band is reduced to that of determining the minimum value of S_{22} with the output termination $-R$ is replaced by R. This is just the familiar matching problem of a one-port circuit dealt with in the first part of this chapter.

For a negative resistance amplifier $|S_{22}|_{\text{max}}$ is replaced by $1/|S_{22}|_{\text{max}}$ in Eq. (13-3). The result is:

$$(\omega_2 - \omega_1)\log_e|S_{22}|_{\text{max}} = \frac{\pi}{RC} \qquad (13\text{-}42)$$

PROBLEMS

13-1 A loading consisting of a parallel network of a 30-Ω resistor and a 25-pF capacitor is to be matched to a generator of internal resistance of 50 Ω over a frequency interval of 300 to 500 MHz. Find the order of the equalizer.

13-2 Determine the equalizer in Prob. 13-1 using Eqs (13-28) to (13-30).

13-3 Repeat Prob. 13-2 by making use of Eqs (13-14), (13-15), (13-18) and (13-19).

13-4 Determine an equalizer to match a shunt network of $R = 28\ \Omega$ and $C = 33$ pF over the frequency interval 0 to 50 MHz.

13-5 Derive g_1 and g_2 for $n = 2$ from first principles in terms of $\sinh(a)$ and $\sinh(b)$.

13-6 Check that Eq. (13-27) agrees with the statement in Eq. (13-32).

13-7 Calculate $|S_{22}|_{\text{max}}$ for the circuit in Fig. 13-8a for $n = 2$ and 4.

13-8 Plot $|S_{22}(j\omega)|$ versus frequency for the following function which applies to the problem in Fig. 13-8.

$$S_{22}(s) = \frac{s^3 + 0.33s^2 + 0.799s + 0.127}{s^3 + 3.72s^2 + 7.664s + 7.819}$$

13-9 Obtain $|S_{21}(j\omega)|$ from $|S_{22}(j\omega)|$ in Prob. 13-8 using the unitary condition and display as a function of ω.

13-10 Making use of the low-pass to band-pass transformation given below display the band-pass response of the problem in Fig. 13-8 using $S_{22}(s)$ and $(j\omega)(s)$ in Probs 13-8 and 13-9.

13-11 Realize equalizers with $n = 2$ and 4 for the problem illustrated in Fig. 13-8a.

13-12 Obtain the low-pass and band-pass frequency responses for the equalizers in Prob. 13-11.

13-13 Construct $|S_{21}(j\omega)|$ and $|S_{22}(j\omega)|$ in Eqs (13-4) and (13-6) in terms of a and b.

13-14 Apply the frequency transformation between the low-pass and band-pass circuit in Prob. 13-10 to the results in Prob. 13-13.

13-15 Using $\sinh(a) = 1.86$, $\sinh(b) = 0.165$, $n = 3$ and the results in Probs 13-13 and 13-14, display the scattering coefficients versus ω. Compare the results with those in Probs 13-8, 13-9 and 13-10.

13-16 Verify the frequency and impedance mappings in Fig. 13-8 using the results in Chapter 12.

13-17 Using Eq. (13-3), obtain the minimum theoretical value of the reflection coefficient for the problem illustrated in Fig. 13-8.

BIBLIOGRAPHY

Levy, R., 'Explicit formulas for Chebyshev impedance matching networks, filters and interstages', *Proc. IEE*, Vol. 111, pp. 1099–1106, June 1964.

CHAPTER 14

ABCD PARAMETERS

INTRODUCTION

Two-port networks are usually described in terms of impedance, admittance and scattering matrices. Another matrix description that is often employed to describe a network is the $ABCD$ one. It has the advantage that the overall $ABCD$ description of a cascade arrangement of a number of two-port circuits is determined by evaluating the matrix product of the individual matrices. It is also of some significance in the area of microwave filter design and analysis. A knowledge of the $ABCD$ description of elementary two-port networks is therefore desirable and is formulated in this chapter. Transfer matrices which may be operated upon in a similar way are also sometimes utilized in the description of a two-port network consisting of a chain of two-port elementary circuits. The entries of this type of matrix are constructed in terms of the scattering variables of the networks. This development is dealt with in the chapter on scattering matrices.

ABCD PARAMETERS

For any linear two-port network there will be a linear relationship between input voltage and current (V_1, I_1) and output voltage and current (V_2, I_2) of the form

$$\begin{bmatrix} V_1 \\ I_1 \end{bmatrix} = \begin{bmatrix} A & B \\ C & D \end{bmatrix} \begin{bmatrix} V_2 \\ I_2 \end{bmatrix} \tag{14-1}$$

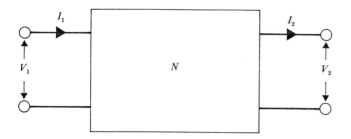

FIGURE 14-1
Voltage/current variables described by *ABCD* parameters.

A, B, C and D are parameters that may be determined from first principles or that may be derived in terms of the entries of the more common immittance matrices of such a network. In the section below, they will be expressed in terms of the open-circuit parameters of the network.

Adopting the port notation for the voltage/current variables in Fig. 14-1 indicates that the output variables of the first network become the input ones for the next circuit. This notation suggests that the overall *ABCD* description of two such networks may be evaluated by forming the matrix product of the *ABCD* matrices of the individual ones, i.e.

$$\begin{bmatrix} A & B \\ C & D \end{bmatrix} = \begin{bmatrix} A_1 & B_1 \\ C_1 & D_1 \end{bmatrix}\begin{bmatrix} A_2 & B_2 \\ C_2 & D_2 \end{bmatrix} \tag{14-2}$$

More generally, the result for the cascade arrangement of n such sections in Fig. 14-2 is

$$\begin{bmatrix} A & B \\ C & D \end{bmatrix} = \begin{bmatrix} A_1 & B_1 \\ C_1 & D_1 \end{bmatrix}\begin{bmatrix} A_2 & B_2 \\ C_2 & D_2 \end{bmatrix} \cdots \begin{bmatrix} A_n & B_n \\ C_n & D_n \end{bmatrix} \tag{14-3}$$

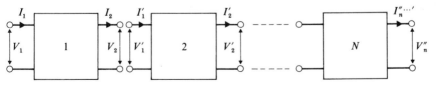

$V_1' = V_2$, $I_1' = I_2$, etc.

FIGURE 14-2
Cascade arrangement of n two-port networks.

ABCD RELATIONSHIPS

The open-circuit parameters of a two-port network with the current convention indicated in Fig. 14-3 are defined in the usual way by

$$\begin{bmatrix} V_1 \\ V_2 \end{bmatrix} = \begin{bmatrix} Z_{11} & Z_{12} \\ Z_{21} & Z_{22} \end{bmatrix} \begin{bmatrix} I_1 \\ I_2 \end{bmatrix} \tag{14-4}$$

Solving the above relationship for V_1 and I_1 for a symmetrical and reciprocal network and noting that the current notation at the output ports in the arrangements in Figs 14-1 and 14-3 are interchanged gives

$$V_1 = \left(\frac{Z_{11}}{Z_{12}}\right) V_2 + \left(\frac{Z_{11}^2}{Z_{12}} - Z_{12}\right) I_2 \tag{14-5a}$$

$$I_1 = \left(\frac{1}{Z_{12}}\right) V_2 + \left(\frac{Z_{11}}{Z_{12}}\right) I_2 \tag{14-5b}$$

The required result between the *ABCD* and open-circuit parameters is obtained by comparing coefficients (14-1) and (14-5):

$$A = \frac{Z_{11}}{Z_{12}} \tag{14-6a}$$

$$B = \frac{Z_{11}^2}{Z_{12}} - Z_{12} \tag{14-6b}$$

$$C = \frac{1}{Z_{12}} \tag{14-6c}$$

$$D = \frac{Z_{11}}{Z_{12}} \tag{14-6d}$$

The derivation of the relationships between the *ABCD* and the corresponding short-circuit parameters is left as an exercise for the reader. The *ABCD* parameters may therefore be evaluated from either a knowledge of the open- or short-circuit

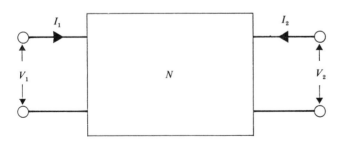

FIGURE 14-3
Voltage/current variables described by open-circuit parameters.

parameters of the circuit. The possibility that a network need not have an impedance or admittance matrix is, however, recalled. Table 14-1 summarizes the solutions of some elementary cells.

Scrutiny of Eqs (14-6) indicates that A and D are dimensionless quantities, C has the dimension of impedance and D that of admittance.

For a reciprocal and symmetrical network the preceding equations indicate that

$$A = D \qquad (14\text{-}7a)$$

$$AD - BC = 1 \qquad (14\text{-}7b)$$

Equations (14-6) indicate that in a reactance network, A and D are pure real numbers and B and C are pure imaginary ones.

TABLE 14-1
ABCD **parameters for some standard two-port circuits**

Component	Circuit	*ABCD* matrix
Series impedance	(a)	$\begin{bmatrix} 1 & Z \\ 0 & 1 \end{bmatrix}$
Shunt admittance	(b)	$\begin{bmatrix} 1 & 0 \\ Y & 1 \end{bmatrix}$
Section of line	(c)	$\begin{bmatrix} \cosh \gamma l & Z_0 \sinh \gamma l \\ \dfrac{\sinh \gamma l}{Z_0} & \cosh \gamma l \end{bmatrix}$ Z_0 = characteristic impedance γ = propagation constant
Ideal transformer	(d)	$\begin{bmatrix} \dfrac{1}{a} & 0 \\ 0 & a \end{bmatrix}$

ABCD PARAMETERS OF SERIES, SHUNT, T AND π NETWORKS

The *ABCD* parameters of series and shunt elements along a transmission line will now be derived. These will then be employed to demonstrate how to obtain some more complicated circuits such as the T and π ones using matrix multiplication.

The development starts by deducing the *ABCD* parameters of the series network in Fig. 14-4*a*. This may be readily done by developing the voltage–current relationships of the network using the appropriate Kirchoff equation:

$$V_1 - V_2 = I_2 Z \qquad (14\text{-}8a)$$

$$I_1 = I_2 \qquad (14\text{-}8b)$$

Rearranging these relationships gives

$$\begin{bmatrix} V_1 \\ I_1 \end{bmatrix} = \begin{bmatrix} 1 & Z \\ 0 & 1 \end{bmatrix} \begin{bmatrix} V_2 \\ I_2 \end{bmatrix} \qquad (14\text{-}9)$$

and

$$A = 1 \qquad (14\text{-}10a)$$

$$B = Z \qquad (14\text{-}10b)$$

$$C = 0 \qquad (14\text{-}10c)$$

$$D = 1 \qquad (14\text{-}10d)$$

Likewise, Kirchoff's equation for the shunt arrangement in Fig. 14-4*b* is

$$I_1 - I_2 = V_2 Y \qquad (14\text{-}11a)$$

$$V_1 = V_2 \qquad (14\text{-}11b)$$

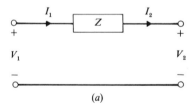

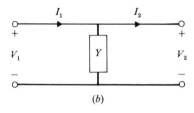

FIGURE 14-4
Voltage–current relationships for (*a*) series element and (*b*) shunt element.

or
$$\begin{bmatrix} V_1 \\ I_1 \end{bmatrix} = \begin{bmatrix} 1 & 0 \\ Y & 1 \end{bmatrix} \begin{bmatrix} V_2 \\ I_2 \end{bmatrix} \tag{14-12}$$

and
$$A = 1 \tag{14-13a}$$
$$B = 0 \tag{14-13b}$$
$$C = Y \tag{14-13c}$$
$$D = 1 \tag{14-13d}$$

The *ABCD* matrices of the standard symmetrical T and π networks will now be derived from a knowledge of the elementary series and shunt elements using simple matrix multiplication. The overall matrix of the T circuit is readily given by

$$\begin{bmatrix} A & B \\ C & D \end{bmatrix} = \begin{bmatrix} 1 & Z \\ 0 & 1 \end{bmatrix} \begin{bmatrix} 1 & 0 \\ Y & 1 \end{bmatrix} \begin{bmatrix} 1 & Z \\ 0 & 1 \end{bmatrix} \tag{14-14}$$

or
$$A = 1 + ZY \tag{14-15a}$$
$$B = Z(2 + ZY) \tag{14-15b}$$
$$C = Y \tag{14-15c}$$
$$D = 1 + ZY \tag{14-15d}$$

The *ABCD* matrix of the π network is determined by

$$\begin{bmatrix} A & B \\ C & D \end{bmatrix} = \begin{bmatrix} 1 & 0 \\ Y & 1 \end{bmatrix} \begin{bmatrix} 1 & Z \\ 0 & 1 \end{bmatrix} \begin{bmatrix} 1 & 0 \\ Y & 1 \end{bmatrix} \tag{14-16}$$

The required result is

$$A = 1 + ZY \tag{14-17a}$$
$$B = Y(2 + ZY) \tag{14-17b}$$
$$C = Z \tag{14-17c}$$
$$D = 1 + ZY \tag{14-17d}$$

It is readily verified that each of these matrices satisfies the symmetry and reciprocity conditions $A = D$ and $AD - BC = 1$.

ABCD PARAMETERS OF DIELECTRIC REGION

The *ABCD* parameters relating the fields on a uniform transmission line supporting TEM mode propagation will now be derived from first principles. The *ABCD* matrix across such a region is defined by

$$\begin{bmatrix} E_y(s) \\ H_z(s) \end{bmatrix} = \begin{bmatrix} A & B \\ C & D \end{bmatrix} \begin{bmatrix} E_y(s + \delta s) \\ H_z(s + \delta s) \end{bmatrix}$$

(14-18)

For simplicity the width of the region is taken as L_1 and its origin is taken at $s = 0$. Thus

$$E_y(0) = AE_y(l_1) + BH_z(l_1)$$

(14-19a)

$$H_z(0) = CE_y(l_1) + DH_z(l_1)$$

(14-19b)

The alternating fields in the dielectric region of a parallel plate waveguide are described with the direction of propagation along the positive z direction by

$$E_y(s) = (R \sin \gamma s + S \cosh \gamma s)$$

(14-20a)

$$H_z(s) = \frac{j\gamma}{\omega \mu_0} (R \cosh \gamma s + S \sinh \gamma s)$$

(14-20b)

where

$$\gamma^2 + \gamma_z^2 + \omega^2 \mu_0 \varepsilon_0 \varepsilon_r = 0$$

(14-21a)

$$\gamma_y = 0$$

(14-21b)

and R and S are arbitrary constants which are to be determined. Forming the fields at $s = 0$ and $s = l_1$ and substituting these into Eqs (14-19) gives

$$S = A(R \sinh \gamma l_1 + S \cosh \gamma l_1) + \frac{jB\gamma}{\omega \mu_0} (R \cosh \gamma l_1 + S \sinh \gamma l_1)$$

(14-22a)

$$\frac{jB\gamma}{\omega \mu_0} R = C(R \sinh \gamma l_1 + S \cosh \gamma l_1) + \frac{jD\gamma}{\omega \mu_0} (R \cosh \gamma l_1 + S \sinh \gamma l_1)$$

(14-22b)

Rearranging gives

$$S\left(1 - A \cosh \gamma l_1 - \frac{jB\gamma}{\omega \mu_0} \sinh \gamma l_1\right) = R\left(A \sinh \gamma l_1 + \frac{jB\gamma}{\omega \mu_0} \cosh \gamma l_1\right)$$

(14-23a)

$$S\left(C \cosh \gamma l_1 + \frac{jD\gamma}{\omega \mu_0} \sinh \gamma l_1\right) = R\left(\frac{j\gamma}{\omega \mu_0} - C \sinh \gamma l_1 - \frac{jD\gamma}{\omega \mu_0} \cosh \gamma l_1\right)$$

(14-23b)

Since these equations are required to hold for any value of S and R, the quantities in parentheses may be set to zero. This gives four equations from which the arbitrary constants $ABCD$ may be derived:

$$1 = A \cosh \gamma l_1 + \frac{jB\gamma}{\omega \mu_0} \sinh \gamma l_1$$

(14-24a)

$$0 = A \sinh \gamma l_1 + \frac{jB\gamma}{\omega \mu_0} \cosh \gamma l_1$$

(14-24b)

$$0 = C \cosh \gamma l_1 + \frac{jD\gamma}{\omega\mu_0} \sinh \gamma l_1 \qquad (14\text{-}24c)$$

$$\frac{j\gamma}{\omega\mu_0} = C \sinh \gamma l_1 + \frac{jD\gamma}{\omega\mu_0} \cosh \gamma l_1 \qquad (14\text{-}24d)$$

These equations may be solved for the *ABCD* parameters by first putting them in matrix form:

$$\begin{bmatrix} 1 & 0 \\ 0 & \dfrac{j\gamma}{\omega\mu_0} \end{bmatrix} = \begin{bmatrix} \cosh \gamma l_1 & \dfrac{j\gamma}{\omega\mu_0} \sinh \gamma l_1 \\ \sinh \gamma l_1 & \dfrac{j\gamma}{\omega\mu_0} \cosh \gamma l_1 \end{bmatrix} \begin{bmatrix} A & C \\ B & D \end{bmatrix} \qquad (14\text{-}25)$$

and then premultiplying both sides by the inverse of the matrix containing the hyperbolic functions. The result is

$$A = \cosh \gamma l_1 \qquad (14\text{-}26a)$$

$$B = \frac{\omega\mu_0}{j\gamma} \sinh \gamma l_1 \qquad (14\text{-}26b)$$

$$C = \frac{j\gamma}{\omega\mu_0} \sinh \gamma l_1 \qquad (14\text{-}26c)$$

$$D = \cosh \gamma l_1 \qquad (14\text{-}26d)$$

and

$$\frac{\omega\mu_0}{j\gamma} \qquad (14\text{-}26e)$$

is the wave impedance Z_0.

These relationships satisfy the identities in Eqs (14-7a, b):

$$AD - BC = 1$$

$$A = D$$

The *ABCD* matrix for air is obtained as a special case of that given above, by putting $\varepsilon_r = 1$.

CIRCUIT RELATIONSHIPS

The input impedance of a cascade of n sections is readily expressed in terms of its overall *ABCD* parameters with the aid of Eq. (14-1):

$$Z_{in} = \frac{V_{in}}{I_{in}} = \frac{AZ_{out} + B}{CZ_{out} + D} \qquad (14\text{-}27)$$

The corresponding reflection coefficient (ρ) at the input terminals is likewise given by

$$\rho = \frac{Z_0 - Z_{\text{in}}}{Z_0 + Z_{\text{in}}} \tag{14-28}$$

Z_{in}, Z_{out} and Z_0 are the input impedance, output impedance and characteristic impedance of the generator network respectively.

The transmission coefficient for such a lossless network is given by the unitary condition by

$$\rho\rho^* + \tau\tau^* = 1 \tag{14-29}$$

In the special case of a lossless network with the source and load resistances equal

$$Z_{\text{in}} = \frac{A + B}{C + D} \tag{14-30}$$

and noting that A and D are pure real numbers and B and C are pure imaginary numbers gives

$$\rho = \frac{(A - D) - (B - C)}{(A + D) + (B + C)} \tag{14-31a}$$

$$\rho^* = \frac{(A - D) - (B - C)^*}{(A + D) + (B + C)^*} \tag{14-31b}$$

Making use of the unitary condition and the reciprocity condition below

$$AD - BC = 1$$

gives

$$\tau\tau^* = \frac{1}{1 + \frac{1}{4}(A - D)^2 - \frac{1}{4}(B - C)^2} \tag{14-32}$$

For the special case of a symmetrical network

$$A = D$$

and

$$\tau\tau^* = \frac{1}{1 - \frac{1}{4}(B - C)^2} \tag{14-33}$$

The general result is described by

$$\tau\tau^* = \frac{1}{\frac{1}{4}\left|A\sqrt{R_L/R_S} + D\sqrt{R_S/R_L} + \sqrt{B/(R_L R_S)} + C\sqrt{R_L R_S}\right|^2} \tag{14-34}$$

where R_S is the source impedance and R_L is the load impedance.

As a simple example of the use of the $ABCD$ notation, it is desired to construct the insertion loss function of the $n = 3$ doubly terminated prototype in Fig. 14-5.

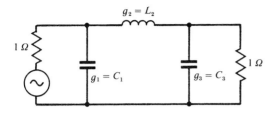

FIGURE 14-5
Doubly terminated $n = 3$ Butterworth circuit.

Here

$$\begin{bmatrix} A & B \\ C & D \end{bmatrix} = \begin{bmatrix} 1 & 0 \\ j\omega & 1 \end{bmatrix} \begin{bmatrix} 1 & j2\omega \\ 0 & 1 \end{bmatrix} \begin{bmatrix} 1 & 0 \\ j\omega & 1 \end{bmatrix}$$

or

$$A = 1 - \omega^2$$

$$B = j2\omega$$

$$C = j(2\omega - 2\omega^2)$$

$$D = 1 - \omega^2$$

and making use of Eqs (14-33) or (14-34) gives the required result:

$$\tau\tau^* = \frac{1}{1 + (\omega^2)^3}$$

in keeping with the Butterworth low-pass generating function.

PROBLEMS

14-1 Deduce the insertion loss function of a degree 3 tee network for which $g_1 = 1$ H, $g_2 = 2$ F, $g_3 = 1$ H.

14-2 Derive the insertion loss functions for Butterworth prototypes for $n = 1, 2, 4$ and 5. The recurrence formula for the element values is

$$g_r = 2 \sin\left(\frac{2r-1}{2n}\pi\right) \qquad \text{for } r = 1, 2, \ldots, n$$

14-3 Derive the insertion loss function in Eq. (14-34) of a reactance network between unequal generator and load resistors.

14-4 Obtain the relationship between the *ABCD* parameters and the short-circuit parameters of a two-port network.

14-5 Obtain the first Kuroda forms in Chapter 18 using the *ABCD* formulation.

14-6 Determine the relationship between the *ABCD* and open-circuited parameters for a reciprocal but non-symmetrical network.

14-7 Write a computer program to calculate the overall $ABCD$ matrix of a network in terms of those of the individual ones.

BIBLIOGRAPHY

Clark, W. P., Hering, K. H., and Charlton, D. A., 'TE-mode solutions for partially ferrite filled rectangular waveguide using $ABCD$ matrices', IEEE, MTT Symposium, 1966.

CHAPTER
15

IMMITTANCE INVERTERS AND CIRCUITS

INTRODUCTION

It is often desirable to replace filters with two kinds of elements by ones with one type spaced by unit elements or immittance inverters. The use of such immittance inverters is illustrated in connection with the development of low-pass, high-pass, band-pass and band-stop prototypes. The high-pass solution may be derived from the low-pass one by substituting the series inductors (shunt capacitors) in the former prototype by series capacitors (shunt inductors) in the latter network with appropriate element values in such a way that the immittance inverters remain unaltered. The band-pass and band-stop solutions are separately summarized. Whether impedance or admittance inverters are used is dictated in part by the type of transmission line employed. The description of some possible immittance inverters are included for completeness.

The use of immittance inverters in the design of microwave structures is illustrated in connection with that of a directly coupled finline band-pass filter.

IMMITTANCE INVERTERS

The low-pass filter prototype, which forms the basis for the high-pass, band-pass and band-stop filters is a Cauer-type ladder network, whose topology is not always

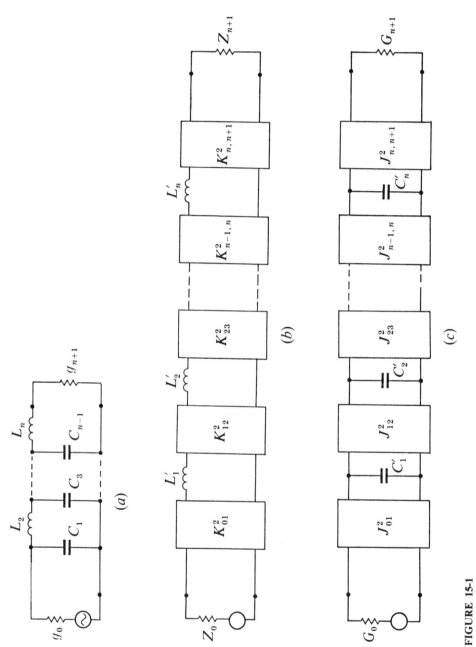

FIGURE 15-1
Cauer low-pass ladder network using (a) shunt and series elements, (b) shunt elements and impedance

the most practical circuit layout at very high frequencies. A more desirable filter architecture would be one involving only shunt or only series elements spaced by immittance inverters. Immittance inverters provide one means of replacing a Cauer-type low-pass ladder network using lumped L and C values by one using only L values and impedance inverters or one using only C values and admittance inverters. This will now be demonstrated in connection with the low-pass lumped element prototype in Fig. 15-1a.

An impedance inverter K_{ij} transforms an impedance $Z_j(s)$ into an impedance $Z_i(s)$:

$$Z_i(s)Z_j(s) = K_{ij}^2 \qquad (15\text{-}1)$$

and an admittance inverter J_{ij} similarly maps an admittance $Y_j(s)$ into $Y_i(s)$ according to

$$Y_i(s)Y_j(s) = J_{ij}^2 \qquad (15\text{-}2)$$

Repeated introduction of either the first or second immittance inverter to the low-pass prototype in Fig. 15-1a maps it to either the topology in Fig. 15-1b or c. The standard topology assumes that the first element of the low-pass prototype is a shunt one. The mapping between the low-pass prototype in Fig. 15-1a and that in Fig. 15-15-1b will now be demonstrated for a degree $n = 3$ network. The derivation starts by expanding the impedance of the network in a first Cauer form:

$$Z_1(s) = \cfrac{1}{sC_1 + \cfrac{1}{sL_2 + \cfrac{1}{sC_3 + 1/g_{n+1}}}} \qquad (15\text{-}3)$$

$Z_1(s)$ is now simulated by an impedance inverter K_{01} terminated by an impedance $Z_1'(s)$:

$$Z_1'(s) = \frac{K_{01}^2}{Z_1(s)} \qquad (15\text{-}4)$$

Scrutiny of the two preceding equations indicates that the required impedance $Z_1'(s)$ at the output terminals of the immittance inverter is

$$Z_1'(s) = K_{01}^2 \left[sC_1 + \cfrac{1}{sL_2 + \cfrac{1}{sC_3 + \cfrac{1}{g_{n+1}}}} \right] \qquad (15\text{-}5)$$

If this impedance is realized as an inductance L' in series with an impedance $Z_2(s)$:

$$Z_1'(s) = sL_1' + Z_2(s) \qquad (15\text{-}6)$$

then

$$L_1' = K_{01}^2 C_1 \tag{15-7}$$

$$Z_2(s) = \cfrac{K_{01}^2}{sL_2 + \cfrac{1}{sC_3 + \cfrac{1}{g_{n+1}}}} \tag{15-8}$$

This operation permits the network in Fig. 15-2a to be put into the form indicated in Fig. 15-2b and completes the first cycle of the synthesis. It is of note that either L_1' or K_{01} in this arrangement is completely arbitrary. The derivation of the required network now proceeds by repeating this cycle. It therefore continues by writing $Z_2(s)$ in terms of an impedance inverter K_{12} and a remainder impedance $Z_2'(s)$ to be determined:

$$Z_2'(s) = \frac{K_{12}^2}{Z_2(s)} \tag{15-9}$$

Eliminating $Z_2(s)$ once again in this equation indicates that $Z_2'(s)$ is given by

$$Z_2'(s) = \frac{K_{12}^2}{K_{01}^2}\left[sL_2 + \cfrac{1}{sC_3 + \cfrac{1}{g_{n+1}}} \right] \tag{15-10}$$

This impedance may be written as

$$Z_2'(s) = sL_1' + Z_3(s) \tag{15-11}$$

where

$$L_2' = \frac{K_{12}^2}{K_{01}^2} L_2 \tag{15-12}$$

$$Z_3(s) = \frac{K_{12}^2}{K_{01}^2}\left[\cfrac{1}{sC_3 + \cfrac{1}{g_{n+1}}} \right] \tag{15-13}$$

The circuit in Fig. 15-2b may therefore be replaced by that in Fig. 15-2c. The derivation now continues by extracting an impedance inverter K_{23} such that

$$Z_3'(s) = \frac{K_{23}^2}{Z_3(s)} \tag{15-14}$$

$Z_3'(s)$ is now fixed by combining the two preceding equations. The result is

$$Z_3'(s) = \frac{K_{01}^2 K_{23}^2}{K_{12}^2}\left[sC_3 + \frac{1}{g_{n+1}} \right] \tag{15-15}$$

Writing this impedance in partial fractions

$$Z_3'(s) = sL_3' + Z_4(s) \tag{15-16}$$

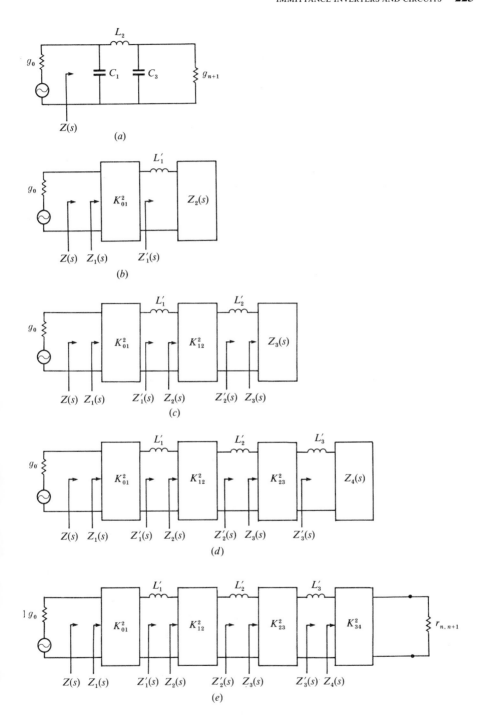

FIGURE 15-2
Derivation of low-pass ladder network using impedance inverters.

indicates that it may be synthesized as a series inductance L_3':

$$L_3' = \frac{K_{01}^2 K_{23}^2}{K_{12}^2} C_3 \qquad (15\text{-}17)$$

and a remainder impedance $Z_4(s)$:

$$Z_4(s) = \frac{K_{01}^2 K_{23}^2}{K_{12}^2} g_{n+1} \qquad (15\text{-}18)$$

The required network after the extraction of this immittance inverter is illustrated in Fig. 15-2d. Since $Z_4(s)$ represents a resistance the synthesis is now complete except for the extraction of one additional impedance inverter K_{34} such that

$$Z_4'(s) = \frac{K_{34}^2}{Z_4(s)} \qquad (15\text{-}19)$$

If $Z_4'(s)$ is taken as r_{n+1} then

$$K_{34}^2 = \frac{K_{01}^2 K_{23}^2}{K_{12}^2} \frac{r_{n+1}}{g_{n+1}} \qquad (15\text{-}20)$$

The required network is indicated in Fig. 15-2e. This network is, however, scaled to a generator impedance g_0 and a cutoff frequency of 1 rad/s. It may now be impedance scaled by recognizing that the element C_1 corresponds to the element g_1 in the low-pass prototype. Impedance scaling this quantity by substituting g_1 by

$$g_1 \left(\frac{g_0}{R_0} \right)$$

indicates that C_1 may be replaced by

$$C_1 \rightarrow g_1 \left(\frac{g_0}{R_0} \right)$$

and that the first inverter K_{01} is fixed according to

$$K_{01} = \sqrt{\frac{R_0 L_1'}{g_0 g_1}} \qquad (15\text{-}21a)$$

Likewise, making use of the fact that L_2 and C_3 corresponds to the elements g_2 and g_3 in the low-pass prototype that impedance scaling, these quantities indicate that the former quantities may be replaced by

$$L_2 \rightarrow g_2 \left(\frac{g_0}{R_0} \right)$$

$$C_3 \rightarrow g_3 \left(\frac{g_0}{R_0} \right)$$

and that the impedance inverters K_{12} and K_{23} are set by

$$K_{12} = \sqrt{\frac{L_1'L_1'}{g_1g_2}}$$

$$K_{23} = \sqrt{\frac{L_2'L_3'}{g_2g_3}}$$

respectively.

The general nature for the inner inverters is therefore

$$K_{j,j+1} = \sqrt{\frac{L_j'L_{j+1}'}{g_jg_{j+1}}} \tag{15-21b}$$

Similar considerations suggest that the outer inverter is described by

$$K_{n,n+1} = \sqrt{\frac{L_n'R_{n+1}'}{g_ng_{n+1}}} \tag{15-21c}$$

It is of note that the inductors $L_1', L_2', \ldots, L_n'$ appearing in the descriptions of the impedance inverters are completely arbitrary. It is also of note that whereas the network has been impedance scaled, its cutoff frequency is still 1 rad/s. Scrutiny of the entries of the impedance inverters indicates that frequency scaling the network to some other frequency leaves the impedance inverters unchanged. The series inductors are, however, modified $(L_j' \rightarrow L_j'/\omega_0)$. The possibility of making the impedance inverters the independent variables and the series inductors the dependent ones is also understood.

It is further observed that the circuit in Fig. 15-2 is not canonical since the impedance inverters do not contribute to the overall amplitude response of the filter but only provide a practical layout of the circuit elements. The design is completed by physically realizing the impedance inverters. Some possible lumped element immittance inverters are discussed in the next section.

The derivation of the topology using admittance inverters is left as an exercise for the reader. The result may be summarized as

$$J_{01} = \sqrt{\frac{G_0C_1'}{g_0g_1}} \tag{15-22a}$$

$$J_{j,j+1} = \sqrt{\frac{C_j'C_{j+1}'}{g_jg_{j+1}}} \tag{15-22b}$$

$$J_{n,n+1} = \sqrt{\frac{C_n'G_{n+1}'}{g_ng_{n+1}}} \tag{15-22c}$$

HIGH-PASS FILTERS USING
IMMITTANCE INVERTERS

The derivation of the high-pass prototype with a cutoff frequency of 1 rad/s using one kind of element and impedance inverters starts by introducing the frequency transformation

$$s \text{ by } \frac{1}{s} \qquad (15\text{-}23)$$

in the impedance description of the network:

$$Z_1(s) = \cfrac{1}{\cfrac{g_1}{s} + \cfrac{1}{\cfrac{g_2}{s} + \cfrac{1}{\cfrac{g_3}{s} + \cfrac{1}{g_4}}}} \qquad (15\text{-}24)$$

Synthesising this impedance in terms of impedance inverters

$$K_{01}, K_{12}, K_{23}, K_{34}$$

and series capacitors

$$C'_1, C'_2, C'_3$$

indicates that

$$K_{02}^2 = \frac{1}{g_1 C'_1}$$

$$K_{12}^2 = \frac{1}{g_1 g_2 C'_1 C'_2}$$

$$K_{23}^2 = \frac{1}{g_2 g_3 C'_2 C'_3}$$

$$K_{34}^2 = \frac{1}{g_3 g_4 C'_3 C'_4}$$

The required result is now obtained by impedance scaling the network:

$$g_1 \rightarrow g_1\left(\frac{g_0}{R_0}\right)$$

$$g_2 \rightarrow g_2\left(\frac{g_0}{R_0}\right)$$

$$g_3 \rightarrow g_3\left(\frac{g_0}{R_0}\right)$$

The result is

$$K_{01}^2 = \frac{R_0}{g_0 g_1 C_1'} \tag{15-25a}$$

$$K_{j,j+1}^2 = \frac{1}{g_j g_{j+1} C_j' C_{j+1}'} \tag{15-25b}$$

$$K_{n,n+1}^2 = \frac{R_{n+1}}{g_n g_{n+1} C_n'} \tag{15-25c}$$

Scrutiny of this result and that associated with the corresponding low-pass prototype indicates that the impedance inverters met in the description of the former circuit are also applicable in this situation provided the series inductors (L_j') of the low-pass prototype are replaced by series capacitors (C_j') in the case of the high-pass one, whose values are given by

$$C_j' = \frac{1}{L_j'} \tag{15-26}$$

The solution to the dual problem utilizing shunt inductors and admittance inverters may be summarized by

$$J_{01}^2 = \frac{G_0}{g_0 g_1 L_1'} \tag{15-27a}$$

$$J_{j,j+1}^2 = \frac{1}{g_j g_{j+1} L_j' L_{j+1}'} \tag{15-27b}$$

$$J_{n,n+1}^2 = \frac{G_{n+1}}{g_n g_{n+1} L_n'} \tag{15-27c}$$

This solution may again be derived from that of the corresponding low-pass one by replacing the shunt capacitors (C_j') in the low-pass prototype by shunt inductors (L_j') whose values are given by

$$L_j' \to \frac{1}{C_j'} \tag{15-28}$$

Figure 15-3 indicates the layouts of these types of circuit.

BAND-PASS FILTERS USING IMMITTANCE INVERTERS

The derivation of the band-pass prototype begins by introducing the low-pass to band-pass frequency transformation

$$s \to \frac{\omega_0}{\text{BW}} \left(\frac{s}{\omega_0} + \frac{\omega_0}{s} \right) \tag{15-29}$$

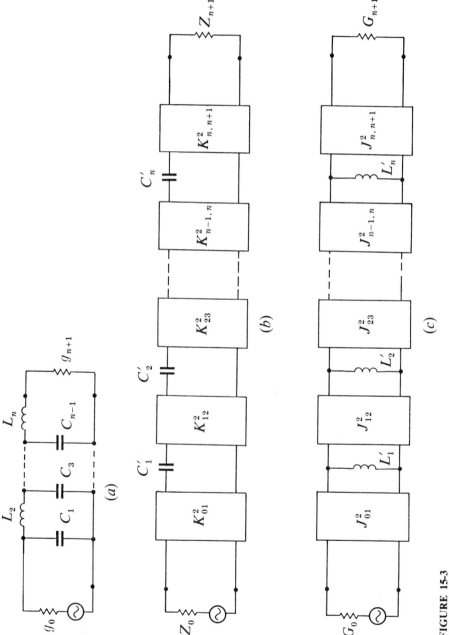

FIGURE 15-3
High-pass ladder networks using immittance inverters.

in the input impedance of the circuit. The result is

$$Z_1(s) = \cfrac{1}{\dfrac{\omega_0}{\text{BW}}\left(\dfrac{s}{\omega_0} + \dfrac{\omega_0}{s}\right)g_1 + \cfrac{1}{\dfrac{\omega_0}{\text{BW}}\left(\dfrac{s}{\omega_0} + \dfrac{\omega_0}{s}\right)g_2 + \cfrac{1}{\dfrac{\omega_0}{\text{BW}}\left(\dfrac{s}{\omega_0} + \dfrac{\omega_0}{s}\right)g_3 + \dfrac{1}{g_4}}}} \qquad (15\text{-}30)$$

This impedance is now synthesized in terms of impedance inverters and series lumped element resonators. If the synthesis starts with the extraction of an impedance inverter K_{01} then the first series lumped element resonator is defined by

$$sL_1' + \frac{1}{sC'} = K_{01}^2\left[\frac{\omega_0}{\text{BW}}\left(\frac{s}{\omega_0} + \frac{\omega_0}{s}\right)\right]g_1$$

which satisfies

$$L_1' = \frac{K_{01}^2 g_1}{\text{BW}} \qquad (15\text{-}31a)$$

$$C_1' = \frac{\text{BW}}{K_{01}^2 g_1 \omega_0^2} \qquad (15\text{-}31b)$$

and $$\omega_0^2 L_1' C_1' = 1 \qquad (15\text{-}31c)$$

The first impedance inverter is therefore deduced from either Eq. (15-31a) or (15-31b) as

$$K_{01}^2 = \frac{L_1'(\text{BW})}{g_1}$$

Impedance scaling this quantity by replacing

$$g_1 \rightarrow g_1\left(\frac{g_0}{R_0}\right)$$

gives the required result:

$$K_{01} = \sqrt{\frac{R_0 x_1 w}{g_0 g_1}} \qquad (15\text{-}32a)$$

where w is a bandwidth parameter

$$w = \frac{\text{BW}}{\omega_0}$$

and x_j is the reactance slope parameter

$$x_j = \frac{\omega_0}{2}\left.\frac{\partial X_j}{\partial \omega}\right|_{\omega = \omega_0} = \omega_0 L_j'$$

The descriptions of the other inverters follow in a like manner:

$$K_{j,j+1} = w\sqrt{\frac{x_j x_{j+1}}{g_j g_{j+1}}} \tag{15-32b}$$

$$K_{n,n+1} = \sqrt{\frac{x_n R_{n+1} w}{g_n g_{n+1}}} \tag{15-32c}$$

R_0 and R_{n+1} are the input and output impedances respectively. Once the immittance inverters are set from a knowledge of L_j', C_j' is calculated from the resonance condition in Eq. (15-31c).

The dual equations for the arrangement in Fig. 15-4 employing admittance inverters are

$$J_{01} = \sqrt{\frac{G_0 b_1 w}{g_0 g_1}} \tag{15-33a}$$

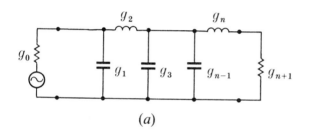

(a)

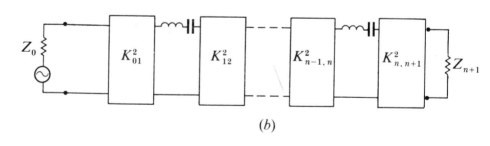

(b)

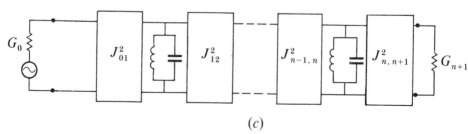

(c)

FIGURE 15-4
Band-pass networks with impedance and admittance inverters.

$$J_{j,j+1} = w\sqrt{\frac{b_j b_{j+1}}{g_j g_{j+1}}} \qquad (15\text{-}33b)$$

$$J_{n,n+1} = \sqrt{\frac{b_n G_{n+1} w}{g_n g_{n+1}}} \qquad (15\text{-}33c)$$

and b_j is the susceptance slope parameter

$$b_j = \frac{\omega_0}{2}\left.\frac{\partial Y_j}{\partial \omega}\right|_{\omega=\omega_0} = \omega_0 C_j' $$

G_0 and G_{n+1} are the input and output conductances respectively. These two situations are illustrated in Fig. 15-4.

BAND-STOP FILTERS USING IMMITTANCE INVERTERS

The derivation of the band-stop relationships proceeds by making use of the appropriate low-pass to band-stop frequency transformation

$$s' \rightarrow \frac{BW}{\omega_0\left(\dfrac{s}{\omega_0} + \dfrac{\omega_0}{s}\right)} \qquad (15\text{-}34)$$

in forming the one-port immittance functions of the circuit. The required result for the topology using impedance inverters is

$$K_{01}^2 = \frac{R_0}{g_0 g_1 w b_1} \qquad (15\text{-}35a)$$

$$K_{j,j+1}^2 = \frac{1}{w g_j g_{j+1} b_j b_{j+1}} \qquad (15\text{-}35b)$$

$$K_{n,n+1}^2 = \frac{R_{n+1}}{g_n g_{n+1} w b_n} \qquad (15\text{-}35c)$$

and for that using admittance ones

$$J_{01}^2 = \frac{G_0}{g_0 g_1 w x_1} \qquad (15\text{-}36a)$$

$$J_{j,j+1}^2 = \frac{1}{w g_j g_{j+1} x_j x_{j+1}} \qquad (15\text{-}36b)$$

$$J_{n,n+1}^2 = \frac{G_{n+1}}{g_n g_{n+1} w x_n} \qquad (15\text{-}36c)$$

These two situations are illustrated in Fig. 15-5.

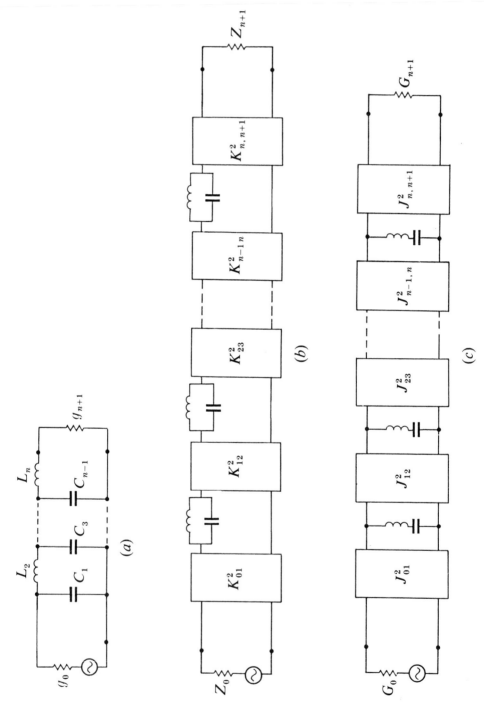

FIGURE 15-5

PRACTICAL IMMITTANCE INVERTERS

One simple impedance inverter already met in this chapter is the quarter-wave long transmission line. Four other practical possibilities that have some merit are illustrated in Fig. 15-6. Although each circuit requires negative elements for its realization these can be absorbed in adjacent elements. The equivalence between the second of these circuits and the ideal impedance inverter consisting of a UE of characteristic impedance K will now be demonstrated. This may be done by establishing a one-to-one equivalence between its $ABCD$ parameters and those of the ideal inverter. The derivation begins by evaluating the $ABCD$ matrix of a uniform transmission line of characteristic impedance Z_0.

$$\begin{bmatrix} A & B \\ C & D \end{bmatrix} = \begin{bmatrix} \cos\theta & jZ_0\sin\theta \\ \dfrac{j\sin\theta}{Z_0} & \cos\theta \end{bmatrix} \tag{15-37}$$

at

$$\theta = 90 \text{ deg} \tag{15-38}$$

The result is

$$\begin{bmatrix} A & B \\ C & D \end{bmatrix} = \begin{bmatrix} 0 & jK \\ \dfrac{j}{K} & 0 \end{bmatrix} \tag{15-39}$$

where

$$K = Z_0 \tag{15-40}$$

The derivation continues by recalling the overall $ABCD$ parameters of a T network consisting of series impedances (Z) and a shunt admittance (Y):

$$A = 1 + ZY \tag{15-41a}$$

$$B = Z(2 + ZY) \tag{15-41b}$$

$$C = Y \tag{15-41c}$$

$$D = 1 + ZY \tag{15-41d}$$

In the situation considered here

$$Z = \frac{1}{-j\omega C} \tag{15-42a}$$

$$Y = j\omega C \tag{15-42b}$$

Evaluating these parameters indicates that

$$\begin{bmatrix} A & B \\ C & D \end{bmatrix} = \begin{bmatrix} 0 & \dfrac{j}{\omega C} \\ j\omega C & 0 \end{bmatrix} \tag{15-43}$$

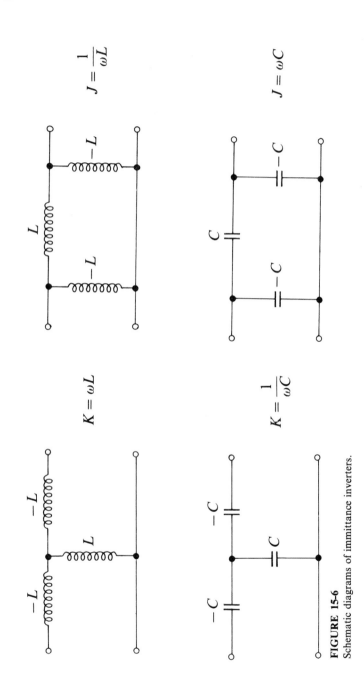

FIGURE 15-6
Schematic diagrams of immittance inverters.

Scrutiny of this matrix and that of the ideal impedance inverter suggests that the two are equivalent provided

$$K = \frac{1}{\omega C} \tag{15-44}$$

The impedance level of the other possibilities in Fig. 15-6 follows.

A much used practical immittance inverter is a T or π circuit embedded between negative lengths of uniform transmission lines. The negative lengths of line are then absorbed in any connecting line. Such T and π circuits are often met in the descriptions of metal posts, irises or similar discontinuities in waveguides or other transmission lines. In practice, the series elements in the T circuit and the shunt ones in the π circuit are often neglected. The required conditions for the T circuits in Fig. 15-7 will now be derived under this assumption. Forming the $ABCD$ matrix of this circuit disregarding the series elements gives

$$\begin{bmatrix} A & B \\ C & D \end{bmatrix} = \begin{bmatrix} \cos\phi - \dfrac{\sin\phi}{2X_p/Z_0} & -jZ_0\left(\sin\phi + \dfrac{\cos\phi - 1}{2X_p/Z_0}\right) \\ \dfrac{j}{Z_0}\left(\sin\phi + \dfrac{\cos\phi + 1}{2X_p/Z_0}\right) & \cos\phi - \dfrac{\sin\phi}{2X_p/Z_0} \end{bmatrix} \tag{15-45}$$

Comparing this matrix with that of the ideal impedance inverter

$$\begin{bmatrix} 0 & \pm jK \\ \pm \dfrac{j}{K} & 0 \end{bmatrix}$$

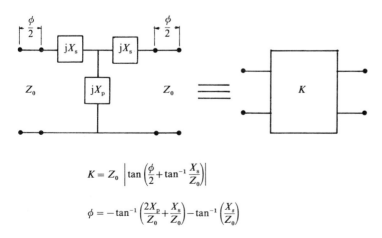

$$K = Z_0 \left| \tan\left(\frac{\phi}{2} + \tan^{-1}\frac{X_s}{Z_0}\right) \right|$$

$$\phi = -\tan^{-1}\left(\frac{2X_p}{Z_0} + \frac{X_s}{Z_0}\right) - \tan^{-1}\left(\frac{X_s}{Z_0}\right)$$

FIGURE 15-7
Schematic diagram of practical impedance inverter.

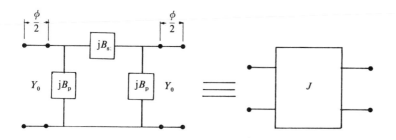

$$J = Y_0 \left| \tan\left(\frac{\phi}{2} + \tan^{-1}\frac{B_p}{Y_0}\right) \right|$$

$$\phi = -\tan^{-1}\left(\frac{2B_s}{Y_0} + \frac{B_p}{Y_0}\right) - \tan^{-1}\left(\frac{B_p}{Y_0}\right)$$

FIGURE 15-8
Schematic diagram of practical admittance inverter.

indicates that

$$\tan \phi = \frac{2X_p}{Z_0} \tag{15-46a}$$

$$\frac{K}{Z_0} = \left|\tan\frac{\phi}{2}\right| \tag{15-46b}$$

Eliminating ϕ between these two relationships also gives

$$\frac{Z_0}{X_p} = \frac{Z_0}{K} - \frac{K}{Z_0} \tag{15-46c}$$

The dual problem is illustrated in Fig. 15-8.

FINLINE BAND-PASS FILTER

A classic directly coupled band-pass filter in finline based on a low-pass lumped element prototype consists of a series of half-wave long cavities spaced by metal or inductive posts. Figure 15-9 indicates one practical construction. One possible equivalent circuit for this type of structure is illustrated in Fig. 15-10. It is obtained by representing each discontinuity by an equivalent inductive T circuit and each cavity by a uniform transmission line. The design of this type of filter proceeds by absorbing the reactive T circuits into ideal impedance inverters as a preamble to forming a one-to-one equivalence between it and a suitable band-pass prototype consisting of series resonators and ideal impedance inverters. The series resonators of the band-pass prototype are in this instance implemented by half-wave long distributed cavity resonators. The impedance inverters of the network are often realized by neglecting the series elements of the metal posts and embedding the

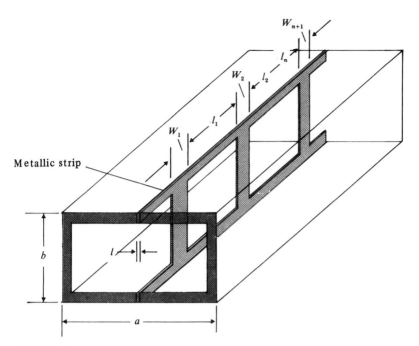

FIGURE 15-9
Schematic diagram of finline band-pass filter.

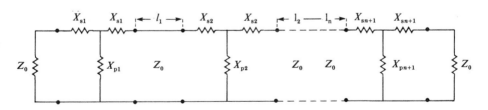

FIGURE 15-10
Equivalent circuit of finline bandpass filter.

shunt ones between suitable negative lengths of uniform transmission lines in the manner discussed in the previous section. These additional lengths of transmission lines are then absorbed in the half-wave long cavities. The equivalent circuit of the finline circuit using this topology is depicted in Fig. 15-11. The design is complete once the K values are specified from the filter specification.

Combining Eqs (15-32) and (15-46a) gives the required design equations:

$$\frac{Z_0}{X_{01}} = Z_0 \left(\frac{g_0 g_1}{x_1 Z_0 w} \right)^{1/2} - \frac{1}{Z_0} \left(\frac{x_1 Z_0 w}{g_0 g_1} \right)^{1/2} \tag{15-47a}$$

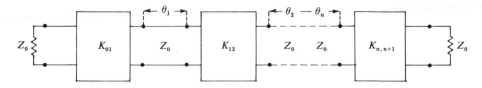

FIGURE 15-11
Equivalent circuit of finline bandpass filter using impedance inverters.

$$\frac{Z_0}{X_{j,j+1}} = Z_0 \left(\frac{g_j g_{j+1}}{x_j x_{j+1} w^2} \right)^{1/2} - \frac{1}{Z_0} \left(\frac{x_j x_{j+1} w^2}{g_j g_{j+1}} \right)^{1/2} \tag{15-47b}$$

$$\frac{Z_0}{X_{n,n+1}} = Z_0 \left(\frac{g_n g_{n+1}}{x_n Z_{n+1} w} \right)^{1/2} - \frac{1}{Z_0} \left(\frac{x_n Z_{n+1} w}{g_n g_{n+1}} \right)^{1/2} \tag{15-47c}$$

The reactance slope parameter for a half-wave section of characteristic impedance Z_j is

$$x_j = \frac{\pi Z_j}{2} \tag{15-48}$$

If the characteristic impedance of each half-wave section is made equal to the characteristic impedance Z_0 then

$$\frac{K_{01}^2}{Z_0^2} = \left(\frac{\pi}{2} \right) \left(\frac{w}{g_0 g_1} \right) \tag{15-49a}$$

$$\frac{K_{j,j+1}^2}{Z_0^2} = \left(\frac{\pi}{2} \right)^2 \left(\frac{w}{g_j g_{j+1}} \right) \tag{15-49b}$$

$$\frac{K_{n,n+1}^2}{Z_0^2} = \left(\frac{\pi}{2} \right) \left(\frac{w}{g_n g_{n+1}} \right) \tag{15-49c}$$

and

$$\frac{Z_0}{X_{01}} = \left(\frac{2g_0 g_1}{\pi w} \right)^{1/2} - \left(\frac{\pi w}{2g_0 g_1} \right)^{1/2} \tag{15-50a}$$

$$\frac{Z_0}{X_{j,j+1}} = \frac{\pi}{2} \left(\frac{g_j g_{j+1}}{w} \right)^{1/2} - \frac{\pi}{2} \left(\frac{w}{g_j g_{j+1}} \right)^{1/2} \tag{15-50b}$$

$$\frac{Z_0}{X_{n,n+1}} = \left(\frac{Z_0}{Z_{n+1}} \right)^{1/2} \left(\frac{2g_n g_{n+1}}{\pi w} \right)^{1/2} - \left(\frac{Z_{n+1}}{Z_0} \right)^{1/2} \left(\frac{\pi w}{2g_n g_{n+1}} \right)^{1/2} \tag{15-50c}$$

The reactance of each step may therefore be evaluated once the immittance inverters have been fixed by the specification of the filter. This fixes the dimensions of the steps.

Once the details of the steps have been determined it only remains to calculate the spacing between the steps. This condition is met by

$$\left(\frac{2\pi}{\lambda_g}\right)l_j = \theta_j + \tfrac{1}{2}(\phi_j + \phi_{j+1}) \tag{15-51a}$$

with

$$\theta_j = \pi \tag{15-51b}$$

In calculating θ_j and θ_{j+1} the exact equivalent circuit of the discontinuity is usually employed:

$$\phi_j = -\tan^{-1}\left(\frac{2X_{pj}}{Z_0} + \frac{X_{sj}}{Z_0}\right) - \tan^{-1}\left(\frac{X_{sj}}{Z_0}\right) \tag{15-51c}$$

The use of these design relationships will now be demonstrated in connection with the design of a seven-section (six-resonator) Chebyshev filter using half-wave lines coupled by immittance inverters. The required specification is

$$f_0 = 1200 \text{ MHz}$$

$$\Delta f = 120 \text{ MHz}$$

$$\text{VSWR} = 1.10$$

$$Z_0 = 50 \ \Omega$$

The element values of the low-pass prototype are here given by

$$g_1 = 0.779\,68$$
$$g_2 = 1.359\,21$$
$$g_3 = 1.688\,00$$
$$g_4 = 1.534\,54$$
$$g_5 = 1.688\,00$$
$$g_6 = 1.359\,21$$
$$g_7 = 0.779\,68$$

and the network is symmetric. The characteristic impedance of the inverters are then evaluated as

$$\frac{Z_0}{K_{01}} = 0.449$$

$$\frac{Z_0}{K_{12}} = 0.1529$$

$$\frac{Z_0}{K_{23}} = 0.1038$$

$$\frac{Z_0}{K_{34}} = 0.0976$$

$$\frac{Z_0}{K_{45}} = 0.1038$$

$$\frac{Z_0}{K_{56}} = 0.1529$$

$$\frac{Z_0}{K_{67}} = 0.449$$

The physical variables are readily calculated using Eqs (15-50).

PROBLEMS

15-1 Demonstrate that the impedance inverters of the low-pass prototype are independent of the cutoff frequency of the filter.

15-2 Design an $n = 4$ Butterworth low-pass prototype with a cutoff frequency of 10 000 rad/s and a generator impedance of 50 Ω using series inductors and impedance inverters only.

15-3 Repeat Prob. 15-2 in the form of a prototype using shunt capacitors and admittance inverters only.

15-4 Lay out an $n = 3$ low-pass Butterworth network using impedance inverters and evaluate its amplitude squared transfer function using the *ABCD* method. Repeat the problem for a circuit employing admittance inverters.

15-5 Repeat Prob. 15-4 in the case of a high-pass filter.

15-6 Repeat Prob. 15-4 in the case of a band-pass filter.

15-7 Repeat Prob. 15-4 in the case of a band-stop filter.

BIBLIOGRAPHY

Cohn, S. B., 'Direct coupled resonator filters', *Proc. IRE*, Vol. 45, pp. 187–196, February 1957.

Levy, R., 'Theory of direct-coupled-cavity filters', *IEEE Trans. on MTT*, Vol. MTT-15, pp. 340–348, June 1967.

CHAPTER
16

THE UNIFORM
TRANSMISSION
LINE

INTRODUCTION

The uniform two-conductor transmission line is an essential element in the realization of practical filter circuits as is already understood. It is usually described in terms of its characteristic impedance (Z_0) and its propagation constant (γ_0). The first parameter specifies the ratio of voltage and current on the line. The second describes its phase constant. For a uniform line the former quantity is fixed by the physical details of the line; the latter quantity corresponds to that of free space. Two possibilities are indicated in Fig. 16-1. The classic transmission line terminated in an arbitrary impedance is of special interest in microwave and filter engineering. If the load impedance is different from that of the characteristic one then both forward and backward waves will exist on the line and the input impedance will be dependent on the plane at which it is measured. The ratio of the reflected to incident waves at any plane on such a line is known as the reflection coefficient. This situation is associated with standing waves of both the voltage and current variables on the line. The ratio of the maximum and minimum values of the voltage on the line is known as the voltage standing wave ratio (VSWR). The relationship between the reflection coefficient and the impedance at any plane on the line is a bilinear transformation. Such a transformation maps circles into circles (a straight line is a degenerate circle). Non-uniform lines such as radial or tapered ones are also of some interest but are outside the remit of this work.

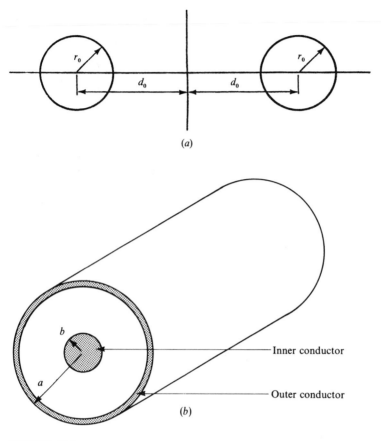

FIGURE 16-1
Schematic diagram of (a) two-wire and (b) coaxial transmission lines.

One possible low-frequency equivalent circuit of a lossless uniform transmission line is a ladder network comprising series inductors and shunt capacitors. Open- and short-circuited lines may also be employed to realize lumped element capacitors, inductors, as well as series and shunt resonators.

TRANSMISSION LINE EQUATIONS

The current flowing in a two-wire line and the potential difference between the two conductors is a function of distance z and time t. The effect of a short length of line δz is depicted in Fig. 16-2.

The series impedance and shunt admittance per unit length of the line in Fig. 16-2 are given by

$$Z = R + j\omega L \qquad (16\text{-}1)$$

$$Y = G + j\omega C \qquad (16\text{-}2)$$

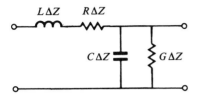

FIGURE 16-2
Voltage and current along small section of two-wire transmission line.

respectively, where $R(\Omega/\text{m})$, $L(\text{H/m})$, $G(\text{S/m})$ and $C(\text{F/m})$ are the resistance, inductance, capacitance and conductance for unit length respectively.

The potential drop across a section line of length δz is

$$(R\,\delta z)I_z + (L\,\delta z)\frac{\partial I_z}{\partial t} = \frac{\partial V_z}{\partial z}\,\delta z \qquad (16\text{-}3a)$$

Similarly, the decrease in the current across the same section is

$$(G\,\delta z)V_z + (C\,\delta z)\frac{\partial V_z}{\partial t} = -\frac{\partial I_z}{\partial z}\,\delta z \qquad (16\text{-}3b)$$

The negative signs in these equations reflect the fact that both the voltage and current on the line decrease with increasing z.

Taking the time variation of V_z and I_z as

$$\exp(\text{j}\omega t) \qquad (16\text{-}4)$$

and assuming that δz is small indicates that the spatial variations of V_z and I_z may be described by

$$\frac{\text{d}V_z}{\text{d}z} = -(R + \text{j}\omega L)I_z \qquad (16\text{-}5a)$$

$$\frac{\text{d}I_z}{\text{d}z} = -(G + \text{j}\omega C)V_z \qquad (16\text{-}5b)$$

These relationships may also be written in terms of Eqs (16-1) and (16-2) as

$$\frac{\text{d}V_z}{\text{d}z} = -ZI_z \qquad (16\text{-}6a)$$

$$\frac{\text{d}I_z}{\text{d}z} = -YV_z \qquad (16\text{-}6b)$$

Differentiating and rearranging these two relationships indicate that both the voltage and current variables of the line satisfy the classic wave equations below:

$$\frac{\text{d}^2V_z}{\text{d}z^2} - \gamma^2 V_z = 0 \qquad (16\text{-}7a)$$

$$\frac{\text{d}^2I_z}{\text{d}z^2} - \gamma^2 I_z = 0 \qquad (16\text{-}7b)$$

where γ is known as the complex propagation constant of the line:

$$\gamma = \sqrt{ZY} \tag{16-8}$$

This quantity is usually written in terms of its real and imaginary parts as

$$\gamma = \alpha + j\beta \tag{16-9}$$

where α is the attenuation of the line per unit length (nepers per metre) and β is the phase constant per unit length (radians per metre).

At very high frequencies or for transmission lines with very small losses

$$j\omega L \gg R \tag{16-10}$$

$$j\omega C \gg G \tag{16-11}$$

and α and β may be approximated by

$$\alpha = 0 \tag{16-12}$$

$$\beta = \omega\sqrt{LC} \tag{16-13}$$

Adopting linear combinations of the forward and backward travelling waves as the solutions for the wave equations describing V_z and I_z in Eqs (16-7a,b) gives

$$V_z = Ae^{-\gamma z} + Be^{\gamma z} \tag{16-14a}$$

$$I_z = \frac{1}{Z_0}(Ae^{-\gamma z} - Be^{\gamma z}) \tag{16-14b}$$

where Z_0 is known as the characteristic impedance of the line:

$$Z_0 = \sqrt{\frac{Z}{Y}} \tag{16-15}$$

The nature of the travelling waves may be understood by introducing the frequency variable $\exp(j\omega t)$ and replacing γ by $j\beta$ in these trial solutions:

$$V_z = A\exp\left[j\omega\left(t - \frac{\beta}{\omega}z\right)\right] + B\exp\left[j\omega\left(t + \frac{\beta}{\omega}z\right)\right] \tag{16-16a}$$

$$I_z = \frac{1}{Z_0}\left\{A\exp\left[j\omega\left(t - \frac{\beta}{\omega}z\right)\right] - B\exp\left[j\omega\left(t + \frac{\beta}{\omega}z\right)\right]\right\} \tag{16-16b}$$

To verify that these equations are linear combinations of forward and backward travelling waves consider the function

$$t - \frac{\beta}{\omega}z \tag{16-17}$$

Since the function must hold at any time it is apparent that

$$t - \frac{\beta}{\omega}z = t + \delta t - \frac{\beta}{\omega}(z + \delta z) \tag{16-18}$$

Scrutiny of this relationship indicates that such a wave travels with a phase velocity

$$v = \frac{\delta z}{\delta t} = \frac{\omega}{\beta} \tag{16-19}$$

so that a term of the form $Af(t - z/v)$ describes a forward travelling wave. A function of the form $Bf(t + z/v)$ represents a backward travelling wave.

Before applying the boundary conditions along the line, in order to evaluate the constants A and B it is helpful to shift the terminal plane from z to that of the termination l. It is also convenient in what follows to take the load terminals rather than the generator ones as the origin.

In order to retain a positive coordinate system the generator terminals then coincide with the plane at $z = -l$. Introducing this transformation in Eqs (16-14a,b) gives

$$V(l) = Ae^{\gamma l} + Be^{-\gamma l} \tag{16-20a}$$

$$I(l) = \frac{1}{Z_0}(Ae^{\gamma l} - Be^{-\gamma l}) \tag{16-20b}$$

The arbitrary constants A and B are now deduced by applying the boundary conditions at $l = 0$:

$$V(0) = V_L = A + B \tag{16-21a}$$

$$I(0) = I_L = \frac{1}{Z_0}(A - B) \tag{16-21b}$$

The two constants are then determined as

$$A = \frac{V_L + I_L Z_0}{2} \tag{16-22a}$$

$$B = \frac{V_L - I_L Z_0}{2} \tag{16-22b}$$

The voltage and current variables along the line are therefore described by

$$V(l) = V_L \cosh \gamma l + I_L Z_0 \sinh \gamma l \tag{16-23a}$$

$$I(l) = I_L \cosh \gamma l + \left(\frac{V_L}{Z_0}\right) \sinh \gamma l \tag{16-23b}$$

and the impedance at any point along such a line is given in terms of the terminating one:

$$Z_L = \frac{V_L}{I_L}$$

by

$$Z(l) = Z_0 \frac{Z_L \cosh \gamma l + Z_0 \sinh \gamma l}{Z_0 \cosh \gamma l + Z_L \sinh \gamma l} \tag{16-24}$$

This is a classic result.

TRANSMISSION LINE IDENTITIES

The uniform transmission line displays some interesting properties which may be readily deduced from the relationship in Eq. (16-24). Some of these will now be enunciated. For the short-circuited line of length l in Fig. 16-3a,

$$Z_L = 0 \qquad (16\text{-}25)$$

and the impedance $Z(l)$ becomes

$$Z(l) = Z_{\text{s.c.}} = Z_0 \tanh \gamma l \qquad (16\text{-}26)$$

For the open-circuited line of length l in Fig. 16-3b,

$$Z_L = \infty \qquad (16\text{-}27)$$

and $Z(l)$ is given by

$$Z(l) = Z_{\text{o.c.}} = Z_0 \coth \gamma l \qquad (16\text{-}28)$$

Combining these two identities provides a method of determining Z_0 from a knowledge of $Z_{\text{o.c.}}$ and $Z_{\text{s.c.}}$:

$$Z_0^2 = Z_{\text{o.c.}} Z_{\text{s.c.}} \qquad (16\text{-}29)$$

If the line is terminated in its characteristic impedance then

$$Z(l) = Z_0 \qquad (16\text{-}30)$$

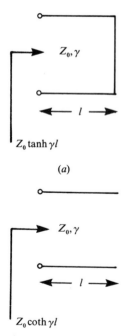

FIGURE 16-3

Input impedance of (a) short-circuited and (b) open-circuited lines.

and the input impedance at any plane coincides with its characteristic impedance Z_0. The impedance of a loss-free line terminated in a load impedance Z_L with

$$\beta l = \frac{\pi}{2} \tag{16-31}$$

is

$$Z(l) = \frac{Z_0^2}{Z_L} \tag{16-32}$$

This property of a quarter-wave long line is used in everyday engineering to match an impedance Z_L to one $Z(l)$.

The impedance of a loss-free line terminated in a load Z_L with

$$\beta l = \pi \tag{16-33}$$

is also of some interest. The load Z_L is in this instance mapped onto itself:

$$Z(l) = Z_L \tag{16-34}$$

Open- and short-circuited transmission lines may also be employed to realize lumped element capacitors and inductors. Series and shunt resonators using such quarter-wave long lines are two other possibilities.

STANDING WAVE RATIO AND REFLECTION COEFFICIENT

A useful quantity in transmission line and filter theory is the voltage (or current) reflection coefficient. The voltage reflection coefficient is defined as

$$\Gamma_v(l) = \frac{\text{reflected voltage}}{\text{incident voltage}} \tag{16-35}$$

or

$$\Gamma_v(l) = \frac{Be^{-\gamma l}}{Ae^{\gamma l}} = \frac{B}{A} e^{-2\gamma l} \tag{16-36}$$

At the plane of the load, $l = 0$, and the reflection coefficient is given by

$$\Gamma_v(l) = \Gamma_v(0) = \frac{B}{A} \tag{16-37}$$

Expressing the arbitrary constants A and B in terms of the original variables in Eqs (16-22a,b) gives the standard result

$$\Gamma_v(0) = \frac{Z_L - Z_0}{Z_L + Z_0} \tag{16-38}$$

A relationship of this type is known as a bilinear transformation.

Since $\Gamma_v(0)$ is in general a complex quantity it may be written as

$$\Gamma_v(0) = |\Gamma_v(0)| e^{j\theta_v} \tag{16-39}$$

$\Gamma_v(l)$ is therefore related to $\Gamma_v(0)$:

$$\Gamma_v(l) = |\Gamma_v(0)| e^{j(\theta_v - 2\beta l)} \tag{16-40}$$

provided

$$\gamma = j\beta$$

The current reflection coefficient is similarly specified by

$$\Gamma_I(l) = \frac{\text{reflected current}}{\text{incident current}} \tag{16-41}$$

or

$$\Gamma_I(l) = -\frac{B}{A} e^{-2\gamma l} = -\Gamma_v(l) \tag{16-42}$$

Since it is in practice more convenient to measure voltage than current, it is usual in everyday engineering to work with the voltage reflection coefficient.

Some useful relationships between Z_L and $\Gamma_v(0)$ are

$$Z_L = 0, \qquad \Gamma_v(0) = -1 \tag{16-43a}$$

$$Z_L = \infty, \qquad \Gamma_v(0) = +1 \tag{16-43b}$$

$$Z_L = Z_0, \qquad \Gamma_v(0) = 0 \tag{16-43c}$$

The voltage and current along the transmission line may also be written in terms of $\Gamma_v(0)$. The result is

$$V(l) = Ae^{\gamma l} + Be^{-\gamma l} = Ae^{\gamma l}[1 + \Gamma_v(l)] \tag{16-44}$$

$$I(l) = \frac{Ae^{\gamma l}}{Z_0} [1 - \Gamma_v(l)] \tag{16-45}$$

and

$$Z(l) = Z_0 \frac{1 + \Gamma_v(l)}{1 - \Gamma_v(l)} \tag{16-46}$$

Another useful parameter encountered in transmission line theory is the voltage standing wave ratio. This quantity is defined by

$$\text{VSWR} = \frac{V_{\max}}{V_{\min}} \tag{16-47}$$

Forming the above voltage variables gives

$$V_{\max} = [1 + |\Gamma_v(0)|] \tag{16-48}$$

$$V_{\min} = [1 - |\Gamma_v(0)|] \tag{16-49}$$

The relationship between the VSWR and the voltage reflection coefficient is therefore given by

$$\text{VSWR} = \frac{1 + |\Gamma_v(0)|}{1 - |\Gamma_v(0)|} \tag{16-50}$$

Some additional useful identities between the VSWR and the reflection coefficient are

$$\Gamma_v(0) = 0, \qquad \text{VSWR} = 1 \qquad (16\text{-}51)$$

$$\Gamma_v(0) = \pm 1, \qquad \text{VSWR} = \infty \qquad (16\text{-}52)$$

It is also of note that at a position of maximum voltage the current is at a minimum and

$$\frac{V_{\max}}{I_{\min}} = Z_0(\text{VSWR}) \qquad (16\text{-}53)$$

and that at a position of minimum voltage the current is at a maximum and

$$\frac{V_{\min}}{I_{\max}} = \frac{Z_0}{\text{VSWR}} \qquad (16\text{-}54)$$

These two relationships indicate that the maximum and minimum values of the impedance on a transmission line are real.

The impedance at the plane $l_{\min}$ at which the voltage is a minimum is

$$\frac{Z_0}{\text{VSWR}} = Z_0 \frac{Z_L + jZ_0 \tan \beta l_{\min}}{Z_0 + jZ_L \tan \beta l_{\min}} \qquad (16\text{-}55)$$

and

$$\frac{Z_L}{Z_0} = \frac{1 - j(\text{VSWR}) \tan \beta l_{\min}}{\text{VSWR} - j \tan \beta l_{\min}} \qquad (16\text{-}56)$$

This relationship permits an unknown load to be determined from an experimental knowledge of the VSWR and $l_{\min}$.

PROBLEMS

16-1 Evaluate the real and imaginary parts (α and β) of the complex propagation constant (γ) for $j\omega L > R$ and $j\omega c > G$.

16-2 Calculate the real and imaginary part of Z_0 under the same circumstances.

16-3 Demonstrate Eqs (16-53) and (16-54).

16-4 Verify Eqs (16-55) and (16-56).

CHAPTER

17

LOW-PASS AND HIGH-PASS FILTER CIRCUITS USING PLANAR ELEMENTS

INTRODUCTION

The existence of some classic equivalences between a lumped element and a uniform transmission line is of importance in the design of UHF circuits. This chapter will introduce some of these as a preamble to describing some possible low- and high-pass filter circuits. Suspended striplines, coaxial lines, slot lines, coplanar waveguides, finlines and waveguides are possibilities employed in practice.

One equivalence is between an electrical short section of uniform distributed transmission line terminated in a short-circuit and a quasi lumped element inductor. Another is between a similar open-circuited line and a quasi lumped element capacitor. One drawback of simulating a capacitor in this way is the difficulty of fabricating an ideal open-circuit boundary condition. One shortcoming of constructing an inductance in this manner is the practical nuisance of having to provide a return path between the inner and outer conductors of the line. If the filter topology requires only series inductors and shunt capacitors then the additional equivalences between short sections of high and low impedance lines and these elements provide one practical planar solution. Series capacitors may also be fabricated by suitably breaking the centre conductor. The provision of a suitable return path cannot, however, be avoided if a shunt inductance is required.

Quarter-wave long open- and short-circuited transmission lines may also be utilized to realize series and parallel resonators. The types of equivalences are satisfied provided the reactance or susceptance slope parameters of the two circuits in the vicinity of the centre frequencies is the same.

UHF LINES AS CIRCUIT ELEMENTS

It will now be demonstrated that a short section of a short-circuited lossless transmission line is equivalent to a lumped element inductor and that a similar open-circuited transmission line is equivalent to a lumped element capacitor. Such lines are therefore often employed to simulate lumped element inductors and capacitors at UHF frequencies. The derivations of the required equivalences start by noting the lossless transmission line relationship below:

$$\gamma_0 = j\beta \qquad (17\text{-}1a)$$

$$\beta = \frac{\omega}{v} \qquad (17\text{-}1b)$$

$$v = \frac{1}{\sqrt{LC}} \qquad (17\text{-}1c)$$

$$Z_0 = \sqrt{\frac{L}{C}} \qquad (17\text{-}1d)$$

L and C are the inductance per unit length and the capacitance per unit length respectively; the other quantities have the usual meaning. It is also recalled that the impedance $[Z(l)]$ on this type of line is related to that of the load (Z_L) by

$$Z(l) = Z_0 \frac{Z_L \cosh \gamma l + Z_0 \sinh \gamma l}{Z_0 \cosh \gamma l + Z_L \sinh \gamma l} \qquad (17\text{-}1e)$$

This equation relates the impedance at one pair of terminals on a line with a characteristic impedance Z_0 to that of the impedance of a load at another pair of terminals.

The equivalence between a short-circuited transmission and a lumped element inductance will be derived first. The input impedance of a short-circuited lossless uniform transmission line in Fig. 17-1a is obtained by letting Z_L equal zero in Eq. (17-1e). The result is

$$Z(l) = Z_{\text{s.c.}} = jZ_0 \tan \beta l \qquad (17\text{-}2a)$$

The small angle approximation of this equation is

$$Z_{\text{s.c.}} = jZ_0 \beta l \qquad (17\text{-}2b)$$

Substituting for β in this equation in terms of ω indicates that

$$Z_{\text{s.c.}} = jZ_0 \left(\frac{\omega}{v}\right) l \qquad (17\text{-}2c)$$

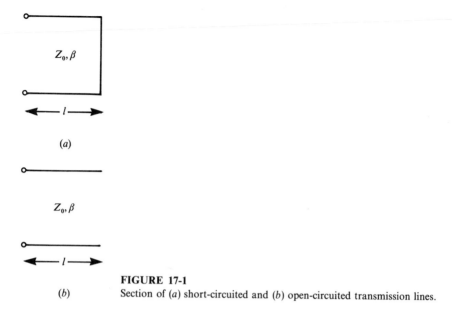

FIGURE 17-1

(b)   Section of (a) short-circuited and (b) open-circuited transmission lines.

and making use of the relationships between Z_0 and v and the constitutive parameters L and C gives

$$Z_{\text{s.c.}} = j\omega(Ll) \qquad (17\text{-}2d)$$

Such a line is therefore equivalent to a lumped element inductance L' given by

$$L' = Ll \qquad (17\text{-}3)$$

This identity is illustrated in Fig. 17-2a.

The equivalence between a short section of open-circuited transmission line and a lumped element capacitor begins by forming the input impedance of the structure in Fig. 17-1b. Specializing Eq. (17-1e) to this situation gives

$$Z(l) = Z_{\text{o.c.}} = -jZ_0 \cot \beta l \qquad (17\text{-}4a)$$

Evaluating this quantity in the case of an electrically short line and writing Z_0 and β in terms of L and C indicates that

$$Z_{\text{o.c.}} = \frac{-j}{\omega Cl} \qquad (17\text{-}4b)$$

A short section of line open-circuited at the output terminals is therefore equivalent to a frequency-independent capacitance C' at its input terminals, which is described by

$$C' = Cl \qquad (17\text{-}5)$$

This correspondence is illustrated in Fig. 17-2b. Figure 17-3 depicts low-pass, band-pass, band-stop and high-pass networks based on these types of circuits.

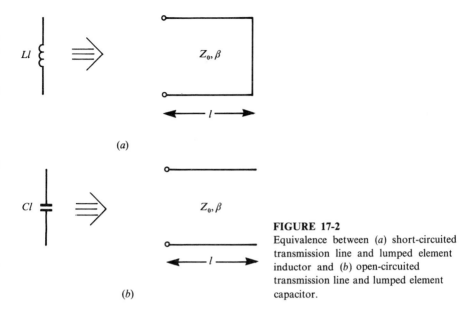

FIGURE 17-2
Equivalence between (*a*) short-circuited transmission line and lumped element inductor and (*b*) open-circuited transmission line and lumped element capacitor.

PLANAR LUMPED ELEMENT SERIES INDUCTORS AND SHUNT CAPACITORS

Of some significance in the design of practical planar filters is the fact that a short section of high impedance line in an otherwise uniform line may be represented by a series lumped element inductor and that a similar length of low impedance line may be characterized by a shunt lumped element capacitor. Figure 17-4*a* and *b* illustrates stripline implementations of the two topologies discussed here. A low-pass filter may therefore, for instance, be readily realized by a cascade arrangement of short sections of high and low impedance transmission lines. These two identities will now be deduced.

The derivation of the first equivalence begins by specializing the standard relationship between the input and output impedances Z_{in} and Z_{out} spaced by a lossless uniform transmission line of characteristic impedance Z_0 and propagation constant γ in Eq. (17-1*e*):

$$Z_{in} = Z_0 \frac{Z_{out} + jZ_0 \tan \beta l}{Z_0 + jZ_{out} \tan \beta l} \tag{17-6}$$

If the length (l) of the connecting transmission line is small compared to the wavelength then

$$\tan \beta l \approx \beta l \tag{17-7}$$

and Z_{in} becomes

$$Z_{in} \approx Z_0 \frac{Z_{out} + jZ_0 \beta l}{Z_0 + jZ_{out}\beta l} \tag{17-8}$$

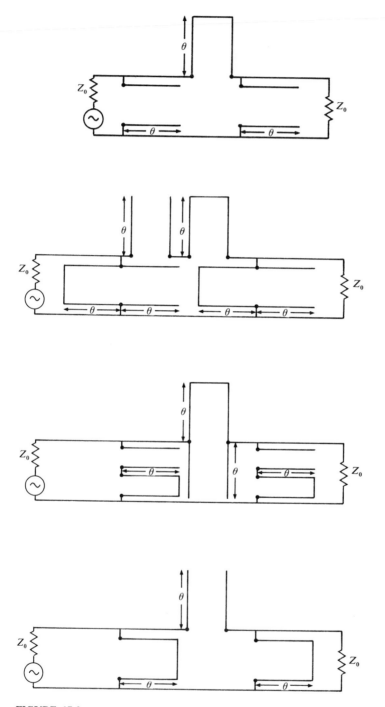

FIGURE 17-3
Topologies of low-pass, band-pass, band-stop and high-pass circuits using uniform transmission lines.

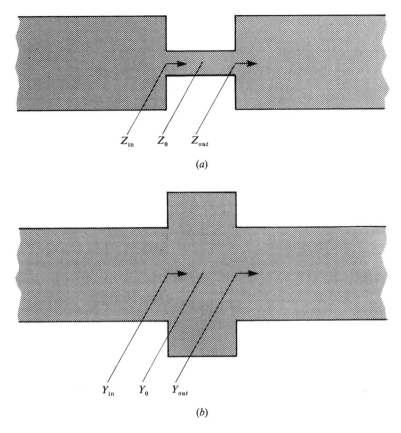

FIGURE 17-4
Uniform transmission line with (a) high and (b) low impedance sections.

The real and imaginary parts R and X of this quantity may be readily expressed by

$$R \approx Z_0 \frac{Z_0 Z_{\mathrm{out}}[1 - (\beta l)^2]}{Z_0^2 + (Z_{\mathrm{out}}\beta l)^2} \tag{17-9}$$

$$X \approx Z_0 \frac{Z_0^2[1 - (Z_{\mathrm{out}}/Z_0)^2]}{Z_0^2 + (Z_{\mathrm{out}}\beta l)^2} \tag{17-10}$$

If the value of the characteristic impedance Z_0 is very large compared to that of Z_{out} and if βl is also very small compared to the wavelength then R and X in the preceding equations may be approximately written as

$$R \approx Z_{\mathrm{out}} \tag{17-11}$$

$$X \approx j\omega(Ll) \tag{17-12}$$

In obtaining this result use is also made of the relationships in Eqs (17-1) between the capacitance and inductance of the line per unit length and its phase constant and characteristic impedance β and Z_0.

The impedance of a short section of high impedance line loaded by a real impedance Z_{out} is therefore equivalent to that of a series inductance

$$L' = Ll \qquad (17\text{-}13)$$

terminated in the same load impedance as asserted.

A similar development indicates that a short section of low impedance line in an otherwise uniform line may be utilized to simulate a shunt capacitance C' given by

$$C' = Cl \qquad (17\text{-}14)$$

DISTRIBUTED LINES AS SERIES AND SHUNT RESONATORS

Distributed transmission lines with open- and short-circuited loads display immittance slope parameters at discrete frequencies that are similar to those encountered with series and shunt lumped element resonators in the vicinity of their resonant frequencies. These lines are therefore often used as cavity resonators in the microwave region. Since it is inappropriate to discuss distributed networks in terms of capacitance and inductance, the equivalence between the two types of circuit is usually expressed in terms of their immittance slope parameters. The two circuits are equivalent provided the respective slope parameters are identical. The equivalence between the admittance functions of the shunt lumped element and distributed short-circuited transmission line in Fig. 17-5a and b will now be deduced.

The input admittance of a short-circuited distributed transmission line is

$$Y_{in} = jB = -jY_0 \cot \beta l \qquad (17\text{-}15a)$$

Expanding the above quantity in the vicinity of $\beta_0 l = \pi/2$ gives

$$Y_{in} \approx \frac{jY_0\pi}{4}\left(\frac{\omega}{\omega_0} - \frac{\omega_0}{\omega}\right) \qquad (17\text{-}15b)$$

In obtaining this relationship use is made of the fact that

$$\beta = \frac{\omega}{v} = \frac{\omega_0}{v}\frac{\omega}{\omega_0} = \beta_0\left[1 + \left(\frac{\omega - \omega_0}{\omega_0}\right)\right] \qquad (17\text{-}16a)$$

This relationship may also be written as

$$\beta \approx \beta_0\left[1 + \frac{1}{2}\left(\frac{\omega}{\omega_0} - \frac{\omega_0}{\omega}\right)\right] \qquad (17\text{-}16b)$$

provided

$$\omega + \omega_0 \approx 2\omega \qquad (17\text{-}16c)$$

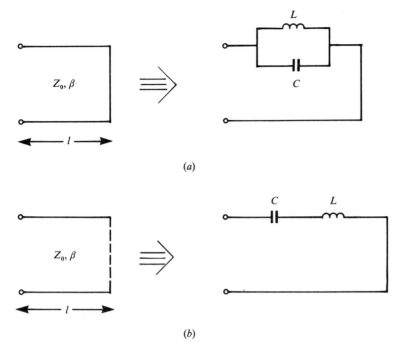

FIGURE 17-5
Equivalence between shunt LC network and (a) short-circuited and (b) open-circuited transmission lines.

The input admittance of a shunt resonant circuit is

$$Y_{in} = jB = j\left(\omega C - \frac{1}{\omega L}\right) \qquad (17\text{-}17a)$$

or

$$Y_{in} = j\omega_0 C\left(\frac{\omega}{\omega_0} - \frac{\omega_0}{\omega}\right) \qquad (17\text{-}17b)$$

It is readily observed that both admittance functions have the same nature in the vicinity of $\omega = \omega_0$. The equivalence between them is satisfied by ensuring that the susceptance slope parameter B'

$$B' = \frac{\omega}{2}\frac{\partial B}{\partial \omega}\bigg|_{\omega=\omega_0} \qquad (17\text{-}18)$$

of each circuit is equal.

Evaluating this quantity for the shunt lumped element network gives

$$B' = \frac{\omega}{2}\left(C + \frac{1}{\omega^2 L}\right)\bigg|_{\omega=\omega_0} = \frac{\omega_0}{2}(2C) = \omega_0 C \qquad (17\text{-}19)$$

Likewise, the susceptance slope parameter of the corresponding distributed circuit is

$$B' = \frac{\beta l}{2} \left. \frac{\partial(-Y_0 \cot \beta l)}{\partial(\beta l)} \right|_{\beta l = \pi/2} = \frac{\pi Y_0}{4} \tag{17-20}$$

The lumped and distributed circuits are therefore equivalent provided

$$\omega_0 C = \frac{\pi Y_0}{4} \tag{17-21}$$

or

$$C = \frac{\pi Y_0}{4\omega_0} \tag{17-22}$$

and

$$L = \frac{1}{\omega_0^2 C} \tag{17-23}$$

This correspondence is illustrated in Fig. 17-5a.

For an open-circuited transmission line with $\beta l = \pi/2$, $Z_{o.c.}$ behaves as a series resonator. The relationship between Z_0 and β and L and C are in this instance established by ensuring that the reactance slope parameters of the two circuits are equal. This quantity is defined by

$$X' = \frac{\omega}{2} \left. \frac{\partial X}{\partial \omega} \right|_{\omega = \omega_0} \tag{17-24}$$

The reactance of a series lumped element network is given by

$$Z = jX = j\omega L - \frac{j}{\omega C} \tag{17-25a}$$

and the reactance slope parameter is

$$X' = \frac{\omega}{2} \left. \left(L + \frac{1}{\omega^2 C} \right) \right|_{\omega = \omega_0} \tag{17-25b}$$

Making use of the fact that at resonance

$$\omega_0^2 L C = 1 \tag{17-25c}$$

gives the required result

$$X' = \omega_0 L \tag{17-25d}$$

The input impedance of the open-circuited distributed circuit is

$$Z(l) = jX = -jZ_0 \cot \beta l \tag{17-26a}$$

and its reactance slope parameter is

$$X' = \frac{\beta l}{2} \left. \frac{\partial(Z_0 \cot \beta l)}{\partial(\beta l)} \right|_{\beta l = \pi/2} = \frac{\pi Z_0}{4} \tag{17-26b}$$

The required relationships between the two circuits is now established by equating the two reactance slope parameters. The result is

$$L = \frac{\pi Z_0}{4\omega_0} \qquad (17\text{-}27a)$$

$$C = \frac{1}{\omega_0^2 L} \qquad (17\text{-}27b)$$

This equivalence is depicted in Fig. 17-5b.

The topologies of band-pass and band-stop filters using open- and short-circuited lines are indicated in Fig. 17-6.

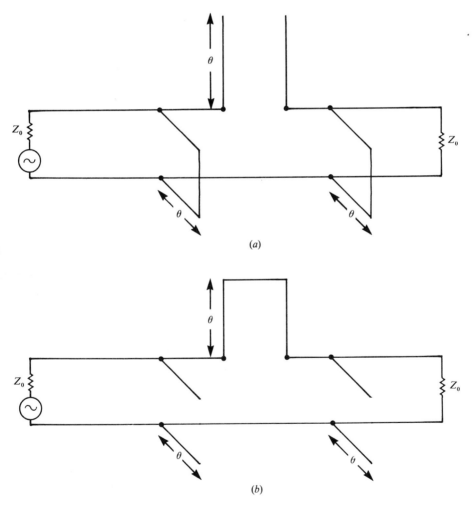

(a)

(b)

FIGURE 17-6
Topologies of band-pass and band-stop filters using resonant lines.

DISSIPATION

The dissipation in any resonator is usually expressed in terms of an unloaded quality factor (Q_u). For a series resonator, this quantity is defined by

$$Q_u = \frac{X'}{R} \qquad (17\text{-}28a)$$

and for a shunt resonator it is given by

$$Q_u = \frac{B'}{G} \qquad (17\text{-}28b)$$

where R and G represent the equivalent series and shunt resistance and conductance respectively.

The impedance of a short-circuited line with dissipation is

$$Z_{s.c.} = jZ_0 \tan(\beta l - j\alpha l) \qquad (17\text{-}29a)$$

Expanding this function in terms of α and β indicates that

$$Z_{s.c.} = jZ_0 \frac{\tan \beta l - j \tan \alpha l}{1 + j \tan \beta l \tan \alpha l} \qquad (17\text{-}29b)$$

and evaluating it in the vicinity of

$$\beta l = \frac{\pi}{2} \qquad (17\text{-}30a)$$

leads to

$$Z_{s.c.} = \frac{Z_0}{\tan \alpha l} \qquad (17\text{-}30b)$$

The equivalent shunt conductance of the resonator is therefore described by

$$G \approx \frac{\alpha l}{Z_0} \qquad (17\text{-}30c)$$

Similarly, the result for an open-circuited line quarter-wave long with

$$\beta l = \frac{\pi}{2} \qquad (17\text{-}31a)$$

is

$$Z_{s.c.} = Z_0 \tan \alpha l \qquad (17\text{-}31b)$$

or

$$R = Z_0 \alpha l \qquad (17\text{-}31c)$$

These equivalences are depicted in Fig. 17-7.

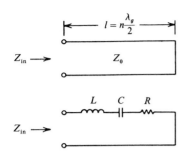

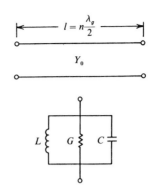

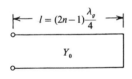

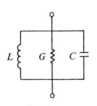

$$R \approx Z_0 \alpha_t l = \frac{n}{2} Z_0 \alpha_t \lambda_g$$

$$X' = \omega_0 L = \frac{1}{\omega_0 C} \approx \frac{n\pi Z_0}{2}\left(\frac{\lambda_g}{\lambda_0}\right)^2$$

$$Z_{\text{in}} = R + jX'\left(\frac{\omega}{\omega_0} - \frac{\omega_0}{\omega}\right)$$

$$Q = \frac{X'}{R} \approx \frac{\pi \lambda_g}{\alpha_t \lambda_0^2}$$

$$n = 1, 2, 3, \ldots$$

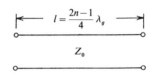

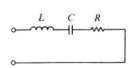

$$G \approx Y_0 \alpha_t l = \frac{n}{2} Y_0 \alpha_t \lambda_g$$

$$B' = \omega_0 C = \frac{1}{\omega_0 L} \approx \frac{n\pi Y_0}{2}\left(\frac{\lambda_g}{\lambda_0}\right)^2$$

$$Y_{\text{in}} = G + jB'\left(\frac{\omega}{\omega_0} - \frac{\omega_0}{\omega}\right)$$

$$Q = \frac{B'}{G} \approx \frac{\pi \lambda_g}{\alpha_t \lambda_0^2}$$

$$n = 1, 2, 3, \ldots$$

$$G = Y_0 \alpha_t l = \frac{2n-1}{4} Y_0 \alpha_t \lambda_g$$

$$B' = \omega_0 C = \frac{1}{\omega_0 L} \approx \frac{2n-1}{4}\pi Y_0 \left(\frac{\lambda_g}{\lambda_0}\right)^2$$

$$Y_{\text{in}} = G + jB'\left(\frac{\omega}{\omega_0} - \frac{\omega_0}{\omega}\right)$$

$$Q = \frac{B'}{G} \approx \frac{\pi \lambda_g}{\alpha_t \lambda_0^2}$$

$$n = 1, 2, 3, \ldots$$

$$R \approx Z_0 \alpha_t l = \frac{2n-1}{4} Z_0 \alpha_t \lambda_g$$

$$X' = \omega_0 L = \frac{1}{\omega_0 C} \approx \frac{2n-1}{4}\pi Z_0 \left(\frac{\lambda_g}{\lambda_0}\right)^2$$

$$Z_{\text{in}} = R + jX'\left(\frac{\omega}{\omega_0} - \frac{\omega_0}{\omega}\right)$$

$$Q = \frac{X'}{R} \approx \frac{\pi \lambda_g}{\alpha_t \lambda_0^2}$$

$$n = 1, 2, 3, \ldots$$

FIGURE 17-7
Open- and short-circuited uniform transmission lines with dissipation.

LOW-PASS AND HIGH-PASS
FILTER CIRCUITS

Low-pass filter circuits with either all pole or finite poles require only series inductors and capacitors connected to the ground plane. Short sections of high and low impedance lines are therefore all that is required for their implementation. Figure 17-8 depicts an all-pole low-pass prototype and its printed circuit realization. Figure 17-9 illustrates a low-pass prototype with finite attenuation poles and the corresponding microwave circuit.

The realization of a high-pass network requires series capacitors and shunt inductors, neither of which are readily realized in pure planar form. The series capacitor can, in principle, be realized by employing an open-circuited transmission line; however, an ideal open-circuit boundary condition is difficult to achieve in practice and this possibility is therefore not adopted if it can be avoided. One way in which the series capacitor is implemented in practice is by breaking the transmission line in the manner indicated in Fig. 17-10. The shunt inductor is in this instance also more difficult to implement. One possible structure in this instance is a short section of short-circuited transmission line short-circuited to the ground plane. Figure 17-11 illustrates the microwave layout of one high-pass filter configuration.

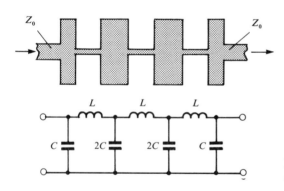

FIGURE 17-8
All-pole low-pass stripline circuit.

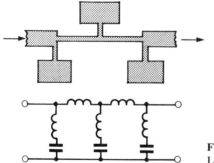

FIGURE 17-9
Low-pass stripline circuit with finite attenuation poles.

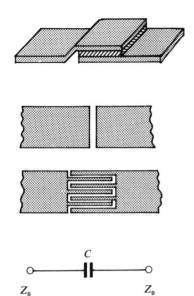

C

Z_0 Z_0

FIGURE 17-10
Schematic diagram of overlap capacitor.

Overlap
capacitor

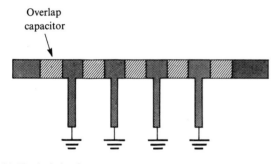

(a) Physical circuit

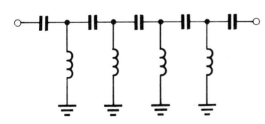

(b) Equivalent circuit;
high-pass filters

FIGURE 17-11
All-pole high-pass stripline circuit.

PROBLEMS

17-1 Calculate the inductance and capacitance of $\lambda_0/32$ long uniform transmission lines with a characteristic impedance of 300 and 5 Ω respectively. Assume that the phase velocity of the line is that of free space. The characteristic impedance of stripline is given by

$$Z_0 = 30\pi \ln\left(1 + \frac{2H}{W}\right)$$

Calculate the ratio $2H/W$ for each situation.

17-2 Lay out a stripline low-pass all-pole $n = 3$ filter with $g_1 = 1$ H, $g_2 = 2$ F, $g_3 = 1$ H to work between 50-Ω striplines at 3 GHz. Indicate the electrical length of the patches utilized.

17-3 Show that a low impedance patch along a uniform transmission line is equivalent to a shunt lumped element capacitor.

CHAPTER

18

SYNTHESIS OF COMMENSURATE LINE NETWORKS

INTRODUCTION

A common approach to the design of a practical distributed circuit is to seek some approximate equivalence between it and a lumped element and distributed circuit. Some possible equivalences have already been discussed in connection with the design of low- and high-pass filters. The exact synthesis of this class of network is, however, separately well established. Indeed, much of the foundation in the classic s plane in connection with the realization of lumped element circuits may be directly applied to immittance functions of commensurate (equal-length) transmission lines by replacing the distributed variable $\tanh \theta$, which appears in the manipulation of transmission line functions, by the complex $t = \Sigma + j\Omega$. The Richards variable t is the basis of modern network theory using distributed networks. Its use permits distributed transmission line functions and lumped resistors to be treated as lumped LCR elements not only for the purpose of realizability but also for synthesis and analysis.

In addition to the basic t-plane distributed inductors and capacitors defined by the Richards transformation, a unit element (UE) may also be introduced. UEs are commensurate sections of transmission line which may be employed to separate circuit elements in high-frequency circuits which would otherwise be located at the same physical point. The Richards theorem states that a UE may always be extracted from a distributed t-plane reactance function and the remainder function is guaranteed to be p.r. and of degree one less than the original one. The characteristic impedance of this UE has the value of the one-port immittance with t replaced by unity. This theorem permits a canonical realization of a reactance function as a cascade of UEs terminated in either an open circuit or a short circuit. The important case of a cascade of UEs terminated in a resistance of R(ohms) is treated separately in Chapter 20.

One way in which it is possible to rearrange the topology of a distributed network is by employing one of the two classic Kuroda identities of the first or second kind, also to be described in this chapter. These identities give the equivalence between a cascade arrangement of a UE and a t-plane reactance and a cascade arrangement of a t-plane reactance and a UE. A feature of the Kuroda identities is that the equivalence between any two circuits using such identities is exact at all frequencies. The principle of this type of network is discussed in terms of $ABCD$ parameters. The use of UEs to space distributed stubs is separately understood. The use of immittance inverters to replace a lumped element filter structure using two kinds of elements by one using only one kind is also now understood.

RICHARDS VARIABLE

The application of modern network theory to the design of microwave distributed networks is based upon the complex frequency transformation introduced by Richards. He showed that distributed networks, composed of commensurate (equal) lengths of transmission line and lumped resistors, could be treated in analysis or synthesis as lumped element LCR networks under the transformation

$$t = \tanh\left(\frac{\pi s}{2\omega_0}\right) \tag{18-1}$$

where s is the usual complex frequency variable

$$s = \sigma + j\omega \tag{18-2}$$

and t is the Richards complex frequency variable

$$t = \Sigma + j\Omega \tag{18-3}$$

For reactance functions $s = j\omega$ and the Richards variable is simply expressed by

$$t = j\Omega \tag{18-4}$$

where

$$\Omega = \tan\left(\frac{\pi\omega}{2\omega_0}\right) = \tan\theta \tag{18-5}$$

ω is the usual real radian frequency variable, ω_0 is the radian frequency at which

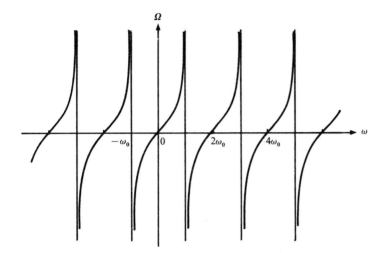

FIGURE 18-1
Mapping between real frequency variable ω and distributed frequency variable Ω showing periodicity.

all line lengths are a quarter-wave long and Ω is the distributed radian frequency variable.

The mapping of the s plane onto the t plane is not one to one, but periodic, corresponding to the periodic nature of the distributed network. Figure 18-1 indicates the relationship between ω and Ω. As ω varies between 0 and ω_0, Ω varies between 0 and ∞. The response of the distributed circuit repeats in frequency intervals of $2\omega_0$. This mapping is theoretically exact at all frequencies. Thus, for a specified response in the ω variable, the corresponding prototype using the Ω variable can be determined using Eq. (18-4), for use in the physical realization of the distributed network.

t-PLANE INDUCTORS AND CAPACITORS

A lumped element LCR network with a realizable impedance $Z(s)$ in the s plane may be converted into one consisting of uniform sections and lumped resistors with an impedance $Z(t)$ in the t plane by replacing the reactance elements by suitable transmission line networks. This transformation is obtained by replacing s by t in $Z(s)$. The impedance of a lumped inductance L in the s plane is

$$Z(s) = Ls \qquad (18\text{-}6)$$

Replacing s by t in this equation gives

$$Z(t) = Lt \qquad (18\text{-}7)$$

Likewise, the impedance of a lumped capacitance C in the s plane is

$$Z(s) = \frac{1}{Cs} \tag{18-8}$$

Replacing s by t yields

$$Z(t) = \frac{1}{Ct} \tag{18-9}$$

The t-plane reactance in Eq. (18-7) is recognized as the input impedance of a short-circuited line of characteristic impedance Z_0 in the s plane equal to the t plane inductor

$$Z_0 = L \tag{18-10}$$

and electrical length θ.

The t-plane reactance in Eq. (18-9) is likewise recognized as the input impedance of an open-circuited line of characteristic impedance Z_0 equal to the reciprocal of the t-plane capacitor

$$Z_0 = \frac{1}{C} \tag{18-11}$$

and electrical length θ.

The equivalence between the lumped and distributed circuits in the s and t planes is depicted in Fig. 18-2. Lumped resistors are invariant under the Richards transformation.

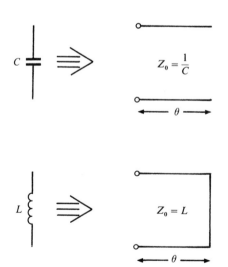

FIGURE 18-2
Element substitution corresponding to the Richards transformation.

POSITIVE REAL COMMENSURATE IMMITTANCE FUNCTIONS

The application of the Richards variable allows the testing procedures developed for lumped element impedance functions to be applied to impedance functions of commensurate transmission line circuits. Similarly, the synthesis of transfer functions in the t plane proceeds in an identical fashion to that of lumped element transfer functions in the s plane. The above statements are readily demonstrated by way of an example:

$$Z(\tanh \theta) = \frac{3 \tanh^3 \theta + 6 \tanh \theta}{5 \tanh^4 \theta + 3 \tanh^2 \theta + 2}$$

To establish whether $Z(\tanh \theta)$ is a positive real function it is merely necessary to replace $\tanh \theta$ by t:

$$Z(t) = \frac{3t^3 + 6t}{5t^4 + 3t^2 + 2}$$

This quantity has now the form associated with lumped element functions and may be tested by ensuring that

$$Z(t) \text{ is real for } t \text{ real} \tag{18-12}$$

$$\text{Re}[Z(t)] > 0 \text{ for } \Sigma > 0 \tag{18-13}$$

The detailed discussion of these two necessary and sufficient tests is, however, outside the remit of this text. Functions that satisfy these two conditions belong to a class of function known as positive real (p.r.) ones.

SYNTHESIS OF t-PLANE INDUCTORS AND CAPACITORS

The synthesis of one-port immittance functions in the t plane is carried out in a like manner to that employed to realize one-port LCR immittances in the s plane, except that it displays t-plane capacitors and inductors which have to be realized in the form of open- and short-circuited UEs. It is recalled that the removal of a pole at either the origin or at infinity reduces the degree of the function by one and that the removal of a pair of conjugate poles at finite frequency reduces it by degree two. The synthesis of this type of circuit will now be illustrated by way of an example in connection with the all-pole t-plane admittance of degree 3 below:

$$Y(t) = \frac{2t^3 + 2t^2 + 2t + 1}{2t^2 + 2t + 1}$$

The realization of this impedance starts by recognizing that it has a pole at infinity which may be extracted in the form of a t-plane capacitor. The value of this capacitance is obtained by evaluating its residue there:

$$C = \lim_{t \to \infty} \frac{Y(t)}{t} = 1$$

The remainder admittance $Y'(t)$ after extraction of this pole

$$Y'(t) = Y(t) - t$$

is of degree one less than that of $Y(t)$ as asserted:

$$Y'(t) = \frac{t+1}{2t^2 + 2t + 1}$$

This admittance has now a zero at infinity but the impedance

$$Z'(t) = \frac{1}{Y'(t)} = \frac{2t^2 + 2t + 1}{t+1}$$

has a pole there. This pole is now extracted from this quantity in the form of a t-plane inductor. The value of this inductance is again equal to its residue:

$$L' = \lim_{t \to \infty} \frac{Z'(t)}{t} = 2$$

The remainder impedance is in this instance given by

$$Z''(t) = Z'(t) - 2t$$

Evaluating this quantity indicates that the remainder impedance

$$Z''(t) = \frac{1}{t+1}$$

is again of degree one less than that of $Z'(t)$.

This impedance has now a zero instead of a pole at infinity but

$$Y''(t) = \frac{1}{Z''(t)} = \frac{t+1}{t}$$

has a pole there that can once more be removed in the form of a t-plane capacitor. Its value is

$$C'' = \lim_{t \to \infty} \frac{Y''(t)}{t} = 1$$

The remainder impedance is again of degree one less than that of the original one. The result is

$$Y'''(t) = Y''(t) - t = 1$$

This last quantity may be realized as a 1-Ω resistor and completes the synthesis procedure.

Figure 18-3 indicates this prototype in terms of t-plane capacitors and inductors and Fig. 18-4 depicts its practical realization in terms of UEs. The response of this circuit in the t plane, unlike that of the same topology in the s plane, is periodic. This is illustrated in Fig. 18-5. The impedance of this type of circuit may be renormalized to a termination of 50 Ω by rescaling the impedances of the UEs by 50.

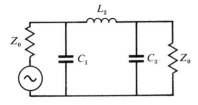

FIGURE 18-3
Realization of $Y(t) = (2t^3 + 2t^2 + 2t + 1)/(2t^2 + 2t + 1)$ in t plane.

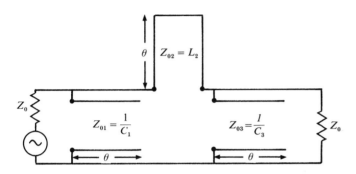

FIGURE 18-4
Commensurate circuit of $Y(t) = (2t^3 + 2t^2 + 2t + 1)/(2t^2 + 2t + 1)$.

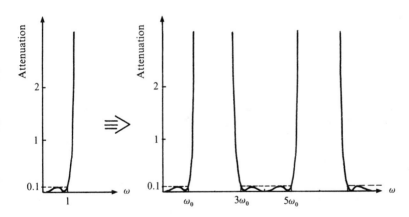

FIGURE 18-5
Frequency transformation between Ω and ω variables for low-pass filter.

RICHARDS THEOREM
(THE UNIT ELEMENT)

In addition to the basic distributed t-plane L and C values defined by the Richards transformation, a unit element (UE) may also be defined. UEs are usually employed to separate the circuit elements in high-frequency circuits which are otherwise located at the same physical point.

The input impedance $Z(t)$ of a transmission line of characteristic impedance Z_0 terminated in an impedance $Z_1(t)$ is given from transmission line theory by

$$Z(t) = Z_0 \frac{Z_1(t) + tZ_0}{tZ_1(t) + Z_0} \qquad (18\text{-}14)$$

where t is the Richards variable.

Solving the above equation for $Z_1(t)$ gives

$$Z_1(t) = Z_0 \frac{Z(t) - tZ_0}{Z_0 - tZ(t)} \qquad (18\text{-}15)$$

The Richards theorem states that a unit element $Z(1)$ may always be extracted from an impedance function and the remainder impedance $Z_1(t)$ is p.r. and in the case of a reactance function of degree one less than $Z(t)$. Putting $t = 1$ in Eq. (18-14) leads to

$$Z(1) = Z_0 \qquad (18\text{-}16)$$

Z_0 is p.r. since $Z(1)$ is p.r. Thus such a unit element may always be realized.

Combining Eqs (18-15) and (18-16) yields

$$Z_1(t) = Z(1) \frac{Z(t) - tZ(1)}{Z(1) - tZ(t)} \qquad (18\text{-}17)$$

Although $Z_1(t)$ appears to be of one degree higher than $Z(t)$, $P_1(t)$ and $Q_1(t)$ have a common factor $(t^2 - 1)$. The demonstration of this feature starts by putting $t = 1$ in Eq. (18-17). This gives

$$Z_1(1) = Z(1) \frac{Z(1) - Z(1)}{Z(1) - Z(1)} \qquad (18\text{-}18)$$

Since both the numerator and denominator polynomials of $Z_1(t)$ are zero for $t = 1$ each contains a common factor $(t - 1)$ which may be cancelled.

The derivation continues by demonstrating that if $Z_1(t)$ is a reactance function then

$$Z_1(t) = -Z_1(-t) \qquad (18\text{-}19)$$

Evaluating this quantity for $t = -1$ enables a common factor $(t + 1)$ to be cancelled as well.

Thus by the repeated extraction of UEs the degree of the reactance function can be reduced to zero and the circuit completely synthesized, the last element having either an open-circuit or short-circuit termination.

The dual admittance relationships to those in Eqs (18-15) and (18-17) are

$$Y(t) = Y_0 \frac{Y(t) + tY_0}{Y_0 + tY(t)} \qquad (18\text{-}20)$$

$$Y_1(t) = Y(1) \frac{Y(t) - tY(1)}{Y(1) - tY(t)} \qquad (18\text{-}21)$$

As an example consider the synthesis of a third-order one-port reactance function incorporating a UE between t-plane capacitors:

$$Y(t) = \frac{t^3 + 3t}{t^2 + 1}$$

$Y(s)$ has a pole at the origin that can be removed in the form of a t-plane capacitor. The result is

$$Y(t) = t + \frac{2t}{t^2 + 1}$$

Thus

$$Y_1(t) = t$$

$$Y_2(t) = \frac{2t}{t^2 + 1}$$

To space the first and last elements by a UE it is now necessary to extract $Y_2(1)$ from $Y_2(t)$:

$$Y_2(1) = 1\,\Omega^{-1}$$

$Y_3(t)$ is now constructed using the relationship in Eq. (18-21):

$$Y_3(t) = \frac{t(t^2 - 1)}{(t^2 - 1)}$$

This remainder admittance exhibits a common factor $t^2 - 1$ in the numerator and denominator polynomials in keeping with the Richards theorem. Cancelling out this common factor gives

$$Y_3(t) = t$$

$Y_3(t)$ has a pole at the origin that can also be extracted as a t-plane capacitor. Figure 18-6 depicts a block diagram of $Y(t)$ with the t-plane capacitor realized by open-circuited UEs. As a second example of the application of the Richards theorem consider the realization of the following one-port reactance function in terms of UEs only:

$$Z(t) = 50\,\frac{t^3 + 3t}{t^2 + 1}$$

The impedance of the first UE is now obtained by replacing t by 1 in $Z(t)$. The result is

$$Z(1) = 100\,\Omega$$

and the remainder impedance is obtained from Eq. (18-17) as

$$Z_1(t) = \frac{100t(t^2 - 1)}{t^4 + t^2 - 2}$$

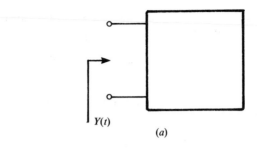

$Y(t)$

(a)

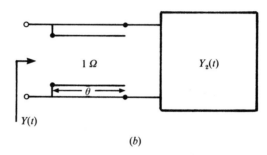

$1\,\Omega$ $Y_2(t)$

θ

$Y(t)$

(b)

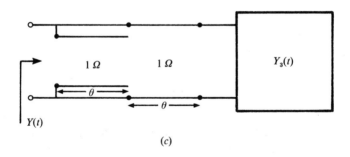

$1\,\Omega$ $1\,\Omega$ $Y_3(t)$

θ

θ

$Y(t)$

(c)

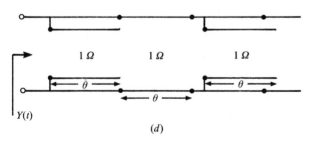

$1\,\Omega$ $1\,\Omega$ $1\,\Omega$

θ θ

θ

$Y(t)$

(d)

FIGURE 18-6
Realization of $Y(t) = (t^3 + 3t)/(t^2 + 1)$ using UE element.

In keeping with the earlier remarks $Z_1(t)$ has a common factor $t^2 - 1$ in both its numerator and denominator polynomials. Eliminating this common factor from the remainder impedance gives

$$Z_1(t) = \frac{100t}{t^2 + 2}$$

The impedance of the second UE is now deduced by replacing t by 1 in $Z_1(t)$. This gives

$$Z_1(1) = \frac{100}{3}\,\Omega$$

The remainder impedance $Z_2(t)$ is now evaluated by making use of Eq. (18-17) as

$$Z_2(t) = \frac{50t(t^2 - 1)}{3(t^2 - 1)}$$

This function is characterized by a common factor $t^2 - 1$. Eliminating this factor gives

$$Z_2(t) = \frac{50t}{3}$$

The impedance of the next UE is

$$Z_3(t) = \frac{50}{3}\,\Omega$$

Evaluating the remainder impedance $Z_3(t)$ yields

$$Z_3(t) = 0$$

The last UE is therefore terminated in a short circuit. Figure 18-7 depicts the step by step realization of $Z(t)$.

PARTIAL EXTRACTION OF UEs

It is also possible to extract a UE with a specified value of characteristic impedance by partially extracting a pole of the immittance function in such a way that the remainder $Z'(t)$ satisfies

$$Z'(1) = 1$$

This will now be illustrated by way of an example by realizing $Z(t)$ below:

$$Z(t) = \frac{t^3 + 4t^2 + 7t + 12}{t^3 + 4t^2 + 7t}$$

in such a way as to realize a UE with a characteristic impedance of unity.

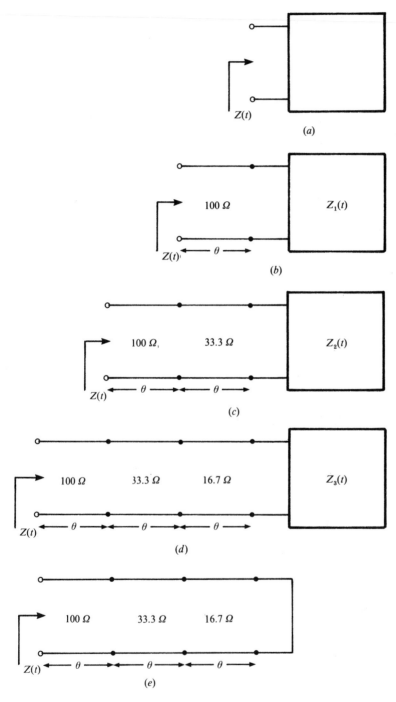

FIGURE 18-7
Realization of $Z(t) = (t^3 + 3t)/(t^2 + 1)$ using UEs only.

This impedance has a pole at the origin which may be partially extracted in a second Cauer form. This extraction is obtained by long division:

$$7t + 4t^2 + t^3 \overline{)\ \ 12 \quad + \quad 7t \quad + \quad 4t^2 \quad + t^3\ } \left(\frac{a}{t}\right)$$
$$\underline{7a \quad + \quad 4at \quad + \quad at^2}$$
$$(12 - 7a) + (7 - 4a)t + (4 - a)t^2 + t^3$$

The result is

$$Z(t) = \frac{a}{t} + \frac{t^3 + (4 - a)t^2 + (7 - 4a)t + (12 - 7a)}{t^3 + 4t^2 + 7t}$$

Evaluating the remainder impedance by replacing t by 1 and equating to unity gives

$$a = 1$$

The realization of $Z(t)$ that permits a UE of unity to be extracted is therefore

$$Z(t) = \frac{1}{t} + \frac{t^3 + 3t^2 + 3t + 5}{t^3 + 4t^2 + 7t}$$

KURODA IDENTITIES

One way in which it is possible to replace a distributed network with two kinds of elements by one using only one kind or rearrange the topology of a network is by having recourse to the Kuroda identities in Figs 18-8 and 18-9. One Kuroda identity of the first kind provides an equivalence between a shunt open-circuited stub and a UE circuit to a UE and series short-circuited stub circuit. A second identity of the first kind gives the equivalence between a series short-circuited stub and a UE and a UE and a shunt open-circuited stub. The two Kuroda identities of the second type interchange stubs of the same kind. The complication of the ideal transformers in the latter identities disappears for symmetrical filters. This may be understood by noting that the transformers can be passed through the circuit from each end, altering the impedance levels of the elements passed over; finally, the symmetrical series of transformers from each half of the filter combine at the centre as two transformers and cancel each other. The Kuroda identities may be deduced by comparing the transfer matrices of the corresponding networks in Fig. 18-10.

The verification of the first identity starts by putting down the overall $ABCD$ matrix of the original circuit:

$$\begin{bmatrix} A & B \\ C & D \end{bmatrix} = \begin{bmatrix} 1 & 0 \\ Y_1 \tanh \theta & 1 \end{bmatrix} \begin{bmatrix} \cosh \theta & Z_0 \sinh \theta \\ \dfrac{\sinh \theta}{Z_0} & \cosh \theta \end{bmatrix} \qquad (18\text{-}22)$$

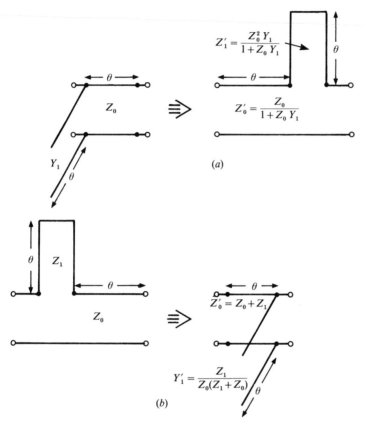

FIGURE 18-8
Kuroda identities of the first type.

and that of the transformed circuit:

$$
\begin{bmatrix} A & B \\ C & D \end{bmatrix} = \begin{bmatrix} \cosh\theta & \left(\dfrac{Z_0}{1+Z_0Y_1}\right)\sinh\theta \\ \left(\dfrac{1+Z_0Y_0}{Z_0}\right)\sinh\theta & \cosh\theta \end{bmatrix} \begin{bmatrix} 1 & \left(\dfrac{Z_0^2Y_1}{1+Z_0Y_1}\right)\tanh\theta \\ 0 & 1 \end{bmatrix}
$$

$$(18\text{-}23)$$

Scrutiny of these two relationships indicates that the overall $ABCD$ matrix of each network is the same provided the characteristic impedance of the UE of the transformed circuit is taken as

$$
Z_0' = \frac{Z_0}{1+Z_0Y_1} \tag{18-24a}
$$

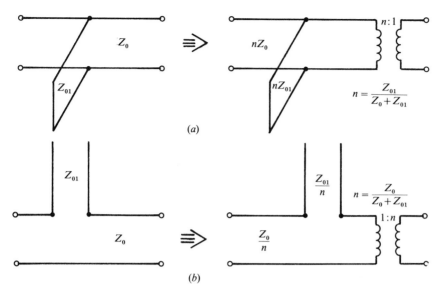

FIGURE 18-9
Kuroda identities of the second type.

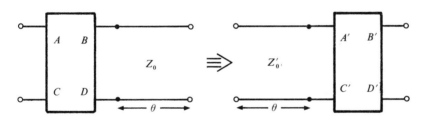

FIGURE 18-10
Schematic diagram showing equivalence between a two-port network and a UE and a UE and a two-port network.

and that the characteristic impedance of the short-circuited UE in series with the first one is taken as

$$Z_1' = \frac{Z_0^2 Y_1}{1 + Z_0 Y_1} \qquad (18\text{-}24b)$$

as asserted.

This result may also be derived by direct synthesis by terminating the original network by a 1-Ω resistance and forming a one-port immittance function in the t plane. The derivation starts by making use of the relationship between the input impedance of a two-port network in the t plane terminated in a 1-Ω resistor and the $ABCD$ description of the circuit:

$$Z(t) = \frac{A(t) + B(t)}{C(t) + D(t)} \tag{18-25}$$

The overall $ABCD$ parameters of the arrangement in Eq. (18-22) in the θ plane is

$$A = \cosh \theta \tag{18-26a}$$

$$B = Z_0 \sinh \theta \tag{18-26b}$$

$$C = Y_1 \sinh \theta + \frac{\sinh \theta}{Z_0} \tag{18-26c}$$

$$D = Z_0 Y_1 \sinh \theta \tanh \theta + \cosh \theta \tag{18-26d}$$

Evaluating these quantities in the t plane indicates that

$$Z(t) = Z_0 \frac{1 + Z_0 t}{Z_0^2 Y_1 t^2 + (Z_0 Y_1 + 1)t + Z_0} \tag{18-27}$$

Extracting a UE from this immittance function by replacing t by 1 gives

$$Z(1) = \frac{Z_0}{1 + Z_0 Y_1} \tag{18-28}$$

in keeping with the result in Eq. (18-24a). The other elements follow. The derivation of the other identities is left as an exercise for the reader. It may be separately demonstrated that it is always possible to map a cascade arrangement of a UE and a p.r. immittance into a p.r. immittance in cascade with a UE.

PROBLEMS

18-1 Realize the following admittance function in terms of shunt stubs and UEs only:

$$Y(t) = \frac{2t^3 + 2t^2 + 2t + 1}{2t^2 + 2t + 1}$$

18-2 Synthesize $Y(t)$ in Prob. 18-1 in terms of shunt and series stubs and use the Kuroda identities to formulate an equivalent circuit using only shunt stubs spaced by commensurate transmission lines.

18-3 Demonstrate the Kuroda identities in Fig. 18-9b, c and d.

18-4 Verify that the transfer parameters A and D are equal to zero for each of the immittance inverters illustrated in this chapter.

18-5 Verify that the product of the transfer parameters B and C is equal to unity for each of the immittance inverters illustrated in this chapter.

18-6 Synthesize $Z(t)$ below:

$$Z(t) = \frac{4t^2 + 6t^2 + 1}{4t^3 + 4t}$$

in a first Cauer form in the t plane.

18-7 Synthesize $Z(t)$ in Prob. 18-6 in terms of UEs only.

18-8 Obtain the correspondence between the topologies in Probs 18-6 and 18-7 by having recourse to the appropriate Kuroda identities.

18-9 Verify the identities in Figs 18-8 and 18-9 using the appropriate Kuroda identities.

18-10 Derive the Kuroda first identities by forming one-port immittance functions of the original circuits and using direct synthesis of the ensuing immittances.

18-11 Repeat Prob. 18-10 in the case of the Kuroda second identities.

18-12 Investigate whether the following functions are p.r.:

$$\frac{\tanh^4 \theta + 10 \tanh^2 \theta + 9}{\tanh^3 \theta + 4 \tanh \theta}$$

$$\frac{(\tanh^2 \theta + 1)(\tanh^2 \theta + 3)(\tanh^2 \theta + 5)}{\tanh \theta (\tanh^2 \theta + 2)(\tanh^2 \theta + 4)}$$

18-13 Synthesize the following impedance function as a cascade of commensurate transmission lines:

$$\frac{\tanh^4 \theta + 10 \tanh^2 \theta + 9}{\tanh^3 \theta + 4 \tanh \theta}$$

BIBLIOGRAPHY

Richards, P. J., 'Resistor-transmission line circuits', *Proc. IRE*, Vol. 34, pp. 217–220, September 1946.

Richards, P. J., 'A special class of functions with positive real parts in a half-plane', *Duke Maths. J.*, Vol. 14, pp. 777–786, September 1947.

CHAPTER

19

EXACT
SYNTHESIS OF
OPTIMUM
QUARTER-WAVE
FILTERS

INTRODUCTION

An important class of commensurate distributed circuits is one comprising a cascade arrangement of quarter-wave lines and open- or short-circuited quarter-wave stubs between resistive loads. The purpose of this chapter is to summarize the Chebyshev and Butterworth high-pass and low-pass amplitude squared insertion loss functions for this class of networks; the low-pass prototype maps into a band-stop one and the high-pass one into a band-pass response. The characteristic functions of a cascade of UEs between real terminations and between a real and a complex STUB-R load are separately summarized. The exact synthesis procedure involves forming the input impedance of the network in the t plane from a knowledge of the magnitude squared transfer function followed by a systematic removal of UEs, t-plane capacitors and inductors until the degree of the input impedance is reduced to zero. This method follows closely the synthesis of a two-port LCR network as a two-port LC network terminated in a 1-Ω resistor. The t-plane capacitors and inductors are then realized by open- and short-circuited stubs of commensurate length. The chapter includes the synthesis of a seven-element filter circuit consisting of two UEs and five stubs. It is again necessary to stress that it is possible to realize any single specification in more than one way, the optimum solution being dependent upon the fabrication problem.

THE BUTTERWORTH APPROXIMATION PROBLEM

The generalized magnitude squared low- and high-pass transfer functions for the Butterworth and Chebyshev approximation problems in the case of a cascade of quarter-wave long transmission lines, stubs and parallel coupled lines have been described by Horton and Wenzel. It is summarized below for the Butterworth case and in the next section for the Chebyshev one. The low-pass solution maps onto the band-stop response and the high-pass one into the band-pass one. The Butterworth characteristic functions for the low- and high-pass circuits are described in the θ plane by

$$K(\theta^2) = \left(\frac{\tan \theta}{\tan \theta_c}\right)^{2m} \left(\frac{\sin \theta}{\sin \theta_c}\right)^{2n} \tag{19-1}$$

and

$$K(\theta^2) = \left(\frac{\tan \theta_c}{\tan \theta}\right)^{2m} \left(\frac{\cos \theta}{\cos \theta_c}\right)^{2n} \tag{19-2}$$

respectively, where m is the number of stubs, n is the number of UEs and θ and θ_c are defined by

$$\theta = \frac{\pi}{2}\left(\frac{\omega}{\omega_0}\right) \tag{19-3}$$

$$\theta_c = \frac{\pi}{2}\left(\frac{\omega_c}{\omega_0}\right) \tag{19-4}$$

The corresponding relationships are described in the t plane by

$$K(t^2) = \left(\frac{t}{t_c}\right)^{2m} \left(\frac{t\sqrt{1-t_c^2}}{t_c\sqrt{1-t^2}}\right)^{2n} \tag{19-5}$$

and

$$K(t^2) = \left(\frac{t_c}{t}\right)^{2m} \left(\frac{\sqrt{1-t_c^2}}{\sqrt{1-t^2}}\right)^{2n} \tag{19-6}$$

where

$$t_c = j \tan \theta_c \tag{19-7a}$$

$$t = j \tan \theta \tag{19-7b}$$

It is also recalled that

$$S_{11}(t)S_{11}^*(t) = \frac{K(t^2)}{1 + K(t^2)} \tag{19-8a}$$

$$S_{21}(t)S_{21}^{*}(t) = \frac{1}{1 + K(t^2)} \tag{19-8b}$$

At $t = t_c$, $K^2(t) = 1$ and $S_{21}(t)S_{21}^{*}(t) = 0.50$, in keeping with the original definition of this type of function.

THE CHEBYSHEV APPROXIMATION PROBLEM

The characteristic functions associated with the low- and high-pass Chebyshev solutions are given in the θ plane by[1,2]

$$K(\theta^2) = \left[T_m\left(\frac{\tan\theta}{\tan\theta_c}\right) T_n\left(\frac{\sin\theta}{\sin\theta_c}\right) - U_m\left(\frac{\tan\theta}{\tan\theta_c}\right) U_n\left(\frac{\sin\theta}{\sin\theta_c}\right) \right]^2 \tag{19-9}$$

and

$$K(\theta^2) = \left[T_m\left(\frac{\tan\theta_c}{\tan\theta}\right) T_n\left(\frac{\cos\theta}{\cos\theta_c}\right) - U_m\left(\frac{\tan\theta_c}{\tan\theta}\right) U_n\left(\frac{\cos\theta}{\cos\theta_c}\right) \right]^2 \tag{19-10}$$

and in the t plane by

$$K(t^2) = \left[T_m\left(\frac{t}{t_c}\right) T_n\left(\frac{t\sqrt{1-t_c^2}}{t_c\sqrt{1-t^2}}\right) - U_m\left(\frac{t}{t_c}\right) U_n\left(\frac{t\sqrt{1-t_c^2}}{t_c\sqrt{1-t^2}}\right) \right]^2 \tag{19-11}$$

and

$$K(t^2) = \left[T_m\left(\frac{t_c}{t}\right) T_n\left(\frac{\sqrt{1-t_c^2}}{\sqrt{1-t^2}}\right) - U_m\left(\frac{t_c}{t}\right) U_n\left(\frac{\sqrt{1-t_c^2}}{\sqrt{1-t^2}}\right) \right]^2 \tag{19-12}$$

where $T_n(x)$ and $U_n(x)$ are the unnormalized Chebyshev polynomials of the first and second kind of order n:

$$T_n(x) = \cos(n \cos^{-1} x) \tag{19-13}$$

$$U_n(x) = \sin(n \cos^{-1} x) \tag{19-14}$$

The Chebyshev functions of the first and second kind are given in polynomial form by

$$T_0(x) = 1 \tag{19-15a}$$

$$T_1(x) = x \tag{19-15b}$$

$$T_2(x) = 2x^2 - 1 \tag{19-15c}$$

$$T_{n+1}(x) = 2xT_n(x) - T_{n-1}(x) \tag{19-15d}$$

and

$$U_0(x) = 0 \tag{19-16a}$$

$$U_1(x) = \sqrt{1-x^2} \tag{19-16b}$$

$$U_2(x) = 2x\sqrt{1-x^2} \tag{19-16c}$$

$$U_{n+1}(x) = 2xU_n(x) - U_{n-1}(x) \tag{19-16d}$$

respectively.

It may also be shown that $T_n(x)$ and $U_n(x)$ are related by

$$-\sqrt{1-x^2}\,U_n(x) = T_{n+1}(x) + T_{n-1}(x) \qquad (19\text{-}17)$$

It is furthermore recalled that the scattering parameters are related to the characteristic function and the ripple level of the circuit by

$$S_{21}(t)S_{21}^*(t) = \frac{1}{1 + \varepsilon^2 K(t^2)} \qquad (19\text{-}18)$$

and

$$S_{11}(t)S_{11}^*(t) = \frac{\varepsilon^2 K(t^2)}{1 + \varepsilon^2 K(t^2)} \qquad (19\text{-}19)$$

The ripple level ε is defined in terms of the VSWR by

$$\varepsilon = \frac{\text{VSWR} - 1}{2\sqrt{\text{VSWR}}} \qquad (19\text{-}20)$$

in the usual way.

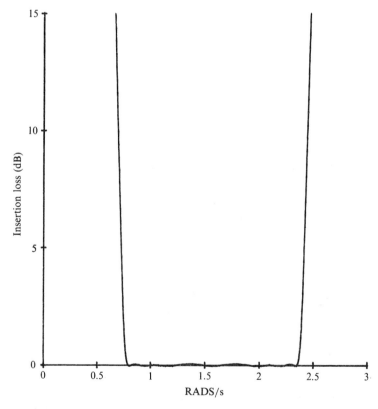

FIGURE 19-1
Frequency response of filter prototype with $n = 4$, $m = 1$.

A modification of the preceding specification that is sometimes of value in controlling the element values of a circuit is obtained by writing Eqs (19-18) and (19-19) as

$$S_{21}(t)S_{21}^*(t) = \frac{1}{1 + K^2 + \varepsilon^2 K(t^2)} \tag{19-21}$$

$$S_{11}(t)S_{11}^*(t) = \frac{K^2 + \varepsilon^2 K(t^2)}{1 + K^2 + \varepsilon^2 K(t^2)} \tag{19-22}$$

The frequency response of this insertion loss function with $n = 4$ and $m = 1$ is illustrated in Fig. 19-1. The maximum and minimum values of the VSWR are given in this instance by

$$(\text{VSWR})_{\text{max}} = (\sqrt{1 + K^2 + \varepsilon^2} + \sqrt{K^2 + \varepsilon^2})^2 \tag{19-23}$$

$$(\text{VSWR})_{\text{min}} = (\sqrt{1 + K^2} + K)^2 \tag{19-24}$$

If the parameter

$$K = 0$$

then

$$(\text{VSWR})_{\text{min}} = 1$$

in keeping with the classic condition.

R AND STUB-R LOADS

One special class of band-pass circuits is that of a number of UEs between unequal loads; another is that of a number of UEs and a single stub between similar terminations. The characteristic function in the θ plane for the first situation is obtained from that in Eq. (19-10) with $m = 0$, and that for the second case is also

$$K(\theta^2) = T_n^2\left(\frac{\cos \theta}{\cos \theta_c}\right) \tag{19-25}$$

The characteristic function of a similar cascade of UEs (n) but embodying a single stub ($m = 1$) is deduced by making use of the identities below:

$$U_1\left(\frac{\tan \theta_c}{\tan \theta}\right) = \frac{\sqrt{1 - (\cos \theta/\cos \theta_c)^2}}{\sin \theta} \tag{19-26a}$$

$$U_n\left(\frac{\cos \theta}{\cos \theta_c}\right) = -\frac{T_{n+1}(\cos \theta/\cos \theta_c) + T_{n-1}(\cos \theta/\cos \theta_c)}{\sqrt{1 - (\cos \theta/\cos \theta_c)^2}} \tag{19-26b}$$

$$T_1\left(\frac{\tan\theta_c}{\tan\theta}\right) = \left(\frac{\sin\theta_c}{\sin\theta}\right)\left(\frac{\cos\theta}{\cos\theta_c}\right) \tag{19-26c}$$

$$T_n\left(\frac{\cos\theta}{\cos\theta_c}\right) = \frac{T_{n+1}(\cos\theta/\cos\theta_c) + T_{n-1}(\cos\theta/\cos\theta_c)}{2(\cos\theta/\cos\theta_c)} \tag{19-26d}$$

The required result in the θ plane is

$$K(\theta^2) = \left[\frac{(1+\sin\theta_c)T_{n+1}(\cos\theta/\cos\theta_c) - (1-\sin\theta_c)T_{n-1}(\cos\theta/\cos\theta_c)}{2\sin\theta}\right]^2 \tag{19-27}$$

The t-plane synthesis of these two classes of network will be dealt with separately in the next chapter.

SEVEN-SECTION EQUAL-RIPPLE FILTER

A seven-section equal-ripple band-pass filter with a 3 to 1 bandwidth and a 0.1-dB ripple in the passband will now be considered in detail as a design example. The filter is to be realized in terms of two unit elements and five t-plane inductors and capacitors. This gives

$$n = 2$$

$$m = 5$$

$$\theta_c = \frac{\pi}{4}$$

$$t_c = j1$$

$$\varepsilon = 0.1526$$

The required characteristic function is then described by

$$K(t^2) = \left[T_5\left(\frac{t_c}{t}\right)T_2\left(\frac{\sqrt{1-t_c^2}}{\sqrt{1-t^2}}\right) - U_5\left(\frac{t_c}{t}\right)U_2\left(\frac{\sqrt{1-t_c^2}}{\sqrt{1-t^2}}\right)\right]^2$$

and the amplitude squared of the reflection coefficient is given by

$$S_{11}(t)S_{11}(-t)$$
$$= \frac{(0.98t^6 + 8.15t^4 + 17.64t^2 + 10.78)^2}{-t^{14} + 2.96t^{12} + 14.96t^{10} + 100.96t^8 + 308.96t^6 + 487.06t^4 + 380.46t^2 + 116.23}$$

The roots of the denominator polynomial of this quantity are now deduced using a root-finding subroutine. The denominator polynomial of $S_{11}(t)$ is constructed from those of the left half-plane as

$$t^7 + 6.16t^6 + 17.47t^5 + 32.36t^4 + 39.13t^3 + 37.57t^2 + 20.73t + 10.78$$

Since the numerator polynomial of $S_{11}(t)S_{11}(-t)$ is a perfect square that of $S_{11}(t)$ is given by

$$0.98t^6 + 8.15t^4 + 17.64t^2 + 10.78$$

A bounded real solution for $S_{11}(t)$ is therefore

$$S_{11}(t) = -\frac{0.98t^6 + 8.15t^4 + 17.64t^2 + 10.78}{t^7 + 6.16t^6 + 17.47t^5 + 32.36t^4 + 39.13t^3 + 37.57t^2 + 20.73t + 10.78}$$

and a realizable solution for $Z(t)$ is

$$Z(t) = \frac{t^7 + 5.18t^6 + 17.47t^5 + 24.21t^4 + 39.13t^3 + 19.93t^2 + 20.73t}{t^7 + 7.13t^6 + 17.47t^5 + 40.51t^4 + 39.13t^3 + 55.21t^2 + 20.73t + 21.56}$$

Since the odd parts of the numerator and denominator polynomials of $Z(t)$ are identical, the network is symmetric. One possible network topology is indicated in Fig. 19-2. The determination of the element values begins with the removal of a unit element by having recourse to the Richards theorem. This gives

$$Z(1) = 0.63$$

and a remainder impedance

$$Z'(t) = \frac{1.84t^6 + 10.20t^5 + 18.84t^4 + 33.65t^3 + 20.08t^2 + 20.98t}{4.64t^6 + 21.09t^5 + 64.83t^4 + 82.37t^3 + 128.05t^2 + 60.52t + 62.96}$$

The remainder impedance $Z'(t)$ is obtained from a knowledge of $Z(t)$ and $Z(1)$ using the Richards theorem:

$$Z'(t) = Z(1)\frac{Z(t) - tZ(1)}{Z(1) - tZ(t)}$$

In obtaining the preceding result a common factor $(t^2 - 1)$ has also been cancelled in both the numerator and denominator polynomials of $Z'(t)$.

A shunt t-plane inductor is next removed in a second Cauer manner by first forming

$$Y'(t) = \frac{1}{Z'(t)}$$

This gives

$$L_1 = 0.33 \text{ H}$$

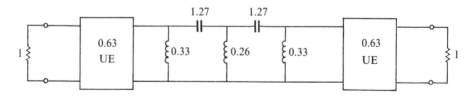

FIGURE 19-2
Schematic diagram of degree 7 filter in t plane.

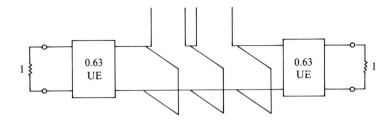

FIGURE 19-3
Schematic diagram of degree 7 filter in θ plane.

and a remainder admittance

$$Y''(t) = \frac{13.98t^5 + 46.87t^4 + 102.76t^3 + 77.15t^2 + 80.25t}{5.54t^5 + 30.73t^4 + 56.77t^3 + 101.42t^2 + 60.52t + 62.96}$$

A t-plane series capacitor of value

$$C_2 = 1.27 \text{ F}$$

and a shunt t-plane inductor of value

$$L_2 = 0.26 \text{ H}$$

are next extracted from the impedance of the remaining network. Since the network is symmetrical the remainder elements are given by inspection in a similar manner. A diagram of the synthesis steps, together with the completely synthesized network, is indicated in Fig. 19-2.

The t-plane inductors (L) are realized by short-circuited lines of characteristic impedance (Z_0):

$$Z_0 = L$$

and the t-plane capacitors (C) are realized by open-circuited lines of characteristic impedance (Z_0):

$$Z_0 = \frac{1}{C}$$

The topology of the required network is indicated in Fig. 19-3.

PROBLEMS

19-1 Verify that the Butterworth low- and high-pass characteristic functions are related by replacing t by $1/t$.

19-2 Plot the frequency responses for the Butterworth approximation problem for $m = 2$, $n = 1, 2, 3$. Repeat for $m = 3$, $n = 1, 2, 3$.

19-3 Repeat Prob. 19-2 for the Chebyshev approximation problem.

19-4 Verify that the characteristic functions in the t plane in Eqs (19-5) and (19-6) reduce to classic s-plane forms when $m = 0$.

19-5 Verify that the characteristic functions in the t plane in Eqs (19-11) and (19-12) reduce to classic s-plane ones when $m = 0$.

19-6 Construct the frequency response of the problem discussed in this chapter.

19-7 Obtain the scattering and impedance eigenvalues of the degree 7 filter discussed in this chapter. Synthesize it in terms of its even impedance eigenvalue.

REFERENCES

1. Norton, M. C. and Wenzel, R. J., 'Optimum quarter-wave TEM filters', *IRE Trans. on MTT*, Vol. MTT-13, pp. 316–327, May 1965, (© 1984 *IEEE*).
2. Carlin, H. J. and Kohler, W., 'Direct synthesis of band-pass transmission line structures', *IEEE Trans. on MTT*, Vol. MTT-13, pp. 283–297, May 1965.

BIBLIOGRAPHY

Levy, R. and Helszajn, J., 'Specific equations for one and two section quarter-wave matching networks for stub-resistor loads, *IEEE Trans. on MTT*, Vol. MTT-30, pp. 55–62, January 1982, (© 1984 *IEEE*).
Riblet, H. J., 'General synthesis of quarter-wave impedance transformers', *IRE Trans. on MTT*, Vol. MTT-5, 1, pp. 36–43, January 1957, (© 1984 *IEEE*).

CHAPTER
20

SYNTHESIS OF
STEPPED
IMPEDANCE
TRANSDUCERS

INTRODUCTION

In Electrical Engineering it is often necessary to match different real impedances or to match a real load to a complex one over some frequency interval. One classic solution to this problem, at microwave frequencies, is to connect the two impedances by one or more quarter-wave long transmission lines (UEs) of appropriate impedance levels. As in the filter problem, an ideal matching network cannot be realized, but transfer functions having Butterworth or Chebyshev amplitude characteristics can be stipulated. This permits the number of UEs and their impedance levels to be fixed in terms of the bandwidth and ripple level (or VSWR) of the network. The purpose of this chapter is to derive the solution in the case of a pure resistive load and also in that of a STUB-R load. The appropriate characteristic functions for these two cases have already been noted.

The classic solution to this type of problem, as in the related filter one, is based upon an insertion loss specification in the θ plane, the use of the unitary condition to deduce a squared amplitude reflection coefficient from a knowledge of that of the transmission one, the mapping of the θ plane into the t plane, the construction of a bounded real reflection coefficient, the use of the bilinear transformation between immittance and reflection coefficient and the synthesis of a one-port immittance function in terms of UEs and t-plane inductors and capacitors that have the topology of the required two-port network.

SCATTERING PARAMETERS OF UE AND UE-STUB MATCHING NETWORKS

Two common circuits met in electrical engineering are that of a number of UEs between unequal real loads and a number of similar UEs between a real impedance and a complex STUB-R load. The amplitude squared scattering parameters associated with these two situations are the same as those met in the related filter problem:

$$S_{21}(\theta)S_{21}^*(\theta) = \frac{1}{1 + K^2 + \varepsilon^2 K(\theta^2)} \tag{20-1}$$

$$S_{11}(\theta)S_{11}^*(\theta) = \frac{K^2 + \varepsilon^2 K(\theta^2)}{1 + K^2 + \varepsilon^2 K(\theta^2)} \tag{20-2}$$

The parameter K may be employed to obtain some degree of control over the element values of the network. The usual ripple parameter is ε. The maximum and minimum values of the VSWRs in the passband are related to K and ε by

$$(\text{VSWR})_{\text{max}} = (\sqrt{1 + K^2 + \varepsilon^2} + \sqrt{K^2 + \varepsilon^2})^2 \tag{20-3}$$

$$(\text{VSWR})_{\text{min}} = (\sqrt{1 + K^2} + K)^2 \tag{20-4}$$

The characteristic function for $m = 0$ and n UEs, which is equal ripple in the passband, has been shown to be

$$K(\theta^2) = T_n^2\left(\frac{\cos\theta}{\cos\theta_c}\right) \tag{20-5}$$

and that for $m = 1$ and n UEs, which is also equal ripple in the passband, has been given by

$$K(\theta^2) = \left[\frac{(1 + \sin\theta_c)T_{n+1}(\cos\theta/\cos\theta_c) - (1 - \sin\theta_c)T_{n-1}(\cos\theta/\cos\theta_c)}{2\sin\theta}\right]^2 \tag{20-6}$$

θ and θ_c are defined by

$$\theta = \frac{\pi}{2}\left(\frac{\omega}{\omega_0}\right) \tag{20-7}$$

$$\theta_c = \frac{\pi}{2}\left(\frac{\omega_c}{\omega_0}\right) \tag{20-8}$$

respectively. θ is also sometimes written in terms of a normalized bandwidth parameter as

$$\theta = \frac{\pi}{2}(1 + \delta) \tag{20-9}$$

where

$$\delta = \frac{\omega + \omega_0}{\omega_0} \tag{20-10}$$

ω_0 is the centre frequency of the transformer, $\omega_{1,2}$ the band edges and ω is the normal frequency variable.

The required characteristic functions in the t plane are obtained from those in the θ plane by having recourse to the Richards transformation

$$t \to j \tan \theta \tag{20-11}$$

as is by now understood.

The topologies of the two circuits discussed here are indicated in Figs 20-1 and 20-2. Figure 20-3 indicates one frequency response.

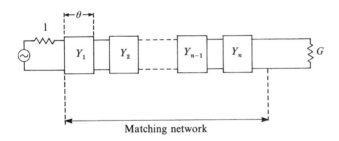

FIGURE 20-1
Schematic diagram of multistep impedance transformer.

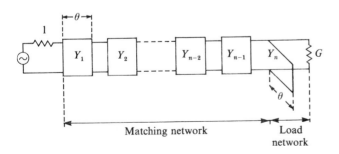

FIGURE 20-2
Schematic diagram of multistep impedance transformer between a load R and a STUB-R load.

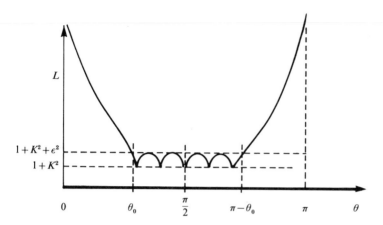

FIGURE 20-3
Frequency response of multistep impedance transformer indicating definitions of ε and k parameters.

EXACT SYNTHESIS OF
STEPPED IMPEDANCE TRANSFORMERS

One important class of matching network between transmission lines with unequal characteristic impedances is the quarter-wave impedance transformer. Such a network may be synthesized by the systematic removal of UEs. This approach, due to Riblet, follows closely that of an *LCR* function as a two-port reactance network terminated in a 1-Ω resistance. The synthesis of such a network with an equal ripple specification will now be developed. Introducing the Richards transformation into the appropriate θ-plane characteristic function gives the required t-plane function below:

$$K(t^2) = T_n^2\left(\frac{1}{\cos\theta_c\sqrt{1-t^2}}\right) \tag{20-12}$$

If K in Eqs (20-1) and (20-2) is taken as zero then

$$S_{21}(t)S_{21}(-t) = \frac{1}{1 + \varepsilon^2 T_n^2[1/(\cos\theta_c\sqrt{1-t^2})]} \tag{20-13}$$

and

$$S_{11}(t)S_{11}(-t) = \frac{\varepsilon^2 T_n^2[1/(\cos\theta_c\sqrt{1-t^2})]}{1 + \varepsilon^2 T_n^2[1/(\cos\theta_c\sqrt{1-t^2})]} \tag{20-14}$$

The synthesis procedure begins by forming $S_{11}(t)$ from a knowledge of $S_{11}(t)S_{11}(-t)$. Since the denominator polynomial of $S_{11}(t)$ is Hurwitz its poles are selected from the LHP ones of $S_{11}(t)S_{11}(-t)$. Since the numerator polynomial need not be Hurwitz its zeros may be selected from either the left or right half-plane roots of the numerator polynomial. A possible bounded real solution for $S_{11}(t)$ is therefore

$$S_{11}(t) = \frac{\pm \alpha \varepsilon T_n[1/(\cos \theta_c \sqrt{1 - t^2})]}{Q(t)} \qquad (20\text{-}15)$$

$Q(t)$ is constructed from the LHP poles of $S_{11}(t)S_{11}(-t)$ except for a multiplication constant α. This constant is chosen such that the specification

$$S_{11} = \frac{Z_{n+1} - Z_0}{Z_{n+1} + Z_0} \qquad (20\text{-}16)$$

is satisfied at $t = 0$.

Once $S_{11}(t)$ is specified, $Z(t)$ is determined from a knowledge of $S_{11}(t)$ in the usual way by making use of the classic bilinear transformation between the two:

$$Z(t) = \frac{1 + S_{11}(t)}{1 - S_{11}(t)} \qquad (20\text{-}17)$$

$Z(t)$ is now synthesized using the development in Chapter 18 by the removal of unit elements of value $Z(1)$ until the order of the function is reduced to zero. The remainder impedance after each extraction is given by the Richards theorem by

$$Z_1(t) = Z(1) \frac{Z(t) - tZ(1)}{Z(1) - tZ(t)} \qquad (20\text{-}18)$$

The order of the network is defined, as in the approximate problem, by Z_{n+1}/Z_0, ε and θ_c, by the fact that at zero frequency S_{11} is set by the mismatch between the generator and load impedances. Combining Eqs (20-14) and (20-16) at $t = 0$ gives

$$\varepsilon T_n\left(\frac{1}{\cos \theta_c}\right) = \frac{Z_{n+1} - Z_0}{2\sqrt{Z_{n+1}Z_0}} \qquad (20\text{-}19)$$

The order of the network is then obtained by writing $T_n(x)$ in the preceding equation in terms of its hyperbolic cosine definition:

$$n = \frac{\cosh^{-1}[(Z_{n+1} - Z_0)/2\varepsilon\sqrt{Z_{n+1}Z_0}]}{\cosh^{-1}(1/\cos \theta_c)} \qquad (20\text{-}20)$$

The frequency response of this type of network is depicted in Fig. 20-4.

As an example consider the design of a transformer consisting of n UEs between normalized generator (Z_0) and load (Z_{n+1}) resistances of 1 and 0.440. The circuit is to have a frequency (f_0) of 3100 MHz and a bandwidth $(f_2 - f_1)$ of 1000 MHz. The VSWR is to be less than 1.05.

Employing (20-3) with $K = 0$ gives $\varepsilon = 0.0244$ and using Eq. (20-8) yields $\cos \theta_c = 0.464$. The order of the filter is then determined by Eq. (20-20) as $n = 2.53$. A suitable integer value for n is therefore 3. Adopting this value of n in Eq. (20-20) yields a value of $\varepsilon = 0.01256$. Evaluating Eq. (20-14) with $n = 3$ and $\varepsilon = 0.01256$ gives, after some algebra,

$S_{11}(t)S_{11}(-t)$

$$= \frac{(0.4222 + 0.081t^2)^2}{(t^3 + 3.0293t^2 + 3.0847t + 1.0858)(-t^3 + 3.0293t^2 - 3.0847t + 1.0858)}$$

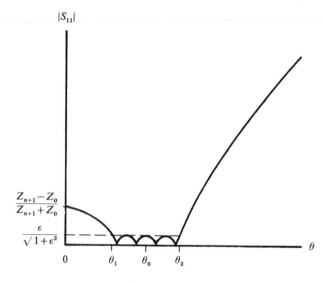

FIGURE 20-4
Frequency response of multistep
impedance transformer
indicating definition of reflection
coefficient for exact synthesis.

One bounded real solution for $S_{11}(t)$ is

$$S_{11}(t) = \frac{-(0.4222 + 0.0811t^2)\alpha}{t^3 + 3.0293t^2 + 3.0847t + 1.0858}$$

The input impedance of the network is then obtained by making use of the bilinear transformation between $Z(t)$ and $S_{11}(t)$. The result is

$$Z(t) = \frac{t^3 + 2.9483t^2 + 3.0847t + 0.6636}{t^3 + 3.1103t^2 + 3.0847t + 1.5080}$$

The two preceding equations satisfy the specification at $t = 0$ with $\alpha = 0.921$, as is observed by constructing Eqs (20-16) and (20-17).

The synthesis of $Z(t)$ begins by extracting a UE using the Richards theorem by replacing t by 1 in $Z(t)$. The result is

$$Z_1(1) = 0.8844$$

The remainder impedance $Z_2(t)$ is then given by resorting to Eq. (20-18) as

$$Z_2(t) = \frac{0.8844t^4 + 1.7522t^3 - 0.2207t^2 - 1.7522t - 0.6636}{1.1303t^4 + 2.3322t^3 + 0.3777t^2 - 2.3322t - 1.5080}$$

In keeping with the Richards theorem (discussed in Chapter 18) the numerator and denominator polynomials of $Z_2(t)$ have a common factor $t^2 - 1$ that may be cancelled. Factoring the numerator and denominator polynomials to identify this factor gives

$$Z_2(t) = \frac{(t^2 - 1)(0.8844t^2 + 1.7522t + 0.6636)}{(t^2 - 1)(1.1303t^2 + 2.3322t + 1.5080)}$$

The second UE is now specified by cancelling the common factor $(t^2 - 1)$ and replacing t by 1 in $Z_2(t)$. This gives

$$Z_2(1) = 0.6638$$

and a remainder impedance $Z_3(t)$:

$$Z_3(t) = \frac{0.7503t^3 + 0.6632t^2 + 0.7512t - 0.6636}{1.3323t^3 + 1.5093t^2 - 1.3322t - 1.5080}$$

Identifying a common term $t^2 - 1$ in both the numerator and denominator polynomials of $Z_3(t)$ yields

$$Z_3(t) = \frac{(t^2 - 1)(0.7503t + 0.6636)}{(t^2 - 1)(1.3323t + 1.5093)}$$

The third UE is then obtained from a knowledge of $Z_3(t)$ as

$$Z_3(1) = 0.497\,65$$

The remainder impedance $Z_4(t)$ is again evaluated from a knowledge of $Z_3(t)$ and $Z_3(1)$ by once more making use of the Richards theorem. The result is

$$Z_4(t) = \frac{0.6629t^2 - 0.6636}{1.5078t^2 - 1.5093}$$

Identifying the common factor $t^2 - 1$ in the preceding equation indicates that

$$Z_4(t) = \frac{0.6636(t^2 - 1)}{1.5093(t^2 - 1)}$$

This last impedance is, subject to rounding-off errors in the calculation, the resistance of the load in the original specification

$$Z_4(t) = R_L = 0.4397$$

Figure 20-5 depicts a schematic diagram showing the step-by-step synthesis of the required transformer configuration.

SYNTHESIS OF QUARTER-WAVE COUPLED STUB-R LOADS

Another important classic problem consists of a cascade of UEs between a real generator impedance and a complex STUB-R load. It is quite common for this class of network to be synthesized from the load terminals and this is the approach used by way of an example in this situation. A simple example of this class of network is one with a single UE and a single stub for which the frequency response is indicated in Fig. 20-6. Since the solution of this problem is of some interest it will be derived in closed form. The characteristic function for this class

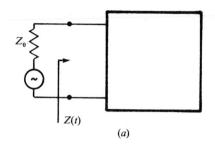

(a)

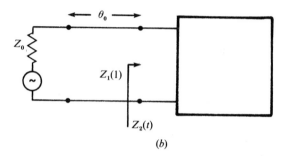

(b)

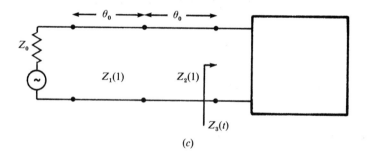

(c)

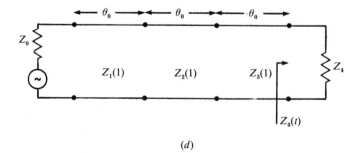

(d)

FIGURE 20-5
Step-by-step synthesis of $n = 3$ quarter-wave impedance transformer.

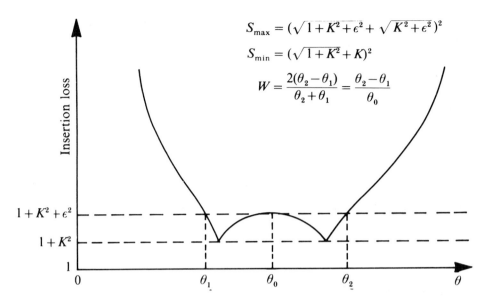

$$S_{max} = (\sqrt{1+K^2+\epsilon^2} + \sqrt{K^2+\epsilon^2})^2$$

$$S_{min} = (\sqrt{1+K^2}+K)^2$$

$$W = \frac{2(\theta_2-\theta_1)}{\theta_2+\theta_1} = \frac{\theta_2-\theta_1}{\theta_0}$$

FIGURE 20-6
Frequency response of degree 2 insertion loss function.

of network is given in the θ plane with $n = 1$ by

$$K(\theta^2) = \left\{ \frac{(1 + \sin \theta_2)[T_2(\cos \theta/\cos \theta_c) - (1 - \sin \theta_c)]}{2 \sin \theta} \right\}^2 \qquad (20\text{-}21)$$

Writing $T_2(\cos \theta/\cos \theta_c)$ in polynomial form:

$$T_2\left(\frac{\cos \theta}{\cos \theta_c}\right) = 2\left(\frac{\cos \theta}{\cos \theta_c}\right)^2 - 1$$

and introducing the Richards transformation

$$t \rightarrow j \tan \theta$$

in the preceding relationship gives

$$K(t^2) = \left[\frac{\beta + t^2}{jt(1 - t^2)^{1/2}} \right]^2 \qquad (20\text{-}22)$$

where

$$\beta = \tan^2 \theta_c + \frac{\tan \theta_c}{\cos \theta_c} \qquad (20\text{-}23)$$

The derivation of a realizable immittance function proceeds by forming the amplitude square of the reflection coefficient. If the network is synthesized from the load terminals then it is necessary to construct $S_{22}(t)S_{22}(-t)$ instead of

$S_{11}(t)S_{11}(-t)$ as has been done up to now. If, furthermore, the parameter K is not neglected, then the required quantity is given by

$$S_{22}(t)S_{22}(-t) = \frac{at^4 + bt^2 + c}{(1 + a)t^4 + (-1 + b)t^2 + c} \tag{20-24}$$

where

$$a = K^2 + \varepsilon^2 \tag{20-25a}$$

$$b = 2\beta\varepsilon^2 - K^2 \tag{20-25b}$$

$$c = \beta^2\varepsilon^2 \tag{20-25c}$$

The derivation now continues by forming $S_{22}(t)$ by factoring $S_{22}(t)S_{22}(-t)$ as

$$S_{22}(t)S_{22}(-t) = \frac{N(t)N(-t)}{D(t)D(-t)} \tag{20-26}$$

One possible solution for $S_{22}(t)$ is then given by

$$S_{22}(t) = \frac{-N(t)}{D(t)} \tag{20-27}$$

where

$$N(t) = (a)t^2 + (2\sqrt{ac} - b)^{1/2}t + \sqrt{c} \tag{20-28}$$

$$D(t) = (a + 1)t^2 + [2\sqrt{(a + 1)c} - b + 1]^{1/2}t + \sqrt{c} \tag{20-29}$$

The negative sign is assigned to $S_{22}(t)$ because at dc $(t = 0)$ the reflection coefficient must be zero to cater for the short-circuited stub.

The normalized admittance at the output terminals is then given from a knowledge of $S_{22}(t)$ by

$$\frac{Y(t)}{G} = \frac{1 - S_{22}(t)}{1 + S_{22}(t)} \tag{20-30}$$

The result in terms of the original variables is

$$\frac{Y(t)}{G} = \frac{d_2t^2 + d_1t + d_0}{n_2t^2 + n_1t} \tag{20-31}$$

where

$$n_2 = \sqrt{a + 1} - \sqrt{a} \tag{20-32a}$$

$$d_2 = \sqrt{a + 1} + \sqrt{a} \tag{20-32b}$$

$$n_1 = [2\sqrt{(a + 1)c} - b + 1]^{1/2} - (2\sqrt{ac} - b)^{1/2} \tag{20-32c}$$

$$d_1 = [2\sqrt{(a + 1)c} - b + 1]^{1/2} + (2\sqrt{ac} - b)^{1/2} \tag{20-32d}$$

$$n_0 = 0 \tag{20-32e}$$

$$d_0 = 2\sqrt{c} \tag{20-32f}$$

Making use of the relationships between the numerator and denominator coefficients $n_{1,2}$ and $d_{0,1,2}$ also indicates that

$$d_1 n_1 - d_0 n_2 = 1 \tag{20-33a}$$

$$n_2 d_2 = 1 \tag{20-33b}$$

The synthesis of this immittance commences by extracting a shunt t-plane inductor in a second Cauer form:

$$L = \frac{G d_0}{n_1}$$

and a remainder admittance

$$Y'(t) = G \frac{n_1 d_2 t + n_1 d_1 - n_2 d_0}{n_1 (n_2 t + n_1)}$$

A UE is now extracted by replacing t by unity in $Y'(t)$. The result is

$$Y'(1) = G \frac{n_1 d_2 + n_1 d_1 - n_2 d_0}{n_1 (n_2 + n_1)}$$

Making use of the relationships in Eqs (20-33a,b) indicates that $Y'(1)$ may also be written as

$$Y'(1) = \frac{G}{n_1 n_2}$$

The remainder admittance $Y''(t)$ is then given by

$$Y''(t) = Y'(1) \frac{Y'(t) - t Y'(1)}{Y'(1) - t Y'(t)}$$

The result is

$$Y''(t) = \frac{G}{n_1^2}$$

Since this admittance coincides with that of the load at the generator terminals

$$\frac{G}{n_1^2} = 1$$

and

$$G = n_1^2$$

The complete result for the circuit in Fig. 20-7 is therefore described by

$$G = n_1^2 \tag{29-34}$$

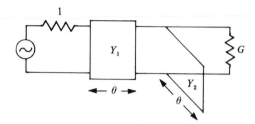

FIGURE 20-7
Topology of degree 2 STUB-R network.

$$Y_1 = \frac{n_1}{n_2} \tag{20-35}$$

$$Y_2 = n_1 d_0 \tag{20-36}$$

It is usual, in tabulating solutions to this type of problem, to replace the stub admittance Y_2 by its equivalent susceptance slope parameter B':

$$B' = \frac{\pi Y_2}{4}$$

This notation has the merit that it permits the Q factor of the load to be also defined without difficulty:

$$Q_L = \frac{B'}{G}$$

A typical family of solutions is tabulated in Table 20-1.

Although this circuit has been synthesized by forming the immittance at the generator terminals it may also be realized from that at the load. This may be done by recognizing that if

$$S_{22}(t) = -\frac{N(t)}{D(t)} \tag{20-37}$$

then

$$S_{11}(t) = \pm \frac{N^*(t)}{D(t)} \tag{20-38}$$

Adopting, once more, the negative solution gives

$$S_{11}(t) = \frac{-(d_2 - n_2)t^2 + (d_1 - n_1)t - d_0}{(d_2 + n_2)t^2 + (d_1 + n_1)t + d_0}$$

The normalized admittance at the input terminals is then given by

$$\frac{Y(t)}{1} = \frac{d_2 t^2 + n_1 t + d_0}{n_2 t^2 + d_1 t}$$

TABLE 20-1
Table of solutions for degree 2 STUB-R network

$R = 1.05$					$R = 1.15$				
$2\delta_0$	b'	g	Q	y_{01}	$2\delta_0$	b'	g	Q	y_{01}
0.100	50.304	16.424	3.063	4.153	0.100	234.000	43.247	5.411	7.052
0.150	15.433	7.846	1.967	2.870					
0.200	6.809	4.844	1.406	2.255	0.150	70.190	19.750	3.554	4.766
0.250	3.672	3.455	1.063	1.905	0.200	30.113	11.530	2.612	3.641
0.300	2.248	2.700	0.833	1.684	0.250	15.738	7.724	2.038	2.980
					0.300	9.328	5.656	1.649	2.550
0.350	1.501	2.245	0.669	1.535	0.350	6.032	4.410	1.368	2.252
0.400	1.067	1.950	0.547	1.431	0.400	4.158	3.601	1.155	2.035
0.450	0.795	1.747	0.455	1.355	0.450	3.010	3.047	0.988	1.872
0.500	0.614	1.603	0.383	1.297	0.500	2.264	2.651	0.854	1.746
0.550	0.488	1.496	0.326	1.253					
0.600	0.397	1.414	0.280	1.219	0.550	1.756	2.358	0.745	1.647
0.660	0.317	1.340	0.237	1.186	0.600	1.397	2.135	0.654	1.567
					0.660	1.091	1.932	0.565	1.491

$R = 1.10$					$R = 1.20$				
$2\delta_0$	b'	g	Q	y_{01}	$2\delta_0$	b'	g	Q	y_{01}
0.100	133.831	30.455	4.396	5.787	0.100	344.387	54.983	6.264	8.123
					0.150	102.999	24.962	4.126	5.473
0.150	40.384	14.070	2.870	3.934	0.200	44.010	14.454	3.045	4.165
0.200	17.457	8.339	2.093	3.029	0.250	22.892	9.591	2.387	3.393
0.250	9.205	5.686	1.619	2.501					
0.300	5.508	4.245	1.297	2.161	0.300	13.498	6.950	1.942	2.888
0.350	3.598	3.377	1.065	1.927	0.350	8.677	5.357	1.620	2.536
0.400	2.505	2.813	0.890	1.759	0.400	5.947	4.324	1.375	2.278
					0.450	4.279	3.616	1.184	2.083
0.450	1.831	2.427	0.754	1.634	0.500	3.200	3.109	1.029	1.932
0.500	1.390	2.151	0.645	1.538	0.550	2.469	2.736	0.903	1.812
0.550	1.088	1.946	0.559	1.463	0.600	1.954	2.451	0.797	1.715
0.600	0.872	1.791	0.487	1.404					
0.660	0.687	1.650	0.417	1.347	0.660	1.517	2.191	0.692	1.622

Source: Levy, R. and Helszajn, J., 'Specific equations for one and two section quarter wave matching networks for stub-resistor loads', *IEEE Trans. on MTT*, Vol. MTT-30, pp. 55–62, January 1982, (© 1984 *IEEE*).

If the first extraction is to be a UE then $Y(1)$ is given by

$$Y(1) = \frac{d_2 + n_1 + d_0}{n_2 + d_1}$$

The required form is now obtained by multiplying this equation top and bottom by n_2 and by making use of the identities in Eqs (20-33a,b).

The required result is

$$Y(1) = \frac{n_1}{n_2}$$

in keeping with the previous derivation. The derivation of the other two elements is left to the reader.

PROBLEMS

20-1 Obtain the order n to match an impedance $Z_0 = 50\,\Omega$ to one of $Z_{n+1} = 25\,\Omega$ with $(f_2 - f_1)/f_0 = 1.03$ and an equiripple VSWR of 1.10.

20-2 Repeat Prob. 20-1 for a VSWR of 1.03.

20-3 Verify the entries $(\text{VSWR})_{max} = 1.20$, $(\text{VSWR})_{min} = 1.07$, $W = 0.20$ in Table 20-1.

20-4 Verify Eqs (20-33a,b).

CHAPTER
21

SCATTERING AND
IMMITTANCE
MATRICES OF
PARALLEL LINE
CIRCUITS

INTRODUCTION

An important planar network is an n-port one consisting of two or more coupled parallel transmission lines. If $n - 2$ of its ports are terminated by electric and magnetic walls in all combinations, then the resultant two-port networks display a number of band-pass, band-stop, all-stop and all-pass circuits that are of some significance in the practice of microwave engineering. In order to proceed with the derivation of these different circuits it is necessary to have the immittance descriptions of such a four-port network at hand. A knowledge of its scattering parameters is of separate interest in the description of the transmission properties of this class of circuit. The scattering and immittance matrices of this type of network with $n = 4$ are therefore formed in this chapter as a preamble to the development of the related two-port filter networks in Chapter 22. The schematic diagram of a stripline version is indicated in Fig. 21-1. If the network has fourfold symmetry the eigenvalue method is appropriate for the description of this class of device. The results are expressed in terms of the parameters of the so-called odd and even modes of the four-port circuit.

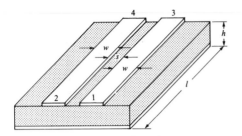

FIGURE 21-1
Schematic diagram of directional coupler.

THE SCATTERING MATRIX OF COUPLED PARALLEL LINES

In order to develop the equivalent circuit of two similar coupled parallel lines it is necessary to have its immittance description. The behaviour of the network is, however, best understood by having recourse to its scattering matrix. One way in which either description may be deduced is by using the even and odd mode approach. Another way, which also reduces to the even and odd mode notations, is to employ the eigenvalue approach already introduced in connection with the description of the symmetrical two-port network. In what follows, the different matrix descriptions of the network will be constructed by having recourse to the eigenvalue method. The derivation of the scattering matrix starts by noting the similarity transformation

$$\bar{A} = \bar{U}\lambda(\bar{U}^*)^{\mathrm{T}} \qquad (21\text{-}1)$$

The matrix $\bar{U}$ is a square matrix with the eigenvectors $\bar{U}_j$ of the network as columns. The eigenvectors may be constructed by applying electric and magnetic walls in all combinations at the symmetry plane of the network. The four possibilities in the case of a symmetrical directional coupler are illustrated in Fig. 21-2. These are fixed from the symmetry of the problem as

$$\bar{U}_1 = \frac{1}{\sqrt{4}}\begin{bmatrix} 1 \\ 1 \\ 1 \\ 1 \end{bmatrix} \qquad (21\text{-}2a)$$

$$\bar{U}_2 = \frac{1}{\sqrt{4}}\begin{bmatrix} 1 \\ 1 \\ -1 \\ -1 \end{bmatrix} \qquad (21\text{-}2b)$$

$$\bar{U}_1 = \frac{1}{\sqrt{4}} \begin{bmatrix} 1 \\ 1 \\ 1 \\ 1 \end{bmatrix}$$

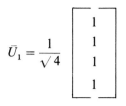

$$\bar{U}_2 = \frac{1}{\sqrt{4}} \begin{bmatrix} 1 \\ 1 \\ -1 \\ -1 \end{bmatrix}$$

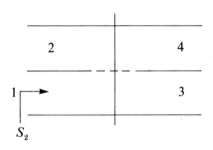

$$\bar{U}_3 = \frac{1}{\sqrt{4}} \begin{bmatrix} 1 \\ -1 \\ 1 \\ -1 \end{bmatrix}$$

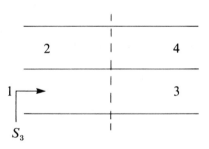

$$\bar{U}_4 = \frac{1}{\sqrt{4}} \begin{bmatrix} 1 \\ -1 \\ -1 \\ 1 \end{bmatrix}$$

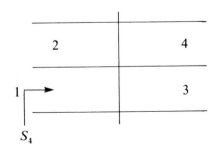

FIGURE 21-2
Eigennetworks of parallel coupled lines.

$$\bar{U}_3 = \frac{1}{\sqrt{4}} \begin{bmatrix} 1 \\ -1 \\ 1 \\ -1 \end{bmatrix} \qquad (21\text{-}2c)$$

$$\bar{U}_4 = \frac{1}{\sqrt{4}} \begin{bmatrix} 1 \\ -1 \\ -1 \\ 1 \end{bmatrix} \qquad (21\text{-}2d)$$

The one-port circuits defined by these eigenvectors are known as the eigennetworks of the device. The one-port reflection coefficients, impedances and admittance variables s_j, z_j and y_j of these circuits are the eigenvalues of the problem. The matrix $\bar{A}$ in the similarity transformation is a real symmetric or Hermitian matrix. If it is taken as the scattering matrix then the diagonal matrix $\bar{\lambda}$ is constructed with the reflection eigenvalues along its main diagonal. If it is one of the immittance matrices of the device it is constructed in terms of the corresponding one-port immittance parameters.

The scattering description of the network will now be formed by specializing the similarity transformation for the situation at hand by replacing $\bar{A}$ by $\bar{S}$ and $\bar{\lambda}$ by $\bar{s}$

$$\bar{S} = \bar{U}\bar{s}(\bar{U}^*)^{\mathsf{T}} \qquad (21\text{-}3a)$$

The matrix $\bar{U}$ is arranged in the form of a square symmetric matrix with the eigenvectors $\bar{U}_j$ of the circuit as columns

$$\bar{U} = \frac{1}{\sqrt{4}} \begin{bmatrix} 1 & 1 & 1 & 1 \\ 1 & 1 & -1 & -1 \\ 1 & -1 & 1 & -1 \\ 1 & -1 & -1 & 1 \end{bmatrix} \qquad (21\text{-}3b)$$

and the conjugate transpose of a real symmetric matrix is obtained by noting that it is the matrix itself

$$(U^*)^{\mathsf{T}} = \frac{1}{\sqrt{4}} \begin{bmatrix} 1 & 1 & 1 & 1 \\ 1 & 1 & -1 & -1 \\ 1 & -1 & 1 & -1 \\ 1 & -1 & -1 & 1 \end{bmatrix} \qquad (21\text{-}3c)$$

The matrix $\bar{s}$ is a diagonal one with the eigenvalues of $\bar{S}$ or the reflection coefficients of the eigennetworks along its main diagonal.

$$\bar{s} = \begin{bmatrix} s_1 & 0 & 0 & 0 \\ 0 & s_2 & 0 & 0 \\ 0 & 0 & s_3 & 0 \\ 0 & 0 & 0 & s_4 \end{bmatrix} \qquad (21\text{-}3d)$$

Assembling these matrices indicates that

$$\bar{S} = \frac{1}{4} \begin{bmatrix} 1 & 1 & 1 & 1 \\ 1 & 1 & -1 & -1 \\ 1 & -1 & 1 & -1 \\ 1 & -1 & -1 & 1 \end{bmatrix} \begin{bmatrix} s_1 & 0 & 0 & 0 \\ 0 & s_2 & 0 & 0 \\ 0 & 0 & s_3 & 0 \\ 0 & 0 & 0 & s_4 \end{bmatrix} \begin{bmatrix} 1 & 1 & 1 & 1 \\ 1 & 1 & -1 & -1 \\ 1 & -1 & 1 & -1 \\ 1 & -1 & -1 & 1 \end{bmatrix} \tag{21-3e}$$

The required relationships are now deduced by expanding the preceding equation. The result is

$$S_{11} = \frac{s_2 + s_2 + s_3 + s_4}{4} \tag{21-4a}$$

$$S_{21} = \frac{s_1 + s_2 - s_3 - s_4}{4} \tag{21-4b}$$

$$S_{31} = \frac{s_1 - s_2 + s_3 - s_4}{4} \tag{21-4c}$$

$$S_{41} = \frac{s_1 - s_2 - s_3 + s_4}{4} \tag{21-4d}$$

Scrutiny of the eigennetworks of the problem suggests that the reflection eigenvalues appearing in the above equations are related by

$$s_2 = -s_1 \tag{21-5}$$

$$s_4 = -s_3 \tag{21-6}$$

Evaluating the scattering parameters using these identities indicates that

$$S_{11} = 0 \tag{21-7a}$$

$$S_{21} = 0 \tag{21-7b}$$

$$S_{31} = \frac{s_1 + s_3}{2} \tag{-7c}$$

$$S_{41} = \frac{s_1 - s_3}{2} \tag{21-7d}$$

in keeping with the classic definition of a symmetrical directional coupler.

The derivation is completed by recognizing that s_1 is associated with the even mode parameters of two coupled waveguides or transmission lines

$$s_1 = s_e \tag{21-8}$$

and that s_4 is associated with the odd mode parameters of the same two lines

$$s_3 = s_o \tag{21-9}$$

The electric fields of these two possibilities are illustrated in Fig. 21-3 in the case of two coupled microstrip lines.

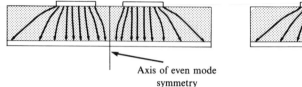 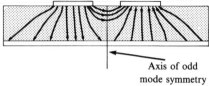

Axis of even mode
symmetry

Axis of odd
mode symmetry

Even mode electric
field distribution

Odd mode electric
field distribution

FIGURE 21-3
Odd and even mode field patterns in coupled microstrip lines.

One possible solution for s_e and s_o in terms of the so-called even and odd phase angles θ_e and θ_o is

$$s_e = \exp - j\theta_e \qquad (21\text{-}10)$$

$$s_o = \exp - j\theta_o \qquad (21\text{-}11)$$

Evaluating the scattering parameters using these two relations indicates

$$S_{11} = 0 \qquad (21\text{-}12a)$$

$$S_{21} = 0 \qquad (21\text{-}12b)$$

$$S_{31} = \frac{\exp - j\theta_e + \exp - j\theta_o}{2} \qquad (21\text{-}12c)$$

$$S_{41} = \frac{\exp - j\theta_e - \exp - j\theta_o}{2} \qquad (21\text{-}12d)$$

and taking out a common term

$$\exp - j\left(\frac{\theta_e + \theta_o}{2}\right)$$

gives the desired result

$$S_{11} = 0 \qquad (21\text{-}13a)$$

$$S_{21} = 0 \qquad (21\text{-}13b)$$

$$S_{31} = \cos\left(\frac{\theta_e - \theta_o}{2}\right) \exp - j\left(\frac{\theta_e + \theta_o}{2}\right) \qquad (21\text{-}13c)$$

$$S_{41} = j \sin\left(\frac{\theta_e - \theta_o}{2}\right) \exp - j\left(\frac{\theta_e + \theta_o}{2}\right) \qquad (21\text{-}13d)$$

An ideal directional coupler is therefore a four-port device having an input port, two mutually isolated output ports and one port isolated from the input one. Also, all its ports are matched. It is furthermore readily shown that the scattering parameters in Eq. (21-13) satisfy the unitary condition.

IMMITTANCE MATRICES OF COUPLED PARALLEL LINES

The detailed derivation of the admittance matrix of the device is identical to that of its scattering description except that the scattering variables are replaced by the short circuit parameters of the network and that the reflection eigenvalues are replaced by the admittance ones. This may be understood by recognizing that the eigennetworks are common to both problems. The open circuit parameters are therefore described by

$$Z_{11} = \frac{z_1 + z_2 + z_3 + z_4}{4} \qquad (21\text{-}14a)$$

$$Z_{12} = \frac{z_1 + z_2 - z_3 - z_4}{4} \qquad (21\text{-}14b)$$

$$Z_{13} = \frac{z_1 - z_2 + z_3 - z_4}{4} \qquad (21\text{-}14c)$$

$$Z_{14} = \frac{z_1 - z_2 - z_3 + z_4}{4} \qquad (21\text{-}14d)$$

The impedance eigenvalues are given in terms of the odd and even mode variables by inspection of the eigennetworks in Fig. 21-2 by

$$z_1 = -jZ_e \cot \frac{\theta}{2} \qquad (21\text{-}15a)$$

$$z_2 = \ \ jZ_e \tan \frac{\theta}{2} \qquad (21\text{-}15b)$$

$$z_3 = -jZ_o \cot \frac{\theta}{2} \qquad (21\text{-}15c)$$

$$z_4 = \ \ jZ_o \tan \frac{\theta}{2} \qquad (21\text{-}15d)$$

The required open-circuit parameters are now constructed by combining the preceding equations. The result is

$$Z_{11} = -j\left(\frac{Z_e + Z_o}{2}\right) \cot \theta \qquad (21\text{-}16a)$$

$$Z_{12} = -j\left(\frac{Z_e - Z_o}{2}\right) \cot \theta \qquad (21\text{-}16b)$$

$$Z_{13} = -j\left(\frac{Z_e + Z_o}{2}\right) \csc \theta \qquad (21\text{-}16c)$$

$$Z_{14} = -j\left(\frac{Z_e - Z_o}{2}\right) \csc \theta \qquad (21\text{-}16d)$$

The derivation of the admittance description of the device proceeds in a similar fashion.

$$Y_{11} = -j\left(\frac{Y_e + Y_o}{2}\right) \cot \theta \qquad (21\text{-}17a)$$

$$Y_{12} = -j\left(\frac{Y_e - Y_o}{2}\right) \cot \theta \qquad (21\text{-}17b)$$

$$Y_{13} = \ j\left(\frac{Y_e + Y_o}{2}\right) \csc \theta \qquad (21\text{-}17c)$$

$$Y_{14} = \ j\left(\frac{Y_e - Y_o}{2}\right) \csc \theta \qquad (21\text{-}17d)$$

IMMITTANCE MATRICES OF COUPLED PARALLEL LINES IN t PLANE

In the synthesis it is necessary to express the open- and short-circuit parameters of the parallel line structure in terms of the t variable

$$t = j \tan \theta \qquad (21\text{-}18)$$

Introducing this substitution into the definition of the open circuit parameters of the network gives

$$Z_{11} = \frac{L_{11}}{t} \qquad (21\text{-}19a)$$

$$Z_{12} = \frac{L_{12}}{t} \qquad (21\text{-}19b)$$

$$Z_{13} = \frac{L_{11}}{t}\sqrt{1 - t^2} \qquad (21\text{-}19c)$$

$$Z_{14} = \frac{L_{12}}{t}\sqrt{1 - t^2} \qquad (21\text{-}19d)$$

where

$$L_{11} = \frac{Z_e + Z_o}{2} \qquad (21\text{-}20)$$

$$L_{12} = \frac{Z_e - Z_o}{2} \tag{21-21}$$

Here L_{11} and L_{12} are linear combinations of the odd and even mode impedances of the two lines.

If Z_{13} and Z_{14} are written in terms of Z_{11} and Z_{12} then

$$Z_{13} = \sqrt{1 - t^2}\, Z_{11} \tag{21-22a}$$

$$Z_{14} = \sqrt{1 - t^2}\, Z_{12} \tag{21-22b}$$

and the impedance matrix of the network takes on a simple symmetric form given by

$$\bar{Z} = \frac{1}{t}\left[\begin{array}{c|c} \bar{L} & \sqrt{1 - t^2}\,\bar{L} \\ \hline \sqrt{1 - t^2}\,\bar{L} & \bar{L} \end{array}\right] \tag{21-23}$$

where

$$\bar{L} = \begin{bmatrix} L_{11} & L_{12} \\ L_{12} & L_{11} \end{bmatrix} \tag{21-24}$$

The derivation of the admittance matrix in the t plane proceeds in a similar fashion. The result is

$$\bar{Y} = \frac{1}{t}\left[\begin{array}{c|c} \bar{C} & -\sqrt{1 - t^2}\,\bar{C} \\ \hline -\sqrt{1 - t^2}\,\bar{C} & \bar{C} \end{array}\right] \tag{21-25}$$

where

$$\bar{C} = \begin{bmatrix} C_{11} & C_{12} \\ C_{12} & C_{11} \end{bmatrix} \tag{21-26}$$

and

$$C_{11} = \frac{Y_e + Y_o}{2} \tag{21-27}$$

$$C_{12} = \frac{Y_e - Y_o}{2} \tag{21-28}$$

C_{11} and C_{12} are in this instance linear combinations of the odd and even mode admittances of the two lines.

It is separately recognized that

$$\bar{C}\bar{L} = \bar{L}\bar{C} = \bar{I} \tag{21-29a}$$

and

$$\bar{Y}\bar{Z} = \bar{Z}\bar{Y} = \bar{I} \tag{21-29b}$$

In the theory of n-coupled lines it is therefore usual to label the $2n$-ports so that the input ports are consecutively numbered from 1 to n and the output ones from $n + 1$ to $2n$.

The $\bar{L}$ and $\bar{C}$ matrices may be realized as two-port all-pass and all-stop networks respectively. The significance of these networks will be discussed separately.

ODD AND EVEN CAPACITANCES OF PARALLEL LINE CIRCUIT

The odd and even admittances in the definition of the $\bar{C}$ matrix may be related to the odd and even mode capacitances on the self and mutual capacitances C'_{11} and C'_{12} of the two coupled lines. The derivation of this property starts by expressing Y_e and Y_o in terms of the even and odd capacitances C_e and C_o and the phase velocity v of the two lines.

$$Y_e = vC_e \tag{21-30a}$$

$$Y_o = vC_o \tag{21-30b}$$

Constructing the even and odd mode capacitances of the two lines in terms of the self and mutual capacitances in Fig. 21-4 by introducing magnetic and electric walls at the plane of symmetry of the circuit indicates that

$$C_e = C'_{11} \tag{21-31a}$$

$$C_o = C'_{11} + 2C'_{12} \tag{21-31b}$$

Another matrix of some interest is the capacitance one defined by

$$\bar{C}'' = \left(\frac{1}{v}\right)\bar{C} \tag{21-32}$$

This matrix may be visualized as relating charge (q) and voltage (v) on the circuit.

The entries of the capacitance matrix are usually given in terms of the even and odd mode capacitances by

$$C''_{11} = \frac{C_e + C_o}{2} \tag{21-33a}$$

$$C''_{12} = \frac{C_e - C_o}{2} \tag{21-33b}$$

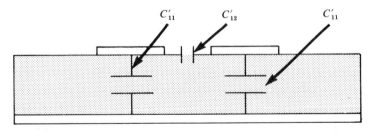

FIGURE 21-4
Capacitance matrix of coupled parallel lines.

or more often in terms of the self and mutual capacitances of the coupled lines by

$$C''_{11} = C'_{11} + C'_{12} \tag{21-34a}$$

$$C''_{12} = -C'_{12} \tag{21-34b}$$

The impedance and admittance descriptions of this type of network are sometimes directly stated in terms of the inductance and capacitance matrices. If this is the case then

$$\bar{Z} = \frac{v}{t} \left[\begin{array}{c|c} \bar{L}'' & \sqrt{1-t^2}\,\bar{L}'' \\ \hline \sqrt{1-t^2}\,\bar{L}'' & \bar{L}'' \end{array} \right] \tag{21-35}$$

and

$$\bar{Y} = \frac{v}{t} \left[\begin{array}{c|c} \bar{C}'' & -\sqrt{1-t^2}\,\bar{C}'' \\ \hline -\sqrt{1-t^2}\,\bar{C}'' & \bar{C}'' \end{array} \right] \tag{21-36}$$

respectively.

The entries of the inductance matrix may be expressed in terms of these variables also by recognizing that

$$\bar{C}''\bar{L}'' = \bar{L}''\bar{C}'' = \left(\frac{1}{v^2}\right)\bar{I} \tag{21-37}$$

Three elementary transformations of the matrix $[C'']$, which may be employed to alter the topology of the network, are the interchange of a pair of rows or columns, the addition of the elements of one row (or column) to the respective ones of another row (or column), and the multiplication of a row or column by a constant factor. The first two operations are not of interest in network theory but the last one is of significant value. This topic is however, outside the remit of this text.

PROBLEMS

21-1 Deduce the impedance matrix of a symmetrical 4-port directional coupler.

21-2 Verify that the scattering parameters of an ideal directional coupler satisfy the unitary condition.

21-3 Determine the scattering parameters of two parallel lines in the t plane.

BIBLIOGRAPHY

Guillemin, E. A., *Synthesis of Passive Networks*, John Wiley, New York, 1957.

Sato, R. and Crystal, E. G., 'Simplified analysis of coupled transmission-line networks', *IEEE Trans. on MTT*, Vol. MTT-18, pp. 122–131, March 1970.

Wenzel, R. J., 'Theoretical and practical applications of capacitance matrix transformation to TEM network design', *IEEE Trans. on MTT*, Vol. MTT-14, pp. 635–647, December 1966.

CHAPTER
22

COUPLED
PARALLEL
LINE
CIRCUITS

INTRODUCTION

A classic planar two-port network is a four-port one consisting of two coupled parallel transmission lines with two of its ports terminated by electric and magnetic walls in all combinations. There are altogether ten possibilities, two of which have a band-pass transmission characteristic and one of which has a band-stop one. The ten possibilities are illustrated in Fig. 22-1. The purpose of this chapter is to summarize the equivalent circuits of some of the different arrangements. The exact derivation using synthesis of one band-stop, two band-pass and one all-stop configuration from a knowledge of each one-port immittance in the t plane is included for completeness. The description of the four-port network has been derived in the previous chapter in terms of its immittance and scattering parameters. A knowledge of the scattering parameters is of separate interest in the description of the transmission properties of these types of circuits.

PARALLEL LINE BAND-STOP
FILTER CIRCUIT

The derivation of equivalent circuits for parallel line filter networks may be readily done in the t plane introduced in Chapter 18. This will now be demonstrated for the band-stop prototype illustrated in Fig. 22-1. The derivation starts by introducing the boundary conditions

$$V_2 = I_4 = 0 \qquad (22\text{-}1)$$

at ports 2 and 3 into the current–voltage relationship of the circuit. This gives

$$
\begin{bmatrix} V_1 \\ 0 \\ V_3 \\ V_4 \end{bmatrix}
=
\begin{bmatrix}
Z_{11} & Z_{21} & Z_{31} & Z_{41} \\
Z_{21} & Z_{11} & Z_{41} & Z_{31} \\
Z_{31} & Z_{41} & Z_{11} & Z_{21} \\
Z_{41} & Z_{31} & Z_{21} & Z_{11}
\end{bmatrix}
\begin{bmatrix} I_1 \\ I_2 \\ I_3 \\ 0 \end{bmatrix}
\qquad (22\text{-}2)
$$

The required relationships between ports 1 and 4 are then described by

$$V_1 = \left(Z_{11} - \frac{Z_{21}^2}{Z_{11}} \right) I_1 + \left(Z_{31} - \frac{Z_{21}Z_{41}}{Z_{11}} \right) I_3 \qquad (22\text{-}3a)$$

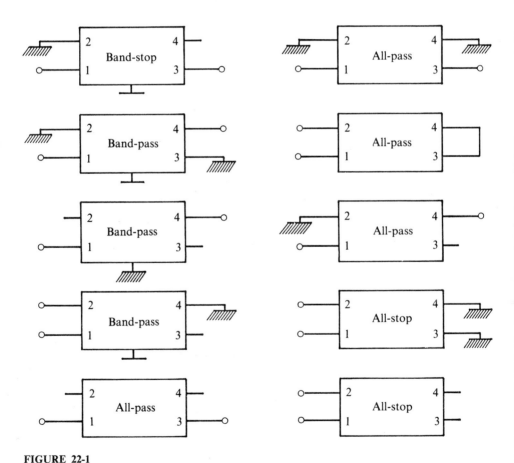

FIGURE 22-1
Two-port networks obtained by open-end short-circuiting different ports of four-port coupled line network.

$$V_3 = \left(Z_{31} - \frac{Z_{21}Z_{41}}{Z_{11}} \right) I_1 + \left(Z_{11} - \frac{Z_{41}^2}{Z_{11}} \right) I_3 \qquad (22\text{-}3b)$$

The open-circuit parameters of the parallel line structure in terms of the t variable are reproduced below

$$Z_{11} = \frac{L_{11}}{t} \qquad (22\text{-}4a)$$

$$Z_{21} = \frac{L_{21}}{t} \qquad (22\text{-}4b)$$

$$Z_{31} = \frac{L_{11}}{t} \sqrt{1 - t^2} \qquad (22\text{-}4c)$$

$$Z_{41} = \frac{L_{21}}{t} \sqrt{1 - t^2} \qquad (22\text{-}4d)$$

where

$$L_{11} = \frac{Z_e + Z_o}{2} \qquad (22\text{-}5)$$

$$L_{21} = \frac{Z_e - Z_o}{2} \qquad (22\text{-}6)$$

L_{11} and L_{21} are connected to the self- and mutual inductances L'_{11} and L'_{21} of the two lines.

The port conditions at ports 1 and 3 are now described by a reduced 2×2 impedance matrix as a preamble to terminating port 3 by a $1\text{-}\Omega$ resistor:

$$Z_{11}(t) = Z_{11} - \frac{Z_{21}^2}{Z_{11}} \qquad (22\text{-}7a)$$

$$Z_{22}(t) = Z_{11} - \frac{Z_{31}^2}{Z_{11}} \qquad (22\text{-}7b)$$

$$Z_{12}(t) = Z_{21}(t) = Z_{31} - \frac{Z_{21}Z_{41}}{Z_{11}} \qquad (22\text{-}7c)$$

Evaluating these parameters in the t plane indicates that

$$Z_{11}(t) = \frac{L_{11}^2 - L_{21}^2}{L_{11}t} \qquad (22\text{-}8a)$$

$$Z_{22}(t) = \frac{L_{11}^2 - L_{21}^2}{L_{11}t} + \frac{L_{21}^2 t}{L_{11}} \qquad (22\text{-}8b)$$

$$Z_{12}(t) = Z_{21}(t) = \frac{L_{11}^2 - L_{21}^2}{L_{11}t} \sqrt{1 - t^2} \qquad (22\text{-}8c)$$

The input impedance of the circuit at port 1 with port 3 terminated in a 1-Ω resistor is now formed in terms of its open-circuit parameters. The result is

$$Z(t) = \frac{Z_{11}(t) + Z_{11}(t)Z_{22}(t) - Z_{21}^2(t)}{1 + Z_{22}(t)} \tag{22-9}$$

Writing this quantity as the ratio of two polynomials in t gives

$$Z(t) = \frac{L_{11}^2 - L_{21}^2 + L_{11}(L_{11}^2 - L_{21}^2)t}{L_{11}^2 - L_{21}^2 + L_{11}t + L_{21}^2 t^2} \tag{22-10}$$

The synthesis of this impedance in the t plane is not unique. One solution is deduced by first extracting a UE with a characteristic impedance $Z(1)$ by replacing t by 1 in $Z(t)$ using the Richards theorem. The result is

$$Z(1) = Z_0 \tag{22-11}$$

where

$$Z_0 = \frac{L_{11}^2 - L_{21}^2}{L_{11}} \tag{22-12}$$

The remainder impedance $Z'(t)$ is guaranteed to be a positive real function and is defined by

$$Z'(t) = Z(1) \frac{tZ(1) - Z(t)}{tZ(t) - Z(1)} \tag{22-13}$$

Evaluating this quantity indicates that the required result is

$$Z'(t) = \frac{L_{21}^2 t}{L_{11}} + 1 \tag{22-14}$$

In obtaining this result a common factor $(t^2 - 1)$ has been cancelled in both the numerator and denominator polynomials of $Z'(t)$ in keeping with the Richards theorem. The remainder impedance $Z'(t)$ may now be realized in a first Foster form as a t-plane inductor:

$$L = \frac{L_{21}^2}{L_{11}} \tag{22-15}$$

in series with a 1-Ω resistor. The required equivalent circuit in Fig. 22-2, in terms of the original variables, has the property of a band-stop circuit. Solving for the odd and even mode parameters of the circuit in terms of Z_0 and L in Eqs (22-12) and (22-15) by making use of Eqs (22-5) and (22-6) gives

$$Z_e = Z_0 + L + \sqrt{LZ_0 + L^2} \tag{22-16}$$

$$Z_o = \frac{Z_e}{1 + 2L/Z_0 + (2/Z_0)\sqrt{LZ_0 + L^2}} \tag{22-17}$$

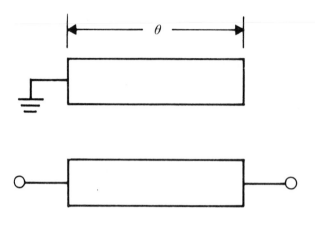

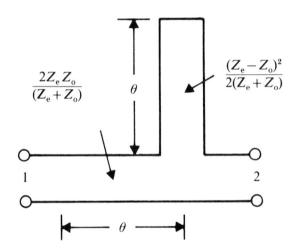

FIGURE 22-2
Two-port band-stop network
obtained by short-circuiting port 2
and open-circuiting port 4 of
parallel coupled lines.

PARALLEL LINE BAND-PASS FILTER CIRCUIT

The configuration of two parallel transmission lines with ports 2 and 3 open-circuited is another arrangement of special interest in the design of practical filter circuits. The boundary conditions are in this instance described by

$$I_2 = I_3 = 0 \tag{22-18}$$

and the open-circuit parameters between the other two ports are given by

$$Z_{11}(t) = Z_{22}(t) = \frac{L_{11}}{t} \tag{22-19}$$

$$Z_{12}(t) = Z_{21}(t) = \frac{L_{21}}{t}\sqrt{1 - t^2} \tag{22-20}$$

L_{11} and L_{21} have again the meaning in Eqs (22-5) and (22-6).

The one-port impedance in the t plane of the network terminated in a 1-Ω resistor is once more obtained by making use of Eq. (22-9). The result is

$$Z(t) = \frac{(L_{11}^2 - L_{21}^2) + L_{11}t + L_{21}^2 t^2}{L_{11}t + t^2} \tag{22-21}$$

The realization of this immittance in terms of UEs and t-plane capacitors and inductors is again not unique. One symmetrical solution is determined by rewriting $Z(t)$ as

$$Z(t) = \frac{L_{11}(L_{11} - L_{21}) + L_{21}(L_{11} - L_{21}) + L_{11}t + L_{21}^2 t^2}{L_{11}t + t^2} \tag{22-22}$$

as a preamble to partially extracting a t-plane capacitor of value

$$C = \frac{1}{L_{11} - L_{21}} \tag{22-23}$$

using a second Cauer form procedure. This permits $Z(t)$ to be written as

$$Z(t) = \frac{L_{11} - L_{21}}{t} + \frac{L_{21}(L_{11} - L_{21}) + L_{21}t + L_{21}^2 t^2}{L_{11}t + t^2} \tag{22-24}$$

The remainder impedance $Z'(t)$ is therefore given by

$$Z'(t) = \frac{L_{21}(L_{11} - L_{21}) + L_{21}t + L_{21}^2 t^2}{L_{11}t + t^2} \tag{22-25}$$

Extracting a UE from this quantity by replacing t by 1 gives

$$Z'(1) = L_{21} \tag{22-26}$$

and a remainder impedance

$$Z''(t) = \frac{Z_{11} - Z_{21}}{t} + 1 \tag{22-27}$$

as is readily verified.

This impedance may be realized in a first Foster form as a t-plane capacitor:

$$C = \frac{1}{L_{11} - L_{21}} \tag{22-28}$$

in series with a 1-Ω resistor. The required equivalent circuit is therefore the symmetric band-pass network indicated in Fig. 22-3. This circuit is constructed by recalling that a t-plane capacitor may be realized by an open-circuit line with a characteristic impedance given by

$$Z_0 = \frac{1}{C} \tag{22-29}$$

Still another filter configuration that is of interest in the design of practical microwave band-pass circuits is the dual case for which ports 2 and 3 are

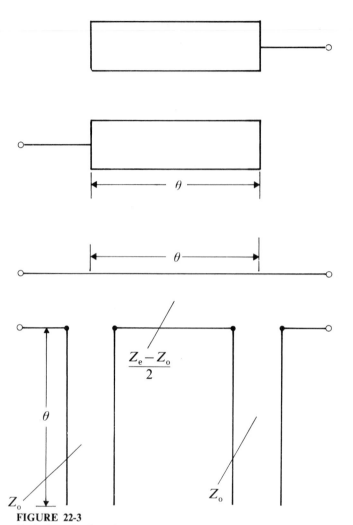

FIGURE 22-3
Equivalent circuit and two-port band-pass network obtained by open-circuiting ports 2 and 4 of parallel coupled lines.

short- instead of open-circuited. The short-circuit parameters of the equivalent two-port network between ports 1 and 4 of the parallel line network are described for this structure by

$$Y_{11}(t) = Y_{22}(t) = \frac{C_{11}}{t} \qquad (22\text{-}30a)$$

$$Y_{12}(t) = Y_{21}(t) = \frac{C_{21}}{t}\sqrt{1 - t^2} \qquad (22\text{-}30b)$$

where

$$C_{11} = \frac{Y_o + Y_e}{2} \qquad (22\text{-}31)$$

$$C_{21} = \frac{Y_o - Y_e}{2} \qquad (22\text{-}32)$$

C_{11} and C_{21} are related to the self- and mutual capacitances of the two coupled lines.

The corresponding one-port admittance of the circuit with the output port terminated by a 1-Ω resistor is

$$Y(t) = \frac{(C_{11}^2 - C_{21}^2) + C_{11}t + C_{21}^2 t^2}{C_{11}t + t^2} \qquad (22\text{-}33)$$

This admittance is readily synthesized in the manner illustrated in Fig. 22-4.

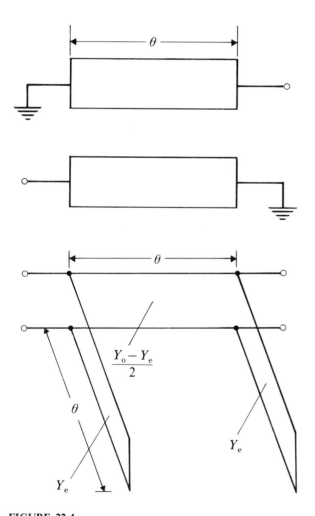

FIGURE 22-4
Two-port band-pass network obtained by short-circuiting ports 2 and 4 of parallel coupled lines.

PARALLEL LINE ALL-STOP FILTER CIRCUIT

Another parallel coupled line arrangement of some interest is the all-stop network in Fig. 22-5 for which the boundary conditions at ports 3 and 4 are

$$V_3 = V_4 = 0 \qquad (22\text{-}34)$$

and the relationship between the voltages and currents at the other two ports are given by

$$I_1 = Y_{11}V_1 + Y_{21}V_2 \qquad (22\text{-}35a)$$

$$I_2 = Y_{21}V_1 + Y_{11}V_2 \qquad (22\text{-}35b)$$

where

$$Y_{11}(t) = Y_{22}(t) = \frac{C_{11}}{t} \qquad (22\text{-}36a)$$

$$Y_{12}(t) = Y_{21}(t) = \frac{C_{21}}{t} \qquad (22\text{-}36b)$$

The one-port immittance of this circuit at port 1 with 2 terminated in a 1-Ω resistor is

$$Y(t) = \frac{C_{11}t + (C_{11}^2 - C_{21}^2)}{t^2 + C_{11}t} \qquad (22\text{-}37)$$

This immittance function may be realized as a symmetrical all-stop network in terms of t-plane inductors only. One symmetrical solution is obtained by rewriting

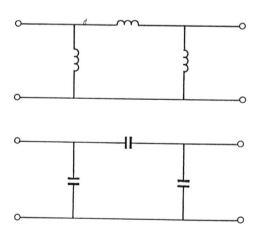

FIGURE 22-5

Equivalent circuit of two-port all-stop network obtained by short-circuiting ports 3 and 4 of parallel coupled lines.

$Y(t)$ as

$$Y(t) = \frac{C_{11}t + C_{21}(C_{11} - C_{21}) + C_{11}(C_{11} - C_{21})}{t^2 + C_{11}t} \tag{22-38}$$

as a preamble to extracting a t-plane inductor in a first Cauer form. The result

$$Y(t) = \frac{C_{11} - C_{21}}{t} + \frac{C_{21}t + C_{21}(C_{11} - C_{21})}{t^2 + C_{11}t} \tag{22-39}$$

is recognized as a shunt t-plane inductor

$$L = \frac{1}{C_{11} - C_{21}} \tag{22-40}$$

and a remainder admittance

$$Y'(t) = \frac{C_{21}t + C_{21}(C_{11} - C_{21})}{t^2 + C_{11}t} \tag{22-41}$$

The synthesis now proceeds by forming $Z'(t)$ from a knowledge of $Y'(t)$:

$$Z'(t) = \frac{1}{Y'(t)} = \frac{t^2 + C_{11}t}{C_{21}t + C_{21}(C_{11} - C_{21})} \tag{22-42}$$

as a preamble to extracting a t-plane inductor in a second Cauer form. This gives

$$Z'(t) = \frac{t}{C_{21}} + \frac{t}{t + (C_{11} - C_{21})} \tag{22-43}$$

$Z'(t)$ may therefore be realized as a series t-plane inductor

$$L = \frac{1}{C_{21}} \tag{22-44}$$

and a remainder impedance

$$Z''(t) = \frac{t}{t + (C_{11} - C_{21})} \tag{22-45}$$

The synthesis procedure is completed by forming $Y''(t)$:

$$Y''(t) = \frac{1}{Z''(t)} = \frac{C_{11} - C_{21}}{t} + 1 \tag{22-46}$$

and realizing it as shunt t-plane inductor

$$L = \frac{1}{C_{11} - C_{21}} \tag{22-47}$$

in parallel with a 1-Ω resistor. The required all-stop network is indicated in Fig. 22-5.

EQUIVALENCE BETWEEN COUPLED PARALLEL LINE CIRCUITS USING THE KURODA IDENTITIES

One important application of the Kuroda identities is in the organization of the architecture of circuits using distributed open- and short-circuited stubs (t-plane capacitors and inductors) and UEs. This will now be illustrated by rearranging the symmetrical equivalent circuits between ports 1 and 4 of a coupled parallel line network with the other two ports terminated in open- or short-circuit conditions. These equivalences are of interest in the analysis of the end sections of coupled parallel line band-pass filter circuits, to be treated in a later chapter. The required equivalence for the situation illustrated in Fig. 22-6 is deduced by having recourse to the second Kuroda identity. The first stub (t-plane capacitor) and UE of the original circuit in Fig. 22-6a are described in terms of the parameters employed in Chapter 21 by

$$Z_{01} = L_{11}L_{21} \tag{22-48a}$$

$$Z_0 = L_{21} \tag{22-48b}$$

The impedance of the UE of the modified circuit in Fig. 22-6b is then given from the Kuroda second identity by

$$\frac{Z_0}{n} = L_{11} \tag{22-49}$$

that of the stub by

$$\frac{Z_{01}}{n} = \frac{L_{11}(L_{11} - L_{21})}{L_{21}} \tag{22-50}$$

and the turns ratio of the ideal transformer by

$$n = \frac{L_{21}}{L_{11}} \tag{22-51}$$

The circuit in Fig. 22-6c is derived from that in Fig. 22-6b by interchanging the positions of the ideal transformer and the first stub and combining the two equivalent t-plane capacitors. The result is

$$C = \frac{L_{11}}{L_{11}^2 - L_{21}^2} \tag{22-52}$$

The required characteristic impedance of the open-circuited stub associated with this t-plane capacitor is

$$Z_0' = \frac{L_{11}^2 - L_{21}^2}{L_{11}} \tag{22-53}$$

This result may be verified by direct synthesis in the t-plane by starting with the

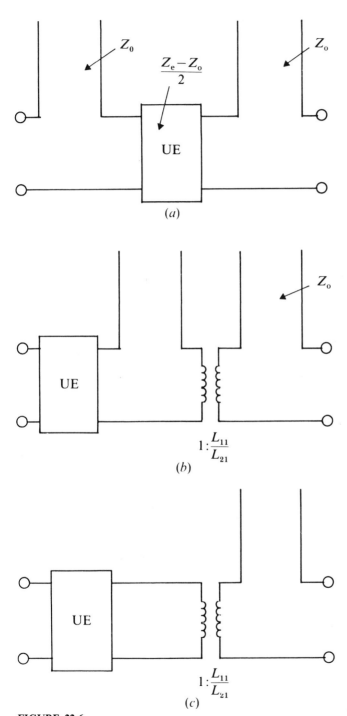

(a)

$1 : \dfrac{L_{11}}{L_{21}}$

(b)

$1 : \dfrac{L_{11}}{L_{21}}$

(c)

FIGURE 22-6
Equivalence between coupled parallel line circuits using the second Kuroda identity (open-circuit case).

one-port impedance description of the network in Eq. (22-9):

$$Z'(t) = \frac{(L_{11}^2 - L_{21}^2) + L_{11}t + L_{21}^2 t^2}{L_{11}t + t^2} \tag{22-54}$$

If the synthesis procedure begins with the extraction of a UE instead of a t-plane capacitor, then

$$Z(1) = L_{11} \tag{22-55}$$

and

$$Z'(t) = \frac{L_{11}(L_{11}^2 - L_{21}^2) + L_{11}^2 t}{L_{21}^2 t} \tag{22-56}$$

If the topology in Fig. 22-6b is to be realized then the synthesis continues by writing $Z'(t)$ as

$$Z'(t) = \frac{L_{11}L_{21}(L_{11} - L_{21}) + L_{11}^2(L_{11} - L_{21}) + L_{11}^2 t}{L_{21}^2 t} \tag{22-57}$$

as a preamble to forming a first Foster form expansion of $Z'(t)$ as a t-plane reactance

$$Z''(t) = \frac{L_{11}(L_{11} - L_{21})}{L_{21}t} \tag{22-58}$$

and a remainder impedance

$$Z'''(t) = \left(\frac{L_{11}}{L_{21}}\right)^2 \left(\frac{L_{11} - L_{21}}{t} + 1\right) \tag{22-59}$$

The first of these impedances is associated with a t-plane capacitor with a capacitance

$$C = \frac{L_{21}}{L_{11}(L_{11} - L_{21})} \tag{22-60}$$

which may be realized by an open-circuited stub with a characteristic impedance

$$Z_0 = \frac{L_{11}(L_{11} - L_{21})}{L_{21}} \tag{22-61}$$

The second one can be realized as an ideal transformer with a turns ratio

$$n = \frac{L_{11}}{L_{21}} \tag{22-62}$$

loaded by a series t-plane capacitor

$$C = \frac{1}{L_{11} - L_{21}} \tag{22-63}$$

and a 1-Ω resistor. The t-plane capacitor is realized as an open-circuited stub with a characteristic impedance

$$Z_0 = L_{11} - L_{21} \tag{22-64}$$

If the topology in Fig. 22-6c is adopted then $Z'(t)$ is written as

$$Z'(t) = \left(\frac{L_{11}}{L_{21}}\right)^2 \left(\frac{L_{11}^2 - L_{21}^2}{L_{11}t} + 1\right) \tag{22-65}$$

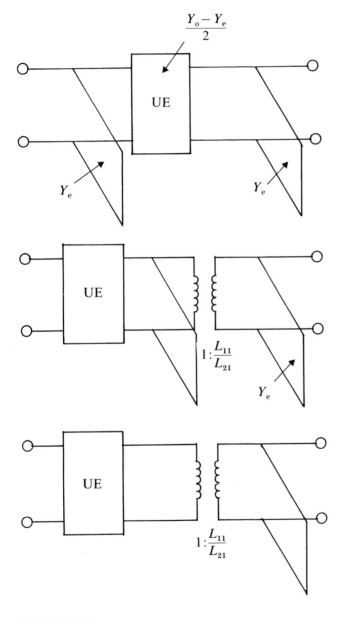

FIGURE 22-7
Equivalences between coupled parallel line circuits using the second Kuroda identity (short-circuit case).

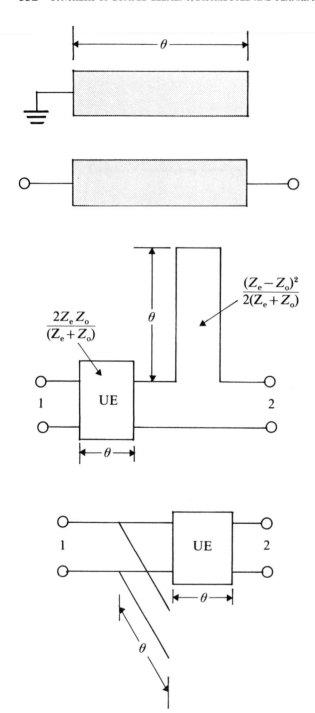

FIGURE 22-8
Equivalent circuits of coupled parallel line band-stop circuit.

and the synthesis proceeds by extracting an ideal transformer having a turns ratio

$$n = \frac{L_{11}}{L_{21}} \tag{22-66}$$

The remainder impedance

$$Z''(t) = \frac{L_{11}^2 - L_{21}^2}{L_{11}t} + 1 \tag{22-67}$$

is then realized in a first Foster form as a series t-plane capacitance

$$\frac{L_{11}}{L_{11}^2 - L_{21}^2} \tag{22-68}$$

in series with a 1-Ω resistor. This defines the characteristic impedance of the last stub.

Figure 22-7 summarizes the solution in the case of two coupled parallel lines with ports 2 and 3 short-circuited. The equivalence between the two possible circuits for the coupled parallel line band-stop circuit is separately understood. The required relationships are illustrated in Fig. 22-8.

COUPLED PARALLEL LINE FILTERS

Two other classic planar filters using parallel line circuits are illustrated in Figs 22-9 and 22-10. These types of circuits may be visualized as half-wave long resonators open- or short-circuited at both ends which are side coupled over quarter-wave intervals. The approximation is obtained by forming a one-to-one correspondence between the immittance matrices of a lumped element prototype using immittance inverters and the t-plane equivalent circuits of the parallel line circuit. Since a high-pass circuit in the t plane maps into a band-pass one in the θ plane, the latter prototype is employed. Figure 22-11 indicates the topology of a low-pass lumped element ladder prototype using ideal immittance inverters, that of the corresponding band-pass one and the equivalent circuit of the parallel line circuit using half-wave long resonators short-circuited at each end.

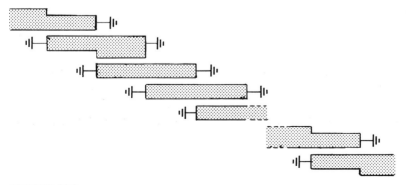

FIGURE 22-9
Planar view of parallel line band-pass filter circuit (short-circuit case).

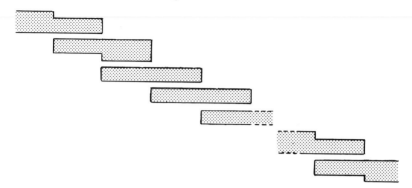

FIGURE 22-10
Planar view of parallel line band-pass filter circuit (open-circuit case).

THE SCATTERING MATRIX

The scattering matrix of the filter circuits described in this chapter may be readily obtained from a knowledge of the immittance parameters by making use of the appropriate relationships in Chapter 1.

The required relationships in terms of the open-circuit parameters are given for a symmetric network by

$$S_{11} = S_{22} = \frac{(Z_{11}^2 - 1) - Z_{21}^2}{(Z_{11} + 1)^2 - Z_{21}^2} \tag{22-69a}$$

$$S_{12} = S_{21} = \frac{2Z_{21}}{(Z_{11} + 1)^2 - Z_{21}^2} \tag{22-69b}$$

Evaluating these quantities in the case of two parallel coupled lines with ports 2 and 4 open-circuited gives

$$S_{11}(t) = S_{22}(t) = \frac{(L_{11}^2 - L_{21}^2) - (1 - L_{21}^2)t^2}{(L_{11}^2 - L_{21}^2) + 2L_{11}t + (1 + L_{21}^2)t^2} \tag{22-70a}$$

and
$$S_{12}(t) = S_{21}(t) = \frac{2L_{21}(\sqrt{1 - t^2})t}{(L_{11}^2 - L_{21}^2) + 2L_{11}t + (1 + L_{21}^2)t^2} \tag{22-70b}$$

In order to analyse a cascade of such circuits it is necessary to form the transmission parameters T_{11}, T_{22}, T_{12} and T_{21} introduced in Chapter 1 or to deduce the appropriate $ABCD$ parameters of the circuit.

The input impedance of the circuit is recovered from the bilinear transformation below:

$$Z(t) = \frac{1 + S_{11}(t)}{1 - S_{11}(t)} \tag{22-71}$$

in the usual way.

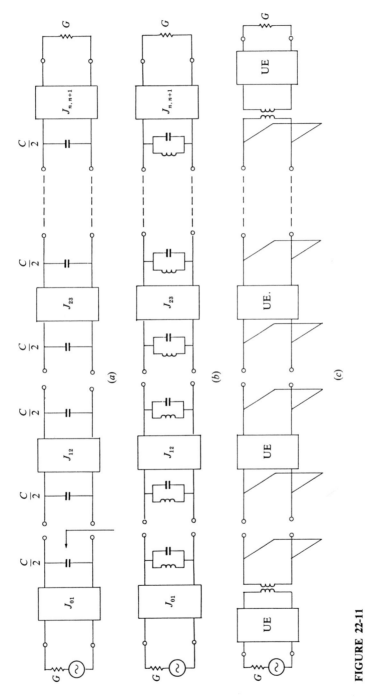

FIGURE 22-11
Partitioned lumped element (a) low-pass prototype and (b) band-pass prototype. (c) Partitioned parallel line prototype (short-circuit case).

PROBLEMS

22-1 Synthesize the $Z_{22}(t)$ in Eq. (22-8) in terms of a t-plane inductor and a UE.

22-2 Deduce the scattering parameters $S_{11}(t)$ and $S_{31}(t)$ between ports 1 and 3 of two parallel lines with ports 2 and 4 short-circuited. Form $S_{31}(t)S_{31}(-t)$ and $Z_{in}(t)$.

22-3 Repeat Prob. 22-2 with ports 2 and 4 open-circuited.

22-4 Verify that Eqs (22-70a,b) satisfy the unitary conditions.

22-5 Obtain the transmission parameters Y_{T11}, T_{22}, T_{12} and T_{21} associated with the scattering parameters in Eqs (22-70a,b) and deduce the scattering matrix of a cascade arrangement of two such circuits.

22-6 Construct $Z(t)$ from $S_{11}(t)$ in Eq. (22-70a) and show that the result is consistent with that in Eq. (22-21).

22-7 Deduce the equivalent circuits of two of the three all-pass networks in Fig. 22-1.

BIBLIOGRAPHY

Cohn, S. B., 'Direct coupled-resonator filters', *Proc. IRE*, Vol. 45, pp. 187–196, February 1957.

CHAPTER
23

PLANAR FILTERS USING PARALLEL LINE CIRCUITS

INTRODUCTION

The main purpose of this chapter is to consider the properties of some planar filter networks using coupled line circuits. The Kuroda identities are first employed to rearrange the topology of filters using ladder prototypes in terms of the architecture encountered in the description of coupled parallel line circuits. This is illustrated in the design of a band-stop filter. The architecture of a band-pass filter using coupled parallel half-wave long line circuits is discussed in Chapter 22. The exact theory of the so-called interdigital band-pass filter is next outlined. This filter consists of an array of quarter-wave long parallel coupled lines that are alternatively open- and short-circuited at each end and for which coupling between non-adjacent lines is neglected. The design of a degree 3 interdigital elliptic filter with coupling between non-adjacent lines is separately described. The usual approach to the design of filter circuits is to utilize a suitable network derived by an exact synthesis procedure. The response of the practical network is then forced to fit that of the ideal prototype as closely as possible.

EXACT SYNTHESIS OF BAND-STOP FILTERS USING PARALLEL LINES

The way in which the Kuroda identities may be used to alter the topology of a ladder network using t-plane capacitors and inductors into one consisting of t-plane inductors and UEs only or t-plane capacitors and UEs only will now be separately illustrated in connection with the design of two band-stop circuits. The first arrangement is suitable for the fabrication of band-stop filters using coupled parallel line circuits. The second topology is suitable for the practical fabrication of band-stop filter using open-circuited stubs spaced by UEs. The derivation of the first topology starts from the knowledge of the low-pass Butterworth characteristic function for $n = 0$ given in the θ plane by

$$K(\theta^2) = \left(\frac{\tan \theta}{\tan \theta_c}\right)^{2m} \tag{23-1}$$

and in the t plane by

$$K(t^2) = \left(\frac{t}{t_c}\right)^{2m} \tag{23-2}$$

The amplitude squared scattering parameters are then given by

$$S_{21}(t)S_{21}^*(t) = \frac{t_c^{2m}}{t^{2m} + t_c^{2m}} \tag{23-3}$$

$$S_{11}(t)S_{11}^*(t) = \frac{t^{2m}}{t^{2m} + t_c^{2m}} \tag{23-4}$$

respectively.

The circular variables in these equations are defined in terms of the midband (ω_o), band-edge (ω_c) and radian (ω) frequencies by

$$\theta = \frac{\pi}{2}\left(\frac{\omega}{\omega_o}\right) \tag{23-5}$$

$$\theta_c = \frac{\pi}{2}\left(\frac{\omega_c}{\omega_o}\right) \tag{23-6}$$

If the degree of the network is taken as 3 and if

$$\theta_c = \frac{\pi}{4}$$

then

$$t_c = j \tan \theta_c = j$$

and

$$S_{21}(t)S_{21}^*(t) = \frac{-1}{t^6 - 1}$$

$$S_{11}(t)S_{11}^*(t) = \frac{t^6}{t^6 - 1}$$

$S_{11}(t)$ is now calculated from a knowledge of $S_{11}(t)S_{11}^*(t)$:

$$S_{11}(t) = \frac{\pm t^3}{t^3 + 2t^2 + 2t + 1}$$

Adopting the negative sign in forming $Y(t)$ gives

$$Y(t) = \frac{2t^3 + 2t^2 + 2t + 1}{2t^2 + 2t + 1}$$

The required network may be synthesized in a first Cauer form. The result in the t plane is indicated in Fig. 23-1. A practical architecture for this network may now be realized by introducing redundant UEs and by making use of the Kuroda identities.

Since the quantity of interest is the reflection coefficient, unit elements may be added at both the input and output terminals of the circuit. UEs that do not modify the response of a circuit are known as redundant UEs. UEs such as are met in the theory of stepped impedance transformers are known as non-redundant ones. Introducing three such elements, two in front and one at the end of the low-pass prototype in Fig. 23-1, gives the network in Fig. 23-2a. The Kuroda identities may now be applied to the shunt capacitors to obtain the topology in Fig. 23-2b. The correspondence between this circuit and the parallel line circuit in Fig. 23-3 is readily understood. The design is complete once the odd and even mode parameters are determined in terms of the t-plane inductors (L) and the impedances (Z_0) of the UEs. The impedance of a t-plane circuit may be scaled by multiplying all normalized resistances and t-plane inductors by Z_0 and dividing all t-plane capacitors by Z_0.

The second possibility may be derived by replacing each UE and t-plane inductor by a t-plane capacitor and a UE. The required topology in the t plane is illustrated in Fig. 23-4 and one practical realization is indicated in Fig. 23-5.

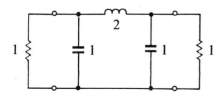

FIGURE 23-1
Butterworth $n = 3$ filter

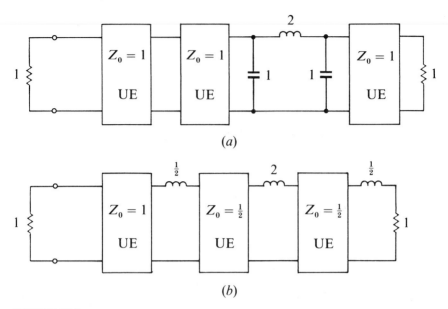

FIGURE 23-2
(a) Butterworth $n = 3$ filter with UEs added. (b) Equivalent circuit after application of the Kuroda identities.

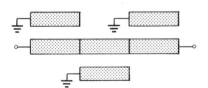

FIGURE 23-3
Parallel $n = 3$ coupled line band-stop filter topology.

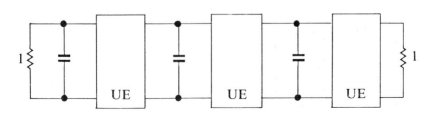

FIGURE 23-4
Butterworth $n = 3$ filter using t-plane capacitors and UEs.

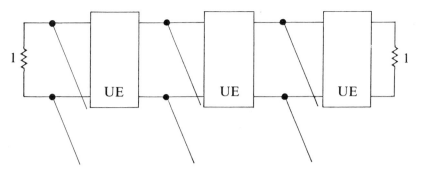

FIGURE 23-5
Band-stop $n = 3$ circuit using stubs and UEs.

THE INTERDIGITAL FILTER

Another planar circuit of some interest is the interdigital filter. It is constructed from an array of quarter-wave long parallel coupled lines by alternately short- and open-circuiting opposite ends of each conductor. The array can begin with either an open- or a short-circuited line. The input and output ports of this type of circuit may be on the same or on the other side. Figures 23-6 and 23-7 indicate the different possibilities. The exact equivalent circuit for this type of circuit may be deduced from a knowledge of its one-port immittance in a like manner to that employed in the case of two parallel lines. The one-port immittance for the circuit is obtained with all the closed ports appropriately terminated by electric or magnetic walls and the output port terminated by a 1-Ω resistor. This approach is somewhat tedious but has been employed to obtain the solutions for symmetric networks with up to éight lines and for asymmetric networks with up to four lines. A simpler solution based on directly operating on the equivalent capacitive matrix description of the circuit has also been described but is outside the remit of this work. Equivalent circuits for interdigital networks with open- and short-circuited lines are summarized in Tables 23-1 and 23-2. Scrutiny of these solutions indicates that the open-circuited topologies have

$$m = 3 \qquad (23\text{-}7a)$$

and
$$n = N - 3 \qquad (23\text{-}7b)$$

and that the short-circuited arrangements have

$$m = 1 \qquad (23\text{-}8a)$$

and
$$n = N - 1 \qquad (23\text{-}8b)$$

where n is the number of UEs, m is the number of t-plane reactances and N is the order of the filter.

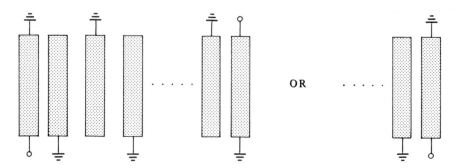

FIGURE 23-6
Schematic diagram of interdigital filter using short-circuited lines.

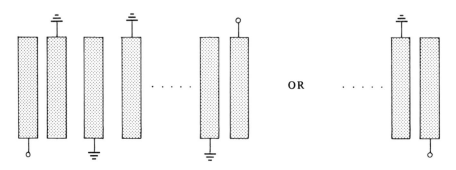

FIGURE 23-7
Schematic diagram of interdigital filter using open-circuited lines.

The high-pass characteristic function for the maximally flat response with $m = 1$ and $n = N - 1$ which maps into the necessary band-pass one is given in the θ plane by

$$K(\theta^2) = \left(\frac{\tan \theta_c}{\tan \theta}\right)^2 \left(\frac{\cos \theta}{\cos \theta_c}\right)^{2(N-1)} \tag{23-9a}$$

and in the t plane by

$$K(t^2) = \left(\frac{t_c}{t}\right)^2 \left(\frac{\sqrt{1 - t_c^2}}{\sqrt{1 - t^2}}\right)^{2(N-1)} \tag{23-9b}$$

The corresponding quantities for the Chebyshev insertion loss function are

$$K(\theta^2) = \left[T_1\left(\frac{\tan \theta_c}{\tan \theta}\right)T_{N-1}\left(\frac{\cos \theta}{\cos \theta_c}\right) - U_1\left(\frac{\tan \theta_c}{\tan \theta}\right)U_{N-1}\left(\frac{\cos \theta}{\cos \theta_c}\right)\right]^2 \tag{23-10a}$$

TABLE 23-1

Equivalent circuits for interdigital networks with open-circuited terminating lines

Number of lines	Interdigital network	s-Plane equivalent circuit
3		C_1 C_2 / L
4		C_1, L_1, Z_1 UE, $C_2 - 1:1$, L_2
5		C_1, Z_1 UE, L, Z_2 UE, C_2
N even		C_1, Z_1 UE, $\frac{Z_{(N-4)}}{2}$, L_1, UE, $\frac{Z_{(N-2)}}{2}$, L_2, UE, $\frac{Z_N}{2}$, Z_{N-3} UE, $C_2 - 1:1$
N odd		C_1, Z_1 UE, $\frac{Z_{(N-3)}}{2}$, L, UE, $\frac{Z_{(N-1)}}{2}$, UE, Z_{N-3} UE, C

Source: Wenzel, R. J., 'Exact theory of interdigital bandpass filters and related coupled circuits', *IEEE Trans. on MTT*, Vol. MTT-13, pp. 559–575, 1965, (© 1984 IEEE).

and

$$K(t^2) = \left[T_1\left(\frac{t_c}{t}\right) T_{N-1}\left(\frac{\sqrt{1-t_c^2}}{\sqrt{1-t^2}}\right) - U_1\left(\frac{t_c}{t}\right) U_{N-1}\left(\frac{\sqrt{1-t_c^2}}{\sqrt{1-t^2}}\right) \right]^2 \tag{23-10b}$$

respectively. The circular variables in these equations have the meaning in Eqs (23-5) and (23-6).

The maximally flat quantities for $m = 3$ and $n = N - 3$ are described by

$$K(\theta^2) = \left(\frac{\tan \theta_c}{\tan \theta}\right)^6 \left(\frac{\cos \theta}{\cos \theta_c}\right)^{2(N-3)} \tag{23-11a}$$

TABLE 23-2
Equivalent circuits for interdigital networks with short-circuited terminating lines

Number of lines	Interdigital network	s-Plane equivalent circuit
3		
4		
5		
N even		
N odd		

Source: as Table 23-1.

$$K(t^2) = \left(\frac{t_c}{t}\right)^6 \left(\frac{\sqrt{1-t_c^2}}{\sqrt{1-t^2}}\right)^{2(N-3)} \tag{23-11b}$$

and the Chebyshev ones by

$$K(\theta^2) = \left[T_3\left(\frac{\tan\theta_c}{\tan\theta}\right) T_{N-1}\left(\frac{\cos\theta}{\cos\theta_c}\right) - U_3\left(\frac{\tan\theta_c}{\tan\theta}\right) U_{N-1}\left(\frac{\cos\theta}{\cos\theta_c}\right) \right]^2 \tag{23-12a}$$

and

$$K(t^2) = \left[T_3\left(\frac{t_c}{t}\right) T_{N-1}\left(\frac{\sqrt{1-t_c^2}}{\sqrt{1-t^2}}\right) - U_3\left(\frac{t_c}{t}\right) U_{N-1}\left(\frac{\sqrt{1-t_c^2}}{\sqrt{1-t^2}}\right) \right]^2 \tag{23-12b}$$

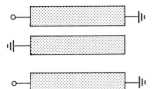

FIGURE 23-8
Topology of $n = 3$, interdigital filter using short-circuited lines.

For the three-strip structure in Fig. 23-8, $m = 1$, $N = 3$ and Eq. (23-9b) gives

$$S_{21}(t)S_{21}^*(t) = \frac{t^2(1 - t^2)^2}{t^6 - 2t^4 + t^2 + t_c^2(1 - t_c^2)^2}$$

$$S_{11}(t)S_{11}^*(t) = \frac{t_c^2(1 - t_c)^2}{t^6 - 2t^4 + t^2 + t_c^2(1 - t_c^2)^2}$$

If

$$\theta_c = \frac{\pi}{4}$$

then

$$t_c = j \tan \theta_c = j$$

and

$$S_{21}(t)S_{21}^*(t) = \frac{t^2(1 - t^2)^2}{t^6 - 2t^4 + t^2 - 4}$$

$$S_{11}(t)S_{11}^*(t) = \frac{-4}{t^6 - 2t^4 + t^2 - 4}$$

The six roots of the denominator polynomial $S_{11}(t)S_{11}^*(t)$ are given using a standard root finding subroutine by

$$0.760\,69 + j0.758\,87$$

$$0.760\,69 - j0.857\,87$$

$$-1.521\,38 + j0$$

$$-0.760\,69 + j0.857\,87$$

$$-0.760\,69 - j0.857\,87$$

$$1.521\,38 + j0$$

$S_{11}(t)$ is now constructed by retaining the LHP roots of the denominator polynomial of $S_{11}(t)S_{11}^*(t)$ and noting that the numerator polynomial is a perfect square. The result is

$$S_{11}(t) = \frac{\pm 2}{t^3 + 3.043t^2 + 3.629t + 2}$$

Adopting the lower sign in order to cater for the boundary condition at the origin due to the single stub indicates that the one-port immittance of the circuit terminated in a 1-Ω resistance is

$$Y(t) = \frac{t^3 + 3.043t^2 + 3.629t + 4}{t^3 + 3.043t^2 + 3.629t}$$

This admittance may now be realized as a symmetrical network consisting of two UEs spaced by a t-plane inductor (Fig. 23-9). The impedance of the first normalized UE is

$$Z = 0.6573$$

the value of the t-plane inductor is

$$L = 4.0 \text{ H}$$

and the impedance of the last UE is

$$Z = 0.6573$$

The derivation of the required network topology of the same-degree interdigital filter terminated in a 1-Ω resistor starts with a knowledge of the boundary conditions:

$$V_2 = V_4 = V_6 = 0 \tag{23-13a}$$

$$I_5 = 0 \tag{23-13b}$$

and the voltage–current relationship:

$$\begin{bmatrix} I_1 \\ I_2 \\ I_3 \\ I_4 \\ 0 \\ I_6 \end{bmatrix} = \frac{1}{t} \left[\begin{array}{c|c} [C] & -[C]\sqrt{1-t^2} \\ \hline -[C]\sqrt{1-t^2} & [C] \end{array} \right] \begin{bmatrix} V_1 \\ 0 \\ V_3 \\ 0 \\ V_5 \\ 0 \end{bmatrix} \tag{23-14}$$

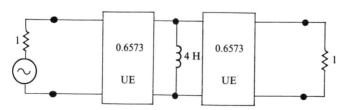

FIGURE 23-9
Topology of $m = 3$, $n = 0$ filter circuit ($t_c = 1$).

If the coupling between non-adjacent lines is neglected $C_{31} = 0$ and $[C]$ is a tridiagonal matrix:

$$[C] = \begin{bmatrix} C_{11} & C_{21} & 0 \\ C_{21} & C_{22} & C_{21} \\ 0 & C_{21} & C_{11} \end{bmatrix} \qquad (23\text{-}15)$$

The derivation then continues by realizing the network in the form of two UEs spaced by a t-plane inductor. This permits a one-to-one correspondence between the network problem and the odd and even variables of the circuit to be established.

The relationship between voltages and currents at ports 1 and 3 of the circuit in Fig. 23-8 are then obtained as

$$I_1 = \frac{C_{11}C_{22} - C_{21}^2(1 - t^2)}{C_{22}t} V_1 - \frac{C_{21}^2(1 - t^2)}{C_{22}t} V_3 \qquad (23\text{-}16a)$$

$$I_3 = -\frac{C_{21}^2(1 - t^2)}{C_{22}t} V_1 + \frac{C_{11}C_{22} - C_{21}^2(1 - t^2)}{C_{22}t} V_3 \qquad (23\text{-}16b)$$

The admittance matrix defined by these two relationships is symmetric, in keeping with the topology of the network. The new short-circuit parameters between the input and output terminals of the required two-port network are therefore described by

$$Y_{11} = Y_{22} = \frac{C_{11}C_{22} - C_{21}^2(1 - t^2)}{C_{22}t} \qquad (23\text{-}17a)$$

$$Y_{12} = Y_{21} = \frac{-C_{21}^2(1 - t^2)}{C_{22}t} \qquad (23\text{-}17b)$$

In order to make a one-to-one correspondence between this circuit and that deduced on the basis of the insertion loss specification it is now necessary to synthesize it in terms of two UEs spaced by a shunt t-plane inductance. Since it is symmetric and since its middle section is a t-plane susceptance rather than a UE it is advantageous to realize it using the even mode eigenvalue of the circuit. Recalling that the odd and even mode eigenvalues are linear combinations of the short-circuit parameters,

$$Y_e = Y_{11} + Y_{21} \qquad (23\text{-}18a)$$

$$Y_o = Y_{11} - Y_{21} \qquad (23\text{-}18b)$$

gives

$$Y_e(t) = \frac{C_{11}C_{22} - 2C_{21}^2(1 - t^2)}{C_{22}t} \qquad (23\text{-}19a)$$

$$Y_o(t) = \frac{C_{11}}{t} \qquad (23\text{-}19b)$$

The synthesis of the even mode eigenvalue starts by extracting a UE by replacing t by 1 in $Y_e(t)$ using the Richards theorem. The result is

$$Y_e(1) = C_{11} \tag{23-20}$$

The first UE is of course also realized by starting with $Y_o(t)$ instead of $Y_e(t)$ or for that matter by starting with Y_{11} in Eq. (23-17b).

The remainder admittance, after cancelling a common factor $1 - t^2$ in both the numerator and denominator polynomials,

$$Y'(t) = \frac{-C_{11}(C_{11}C_{22} - 2C_{21}^2)}{2C_{21}^2 t} \tag{23-21}$$

is readily recognized as a shunt t-plane inductance given by

$$L = \frac{-2C_{21}^2}{C_{11}(C_{11}C_{22} - 2C_{21}^2)} \tag{23-22}$$

The other half of the network is readily obtained from symmetry considerations. The design is completed by satisfying

$$C_{11} = 0.6573 \tag{23-23a}$$

$$\frac{C_{21}^2}{C_{11}(C_{11}C_{22} - 2C_{21}^2)} = 4.0 \text{ H} \tag{23-23b}$$

and expressing the short-circuit parameters of the two-port circuit in terms of those of the interdigital structure defined by the tridiagonal $[C]$ matrix.

The circuit development of the topology of the three-strip structure in Fig. 23-10 proceeds in a like manner to that utilized in deducing that in Fig. 23-8. It starts by realizing the insertion loss specification as a two-port network with the topology in Fig. 23-11 by making use of Eq. (23-11b) with $m = 3$ and $N = 3$. This gives

$$K(t^2) = \left(\frac{t_c}{t}\right)^2 \tag{23-24a}$$

and

$$S_{21}(t)S_{21}^*(t) = \frac{t^6}{t^6 + t_c^6} \tag{23-24b}$$

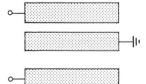

FIGURE 23-10
Topology of $n = 3$ interdigital filter with open-circuited lines.

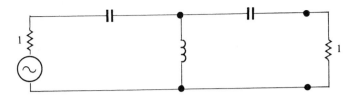

FIGURE 23-11
Topology of $m = 3$, $n = 0$ filter circuit ($t_c = 1$).

$$S_{11}(t)S_{11}^*(t) = \frac{t_c^6}{t^6 + t_c^6} \qquad (23\text{-}24c)$$

This circuit is readily synthesized in terms of t-plane reactances only in the manner indicated in Fig. 23-11. The derivation is completed by realizing the equivalent two-port circuit in Fig. 23-10 in the same topology.

ELLIPTIC FILTERS USING PARALLEL LINE CIRCUITS

Elliptic filter circuits in planar form are also of some practical interest in microwave engineering. One possible planar arrangement of a series resonator in shunt with the circuit which has already been met in this text consists of a cascade arrangement of high and low impedance transmission lines. Another possibility is an n-wire circuit with coupling between every wire. The structure under discussion is indicated in Fig. 23-12a in the case of $n = 3$ and its equivalent circuit in Fig. 23-12b.

The derivation of the equivalent circuit starts with a description of the open-circuit parameters of the network:

$$\bar{Z} = \frac{1}{t}\left[\begin{array}{c|c} [L] & [L]\sqrt{1-t^2} \\ \hline [L]\sqrt{1-t^2} & [L] \end{array}\right] \qquad (23\text{-}25a)$$

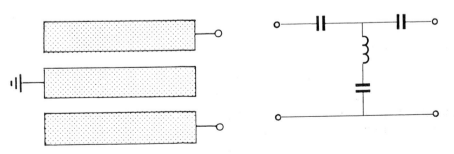

FIGURE 23-12
(a) A three-wire parallel line elliptic section. (b) Equivalent circuit of elliptic section.

where

$$\bar{L} = \begin{bmatrix} L_{11} & L_{21} & L_{31} \\ L_{21} & L_{22} & L_{21} \\ L_{31} & L_{21} & L_{11} \end{bmatrix} \tag{23-25b}$$

and the boundary conditions

$$I_2 = I_4 = I_6 = V_5 = 0 \tag{23-25c}$$

The open-circuit parameters between ports 1 and 3 are then given by

$$Z_{11}(t) = Z_{22}(t) = \frac{L_{11}L_{22} - L_{21}^2(1 - t^2)}{L_{22}t} \tag{23-26a}$$

$$Z_{12}(t) = Z_{21}(t) = \frac{L_{22}L_{31} - L_{21}^2(1 - t^2)}{L_{22}t} \tag{23-26b}$$

Since the network is symmetrical it is possible to synthesize it from a knowledge of its immittance eigenvalues below:

$$z_e(t) = Z_{11}(t) + Z_{21}(t) \tag{23-27a}$$

$$z_o(t) = Z_{11}(t) - Z_{21}(t) \tag{23-27b}$$

This gives

$$z_e(t) = \frac{L_{22}(L_{11} + L_{31}) - 2L_{21}^2(1 - t^2)}{L_{22}t} \tag{23-28a}$$

$$z_o(t) = \frac{L_{11} - L_{31}}{t} \tag{23-28b}$$

The realization of the required network now proceeds by synthesizing $z_e(t)$. Forming a partial fraction expansion of $z_e(t)$ in order to display the eigenvalue $z_o(t)$ indicates that

$$z_e(t) = \frac{L_{11} - L_{31}}{t} + \frac{2L_{22}L_{31} - 2L_{21}^2(1 - t^2)}{L_{22}t} \tag{23-29}$$

This immittance may be realized as a series t-plane capacitor

$$C = \frac{1}{L_{11} - L_{31}} \tag{23-30}$$

and a remainder admittance

$$z_e'(t) = \frac{2L_{22}L_{31} - 2L_{21}(1 - t^2)}{L_{22}t} \tag{23-31}$$

Since it is desirable to realize $z_e(t)$ as a ladder network the synthesis proceeds by forming $y_e'(t)$ from a knowledge of $z_e'(t)$:

$$y_e' = \frac{\frac{1}{2}L_{22}t}{L_{22}L_{31} - L_{21}^2(1 - t^2)} \tag{23-32}$$

If the numerator and denominator polynomials are both divided by $L_{22}t$ then

$$y'_e(t) = \frac{\frac{1}{2}}{(L_{22}L_{31} - L_{21}^2)/L_{22}t + L_{21}^2t/L_{22}} \qquad (23\text{-}33)$$

and $y'_e(t)$ is recognized as the impedance of a t-plane series LC resonator with L and C given by

$$C = \frac{L_{22}}{2(L_{22}L_{31} - L_{21}^2)} \qquad (23\text{-}34a)$$

and

$$L = \frac{2L_{21}^2}{L_{22}} \qquad (23\text{-}34b)$$

The resonant frequency of the attenuation pole is readily obtained as

$$\omega_0^2 LC = \frac{\omega_0^2 L_{21}^2}{L_{22}L_{31} - L_{21}^2} \qquad (23\text{-}34c)$$

The necessary equivalent circuit is indicated in Fig. 23-12b.

If the entries of the capacitance matrix are taken as the primary variables of the circuit then those of the inductive one can be derived by recalling that the two descriptions are related by

$$[L] = [C]^{-1} \qquad (23\text{-}35)$$

The relationship indicates that the condition $C_{31} = 0$ does not imply $L_{31} = 0$. If the coupling between the non-adjacent lines is removed then

$$C_{31} = 0 \qquad (23\text{-}36a)$$

and

$$L_{22}L_{31} - L_{21}^2 = 0 \qquad (23\text{-}36b)$$

If this condition is met then the capacitance of the series resonator of the elliptic filter vanishes and the equivalent circuit reduces to that indicated in Fig. 23-11.

BIBLIOGRAPHY

Cohn, S. B., 'Parallel-coupled transmission line resonator filters', *IRE Trans. on MTT*, Vol. MTT-6, pp. 223–231, April 1958.

Cristal, E. G., 'New design equations for a class of microwave filters', *IEEE Trans. on MTT*, Vol. MTT-19, pp. 486–490, May 1971.

Matthaei, G. L., 'Interdigital bandpass filters', *IEEE Trans. on MTT*, Vol. MTT-10, pp. 479–491, November 1962.

Ozaki, H. and Ishii, J., 'Synthesis of a class of strip-line filters', *IRE Trans. on Circuit Theory*, Vol. CT-5, pp. 104–109, June 1958.

Schiffman, B. M. and Matthaei, G. L., 'Exact design of band-stop microwave filters', *IEEE Trans. on MTT*, Vol. MTT-12, pp. 6–15, January 1964.

Wenzel, R. J., 'Exact theory of interdigital bandpass filters and related coupled circuits', *IEEE Trans. on MTT*, Vol. MTT-13, pp. 559–575, 1965, (© 1984 *IEEE*).

Yamamoto, S., Azakami, Z. and Stakura, K., 'Coupled strip transmission line with three center conductors', *IEEE Trans. on MTT*, Vol. MTT-14, pp. 446–456, October 1966.

CHAPTER
24

TEM AND
QUASI TEM
TRANSMISSION
LINES

INTRODUCTION

At low frequencies, where electric potential and current are perhaps more meaningful concepts than electric and magnetic fields, electrical energy is transported by current-carrying wires; the coaxial and the two wire lines are two examples. This class of line supports a TEM solution with no low frequency cutoff condition. Widely used semi-planar forms of these two lines are the microstrip and stripline geometries. Other variations of these lines, which have some merit at still higher frequencies, are the slot line, the coplanar line, the finline and the inverted microstrip. While, strictly speaking, these lines support quasi TEM rather than TEM solutions in the microwave region this has not prevented the wide use of such lines in modern assemblies and equipment. This is in no small part due to the ease with which it is possible to mount series and shunt elements along these lines without the need to drill holes in the substrate. Quartz and alumina are two substrate materials employed in practice.

PROPAGATION IN COAXIAL LINE

One widely used transmission line that is readily amenable to an approximate closed-form solution is the coaxial line illustrated in Fig. 24-1 having an inner radius b and an outer radius a. Such a structure can support a purely transverse solution (TEM wave) for which

$$E_z = H_z = 0 \qquad (24\text{-}1a)$$

The simplest permissible solution in this type of transmission line that satisfies the boundary conditions at $r = a$ and b is one with

$$E_\theta = H_r = 0 \qquad (24\text{-}1b)$$

and

$$E_r \neq 0, \; H_\theta \neq 0 \qquad (24\text{-}1c)$$

The description of the two components H_θ and E_r may be deduced by having recourse to Maxwell's equations. These equations are given in differential form for a charge-free region ($\rho = 0$) by

$$\nabla X \bar{E} = -\mu_0 \frac{\partial \bar{H}}{\partial t} \qquad (24\text{-}2a)$$

$$\nabla X \bar{H} = \varepsilon_0 \frac{\partial \bar{E}}{\partial t} \qquad (24\text{-}2b)$$

$$\nabla \cdot \bar{D} = 0 \qquad (24\text{-}2c)$$

$$\nabla \cdot \bar{B} = 0 \qquad (24\text{-}2d)$$

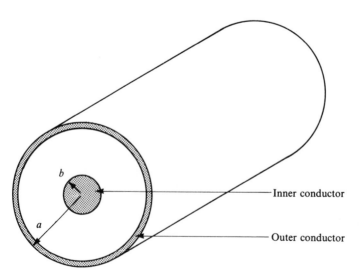

Inner conductor

Outer conductor

FIGURE 24-1
Schematic diagram of coaxial transmission line.

In transmission theory it is usual to assume that all the fields vary in the positive z direction as

$$\exp(-\gamma z) \tag{24-3a}$$

so that it is only necessary to determine the variations of the fields in the transverse plane.

γ is a complex quantity known as the propagation constant which satisfies the wave equation to be deduced:

$$\gamma = \alpha + j\beta \tag{24-3b}$$

α is the attenuation constant of the line per unit length (decibels per metre) and β is the phase constant per unit length (radians per metre).

It is also understood that all the field quantities vary with time as

$$\exp(j\omega t) \tag{24-3c}$$

where ω is the radian frequency (radians per second).

The constitutive parameters μ_0 and ε_0 are given by

$$\mu_0 = 4\pi \times 10^{-7} \text{ H/m}$$

$$\varepsilon_0 = \frac{1}{36\pi} \times 10^{-9} \text{ F/m}$$

The two curl equations in circular variables give

$$\frac{\partial E_r}{\partial z} = -j\omega\mu_0 H_\theta \tag{24-4}$$

and

$$\frac{\partial H_\theta}{\partial z} = -j\omega\varepsilon_0 E_r \tag{24-5}$$

$$\frac{\partial(r H_\theta)}{\partial r} = 0 \tag{24-6}$$

respectively.

The two divergence equations with the boundary conditions in Eq. (24-1) indicate that

$$\left(\frac{1}{r}\right)\frac{\partial(r E_r)}{\partial r} = 0 \tag{24-7}$$

$$\left(\frac{1}{r}\right)\frac{\partial H_\theta}{\partial \theta} = 0 \tag{24-8}$$

The solution for either E_r or H_θ is now derived by forming the wave equation for either quantity by making use of Eqs (24-4) and (24-5). The results are

$$\left(\frac{\partial^2}{\partial z^2} + \omega^2 \mu_0 \varepsilon_0\right) E_r = 0 \tag{24-9}$$

and
$$\left(\frac{\partial^2}{\partial z^2} + \omega^2 \mu_0 \varepsilon_0\right) H_\theta = 0 \tag{24-10}$$

One possible product solution in a semi-infinite line for H_θ which is independent of θ and which satisfies Eq. (24-6) with

$$rH_\theta \tag{24-11}$$

as a constant is

$$H_\theta = \frac{I}{2\pi r} \exp(-\gamma z) \tag{24-12}$$

provided

$$\gamma^2 = \omega^2 \mu_0 \varepsilon_0 \tag{24-13}$$

I is the total current in the conductors.

E_r is now obtained in terms of H_θ by making use of Eq. (24-5):

$$E_r = \frac{\gamma}{j\omega\varepsilon_0} H_\theta \tag{24-14}$$

Product solutions of pure functions of R and Z which are independent of θ therefore satisfy both the curl and divergence equations. The wave impedance defined by Eq. (24-14) corresponds to that of free space:

$$\eta_0 = \frac{\gamma}{j\omega\varepsilon_0} = \sqrt{\frac{\mu_0}{\varepsilon_0}} = 120\pi \ \Omega \tag{24-15}$$

For completeness, the phase velocity is defined by Eqs (24-13) and (24-15) as

$$v = \frac{\omega}{\beta} = \frac{1}{\sqrt{\mu_0\varepsilon_0}} = 3 \times 10^9 \ \text{m/s} \tag{24-16}$$

in keeping with that of the velocity of light.

More generally, in the presence of a dielectric or magnetic filler with a complex relative constitutive parameter ε_r or μ_r,

$$\varepsilon_0 \rightarrow \varepsilon_0(\varepsilon_r' - j\varepsilon_r'') \tag{24-17a}$$

$$\mu_0 \rightarrow \mu_0(\mu_r' - j\mu_r'') \tag{24-17b}$$

and both the impedance and propagation constants are complex quantities.

Figure 24-2 illustrates the field patterns in this type of transmission line.

At low frequencies it is customary to work in terms of voltage and current instead of electric and magnetic fields; it is also usual to work in terms of the characteristic instead of the wave impedance. The voltage across the line is defined by

$$V = \int_a^b E_r \, dr \tag{24-18}$$

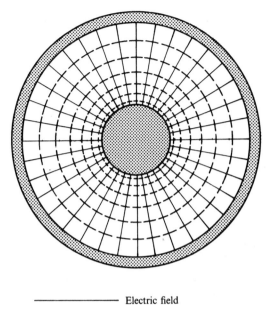

———————— Electric field

– – – – – – – Magnetic field

FIGURE 24-2
Electric and magnetic fields in coaxial line.

Writing E_r in terms of Eqs (24-12) and (24-14) gives

$$E_r = \frac{\eta_0 I}{2\pi r} \qquad (24\text{-}19)$$

and forming V gives the required result:

$$V = \frac{\eta_0 I}{2\pi} \ln\left(\frac{b}{a}\right) \qquad (24\text{-}20)$$

This last equation also defines the characteristic impedance of the line as

$$Z_0 = \frac{V}{I} = \frac{\eta_0}{2\pi} \ln\left(\frac{b}{a}\right) \qquad (24\text{-}21)$$

and I as

$$I = \frac{V}{Z_0} \qquad (24\text{-}22)$$

The characteristic impedance of the line may therefore be varied by adjusting the radii a and b.

TELEGRAPHIST EQUATIONS

It will now be demonstrated that the coupled equations between E_r and H_θ in Eqs (24-4) and (24-5) reduce to a standard form in terms of the voltage and current variables V and I.

Writing H_θ in Eq. (24-12) in terms of V in Eq. (24-20) and noting Eq. (24-4) gives

$$H_\theta = \frac{V}{\eta_0 r \, \ln(b/a)} \tag{24-23a}$$

$$E_r = \eta_0 H_\theta = \frac{V}{r \, \ln(b/a)} \tag{24-23b}$$

Substituting these relationships into Eq. (24-4) and (24-5) gives a new pair of coupled equations in V and I instead of E_r and H_θ:

$$\frac{\partial V}{dz} = -j\omega\mu_0 \frac{\ln(b/a)}{2\pi} I \tag{24-24a}$$

$$\frac{\partial I}{dz} = -j\omega\varepsilon_0 \frac{2\pi}{\ln(b/a)} V \tag{24-24b}$$

The preceding equations are recognized as the standard coupled transmission line or telegraphist equations already encountered elsewhere:

$$\frac{\partial V}{\partial z} = -ZI \tag{24-25a}$$

$$\frac{\partial I}{\partial z} = -YV \tag{24-25b}$$

Z and Y are defined in terms of the inductance and capacitance per unit length of the line

$$Z = j\omega L \tag{24-26a}$$

$$Y = j\omega C \tag{24-26b}$$

In the presence of dissipation Z and Y become

$$Z = R + j\omega L \tag{24-27a}$$

$$Y = G + j\omega C \tag{24-27b}$$

R and G represent the dissipation per unit length of the inductance and capacitance respectively.

It is usual to write the relationship for V and I in Eqs (24-25a,b) in the form met for E_r and H_θ in Eqs (24-9) and (24-10):

$$\left(\frac{\partial^2}{\partial z^2} - ZY\right)V = 0 \tag{24-28a}$$

$$\left(\frac{\partial^2}{\partial z^2} - ZY\right)I = 0 \tag{24-28b}$$

One solution for a semi-infinite line is

$$V = A \exp(-\gamma z) \tag{24-29a}$$

$$I = \frac{A}{Z_0} \exp(-\gamma z) \tag{24-29b}$$

where Z_0 is the characteristic impedance of the line.

The coupled equations are satisfied provided the propagation constant is

$$\gamma = \sqrt{ZY} \tag{24-30}$$

The characteristic impedance may be deduced by differentiating Eq. (24-25b) and making use of Eq. (24-29b). The result is

$$Z_0 = \sqrt{\frac{Z}{Y}} \tag{24-31}$$

In terms of the distributed elements L and C,

$$\gamma = j\beta = j\omega\sqrt{LC} \tag{24-32a}$$

$$Z_0 = \sqrt{\frac{L}{C}} \tag{24-32b}$$

$$v = \frac{1}{\sqrt{LC}} = 3 \times 10^9 \text{ m/s} \tag{24-32c}$$

This result indicates that a knowledge of either the capacitance or inductance of the line and a statement about its phase velocity is sufficient to describe a TEM transmission line.

Scrutiny of the coupled equations indicates that the inductance (henrys per metre) and capacitance (farads per metre) of the line are given respectively by

$$C = \frac{2\pi\varepsilon_0}{\ln(b/a)} \quad \text{F/m} \tag{24-33}$$

$$L = \frac{\mu_0 \ln(b/a)}{2\pi} \quad \text{H/m} \tag{24-34}$$

PARALLEL WIRE TRANSMISSION LINE

Another classic TEM transmission line is the twin or parallel wire arrangement. Although this geometry is amenable to a closed-form solution its derivation requires an appreciation of conformal mapping which is outside the remit of this work. The result will therefore be noted:

$$Z_0 = 120 \cosh^{-1}\left(\frac{d_0}{r_0}\right) \tag{24-35}$$

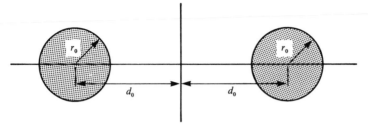

FIGURE 24-3
Schematic diagram of parallel wire transmission line.

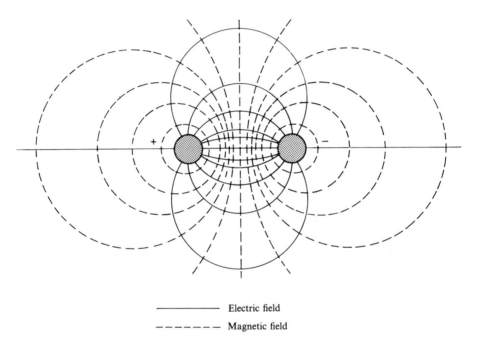

———————— Electric field

— — — — — — Magnetic field

FIGURE 24-4
Field patterns in parallel wire transmission line.

The schematic diagram of this arrangement is illustrated in Fig. 24-3 and the field patterns in Fig. 24-4.

THE STRIPLINE LINE

A semi-planar variation of the coaxial line that is widely used in the microwave region is the stripline one, illustrated in Fig. 24-5. The derivation of its characteristic impedance starts by forming the magnetic field at the position y, in the

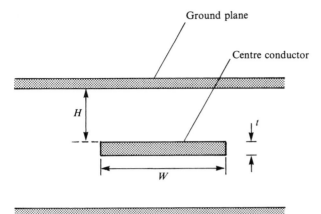

FIGURE 24-5
Schematic diagram of stripline
transmission line.

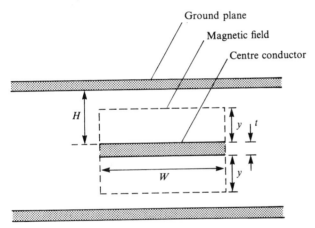

FIGURE 24-6
Definition of magnetic field
in stripline transmission line.

configuration in Fig. 24-6, due to a current I in the centre conductor:

$$h = \frac{I}{2W + 2t + 4y} \tag{24-36}$$

The contributions from each ground plane cancel, since the magnetic field from a uniform infinite current sheet does not depend on the distance from it.

The average field is

$$h = \frac{1}{H} \int_0^H \frac{I \, dy}{2W + 2t + 4y} \tag{24-37}$$

Integrating this quantity leads to

$$h = \frac{1}{4H} \ln\left(\frac{W + t + 2H}{W + t}\right) \tag{24-38}$$

The voltage on the line is given by

$$V = E \, dy = EH \tag{24-39}$$

The impedance of the line is now obtained by combining Eqs (24-38) and (24-39). The result is

$$Z_0 = 30\pi \ln\left(\frac{W + t + 2H}{W + t}\right) \tag{24-40}$$

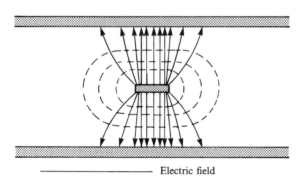

——————— Electric field

– – – – – – – – Magnetic field

FIGURE 24-7
Field patterns in stripline transmission line.

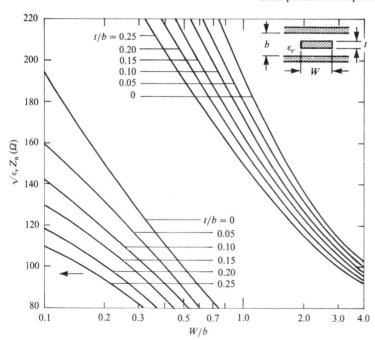

FIGURE 24-8
Characteristic impedance of stripline transmission line.

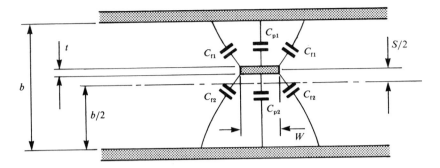

FIGURE 24-9
Capacitance of stripline transmission line.

where W, H and t are the linear dimensions of the stripline in Fig. 24-5. Figure 24-7 shows the field patterns in this line; Fig. 24-8 depicts one graphical solution for Z_0.

The characteristic impedance of a TEM line may also be deduced from a knowledge of the total capacitance of the line and the phase velocity along the line. Figure 24-9 shows the parallel plate and fringing capacitances in this type of geometry. The parallel plate capacitance is readily given by

$$C_p = \frac{\varepsilon_0 \varepsilon_r W}{H} \qquad \text{F/m} \tag{24-41}$$

and that of a single edge may be shown to be approximately determined by

$$C_f = \frac{\varepsilon_0 \varepsilon_r}{H} \ln 2 \qquad \text{F/m} \tag{24-42}$$

The calculation of the impedance of the line based on this approach is left as an exercise for the reader.

THE MICROSTRIP LINE

A classic planar transmission line closely related to the two-wire geometry is the microstrip arrangement. Figure 24-10 illustrates its layout. Its characteristic impedance may be approximately deduced by assuming that the fields along such a line are quasi TEM and that the metallization is infinitely thin. Making use of these assumptions indicates that the characteristic impedance of the line may be expressed in terms of the free-space phase velocity of the substrate and a knowledge of either its inductance (L) or capacitance (C) per unit length:

$$Z_0 = \frac{1}{vC} \qquad \Omega \tag{24-43}$$

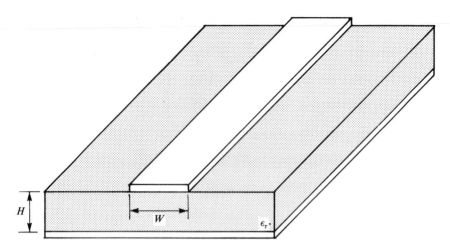

FIGURE 24-10
Schematic diagram of microstrip line.

or
$$Z_0 = vL \quad \Omega \tag{24-44}$$

The capacitance of the geometry per unit length is given by
$$C = \frac{\varepsilon_0 \varepsilon_r W}{H} \quad \text{F/m} \tag{24-45}$$

and the phase velocity of the medium is
$$v = \frac{c}{\sqrt{\varepsilon_r}} \quad \text{m/s} \tag{24-46}$$

ε_r is the relative dielectric constant of the substrate, H is its thickness (metres) and W is the width of the centre conductor (metres).

Combining the preceding relationships readily gives the required result:
$$Z_0 = \frac{\eta_0 H}{\sqrt{\varepsilon_r} W} \tag{24-47}$$

This expression is valid for low impedance lines; for high impedance ones the effect of the fringing fields at the edges of the centre conductor, described by Eq. (24-42), must be taken into account.

The field pattern in this type of line is indicated in Fig. 24-11.

If the fringing is significant then the dielectric constant of the substrate is no longer appropriate to describe the phase constant of the line described by Eq. (24-46); it is then expedient to define an effective dielectric constant (ε_{eff}) in order to accommodate this effect:
$$v = \frac{C}{\sqrt{\varepsilon_{\text{eff}}}} \tag{24-48}$$

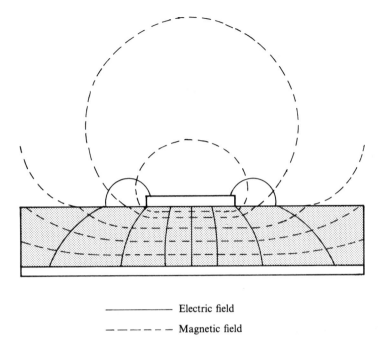

——————— Electric field

— — — — — Magnetic field

FIGURE 24-11
Field patterns in microstrip line.

The effective dielectric constant may be derived by forming the ratio of the capacitance of the line with the dielectric filler and that without it:

$$\varepsilon_{\text{eff}} = \frac{C}{C'} \qquad (24\text{-}49)$$

where C is the capacitance of the line with the dielectric filler and C' is that without it.

This result may be readily demonstrated by investigating the parameters of the two arrangements. Those of the free-space line are

$$Z'_0 = \sqrt{\frac{L}{C}} \qquad (24\text{-}50a)$$

$$Z'_0 = cL \qquad (24\text{-}50b)$$

$$Z'_0 = \frac{1}{cC'} \qquad (24\text{-}50c)$$

and those of the loaded line are given by similar relationships with C' replaced by C and c replaced by v. Combining these equations and noting that the inductance is invariant under the two situations gives

$$L = \frac{1}{c^2 C'} = \frac{1}{v^2 C} \qquad (24\text{-}51)$$

which satisfies Eqs (24-48) and (24-49) as asserted.

For low impedance lines with wide strips

$$\varepsilon_{\text{eff}} \rightarrow \varepsilon_{\text{r}} \qquad (24\text{-}52a)$$

and for high impedance lines with narrow strips

$$\varepsilon_{\text{eff}} \rightarrow \frac{\varepsilon_{\text{r}} + 1}{2} \qquad (24\text{-}52b)$$

This interval is sometimes expressed in terms of a filling factor (q):

$$\varepsilon_{\text{eff}} = 1 + q(\varepsilon_{\text{r}} - 1) \qquad (24\text{-}52c)$$

which lies between

$$\tfrac{1}{2} \leqq q \leqq 1 \qquad (24\text{-}52d)$$

THE COPLANAR WAVEGUIDE

The coplanar waveguide consists of a centre strip with two ground planes located parallel to and in the plane of the strip. Figure 24-12 depicts the schematic diagram of this transmission line and Fig. 24-13 the electric and magnetic fields in the quasi-static situation. This type of structure allows both series and shunt elements to be integrated into the line without the need to drill holes into the substrate. Strictly speaking, the magnetic field has a longitudinal component so that this field is in practice elliptically polarized. The quasi-static model of the coplanar waveguide may once more be established in terms of its quasi-static capacitance. This capacitance may, in this instance, be derived by a technique known as conformal mapping. A conformal mapping fixes a complex variable

$$W = u + \mathrm{j}v \qquad (24\text{-}53a)$$

in the W plane corresponding to one in the z plane:

$$z = x + \mathrm{j}y \qquad (24\text{-}53b)$$

for a given function

$$W = f(W) \qquad (25\text{-}53c)$$

Using this method the dielectric half-plane z in Fig. 24-14 may be transformed into the interior of a rectangle in the W plane, the capacitance of which can be calculated without difficulty from a knowledge of its dimensions. This transformation may be written as

$$\frac{\mathrm{d}W}{\mathrm{d}Z} = \frac{A}{(Z^2 - a_1^2)^{1/2}(Z^2 - b_1^2)^{1/2}} \qquad (24\text{-}54)$$

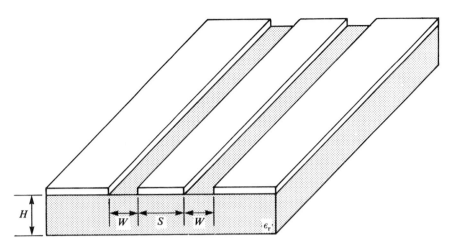

FIGURE 24-12
Schematic diagram of coplanar waveguide.

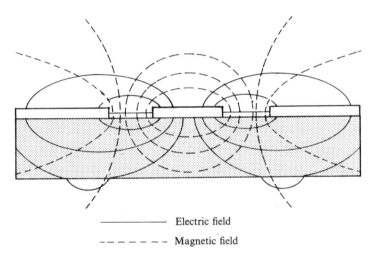

——————— Electric field

– – – – – – Magnetic field

FIGURE 24-13
Field patterns in coplanar waveguide.

The ratio u/v is then deduced by integrating the preceding function:

$$W = u + jv = \int_0^{b_1} \frac{a \, dZ}{(Z^2 - a_1^2)^{1/2}(Z^2 - b_1^2)^{1/2}} \tag{24-55}$$

This is a standard integral whose solution is given in terms of the complete elliptical integral of the first kind $K(k)$:

$$\frac{u}{v} = \frac{K(k)}{K'(k)} \tag{24-56}$$

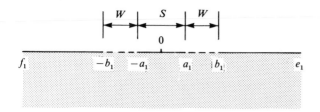

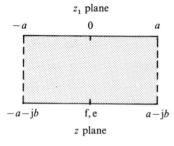

FIGURE 24-14
Conformal transformation between coplanar waveguide and equivalent parallel plate capacitor.

where

$$k = \frac{a_1}{b_1} \tag{24-57}$$

One identity that may be used to evaluate $K'(k)$ is

$$K'(k) = K(k') \tag{24-58}$$

where

$$K' = (1 - k^2)^{1/2} \tag{24-59}$$

The capacitance of the dielectric half-space is now given by

$$C_r = \varepsilon_0 \varepsilon_r \frac{K(k)}{K'(k)} \quad \text{F/m} \tag{24-60}$$

and that of the free-space half-space by

$$C_o = \varepsilon_o \frac{K(k)}{K'(k)} \quad \text{F/m} \tag{24-61}$$

The effective dielectric constant of the overall network is approximately the average of the two half-spaces:

$$\varepsilon_{\text{eff}} = \varepsilon_o \frac{\varepsilon_r + 1}{2} \tag{24-62}$$

and the phase velocity is therefore described in terms of that of the free space (c) by

$$v = \frac{c}{\sqrt{\varepsilon_{\text{eff}}}} \quad \text{m/s} \tag{24-63}$$

Writing the characteristic impedance in terms of the capacitance and phase velocity gives the required result:

$$Z_0 = \frac{\eta_o}{\sqrt{\varepsilon_{\text{eff}}}} \frac{K'(k)}{K(k)} \quad \Omega \qquad (24\text{-}64)$$

THE SLOTLINE

Another planar transmission line is the slotline one. It consists of a dielectric substrate with a narrow slot etched in the metallization of the substrate. The other surface of the substrate is without any metallization. Short-circuits and series stubs are two elements that are, for instance, readily incorporated in this type of line. Figure 24-15a illustrates the schematic diagram of this sort of line: Fig. 24-15b indicates its electric and magnetic field patterns. Since the structure is essentially a two-conductor transmission line it has no low frequency cutoff. The field pattern is, however, a quasi TE one. The detailed derivation of the parameters of this line is outside the remit of this introductory text. However, one possibility is to construct a one-to-one equivalence between the slotline and a semi-elliptical dielectric loaded ridge waveguide. Figure 24-16 gives the required equivalence. The assumption employed here assumes that the fringing field inside the dielectric region is not affected by the details of its contour.

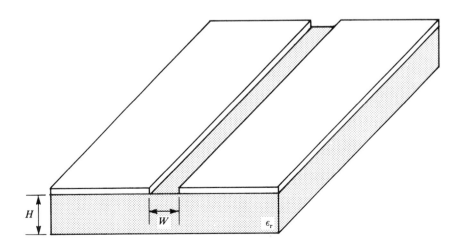

FIGURE 24-15
(a) Schematic diagram of slotline.

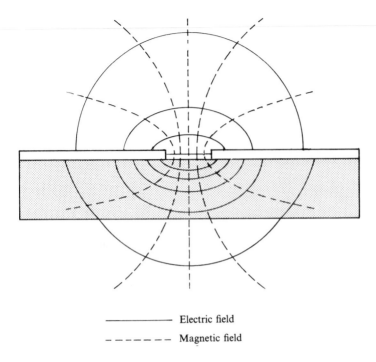

——————— Electric field

– – – – – – Magnetic field

FIGURE 24-15
(b) Field patterns in slotline.

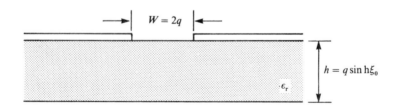

$W = 2q$

$h = q \sin h\xi_0$

ϵ_r

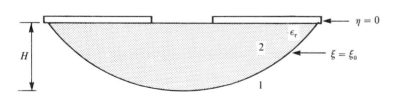

$\eta = 0$

ϵ_r

2

$\xi = \xi_0$

H

1

FIGURE 24-16
Elliptical waveguide model of slotline.

Figure 24-17 illustrates some shielded variations.

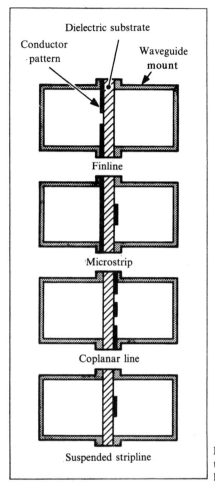

Conductor pattern

Dielectric substrate

Waveguide mount

Finline

Microstrip

Coplanar line

Suspended stripline

E-plane transmission lines

FIGURE 24-17
Schematic diagrams of shielded slotline and coplanar waveguides.

PROBLEMS

24-1 Calculate the inductance of the microstrip line from a knowledge of its capacitance and its phase velocity.

24-2 Derive the impedance of the stripline in terms of its capacitance and phase velocity.

24-3 Determine the ratio b/a of a 50-Ω coaxial line.

24-4 Calculate the ratio W/H of a 50-Ω microstrip line using an alumina substrate ($\varepsilon_r = 9.4$). Repeat for a quartz substrate ($\varepsilon_r = 4.7$).

24-5 Show that the capacitance per unit length of a low impedance stripline transmission line is twice that of the microstrip arrangement.

24-6 It may be shown that the fringing capacitance of a single microstrip edge is

$$C_f = \frac{2\varepsilon_0 \varepsilon_r}{\pi} \ln 2 \quad \text{F/m}$$

Obtain Z_0.

FURTHER READING

Bader, W., 'Kopplungsfreie Kettenschalt ungen', *Telegr. Fernspr. Tech.*, Vol. 31, pp. 177–189, July 1972.

Balabanian, N., *Network Synthesis*, Prentice-Hall, Englewood Cliffs, N.J., 1958.

Bode, H. W., *Network Analysis and Feedback Amplifier Design*, D. Van Nostrand, Princeton, N.J., 1945.

Cauer, W., 'Ein Interpolations Problem mit Funktionen mit Positiven Realteil', *Mathematics Zeitschrift*, Vol. 38, pp. 1–44, 1933.

Chirlian, P. M., *Integrated and Active Network Analysis and Synthesis*, Prentice-Hall, Englewood Cliffs, N.J., 1967.

Christian, E. and Eisenmann, E., *Filter Design Tables and Graphs*, John Wiley, New York, 1966.

Daniels, Richard W., *Approximation Methods for Electronic Filter Design*, McGraw-Hill, New York, 1974.

Darlington, S., 'Synthesis of reactance 4-poles which produce prescribed insertion loss characteristics', *J. Math Phys.*, Vol. 18, pp. 257–353, 1939.

Guillemin, E. A., *Synthesis of Passive Networks*, John Wiley, New York, 1957.

Huelsman, L. P., *Theory and Design of Active RC Circuits*, McGraw-Hill, New York, 1963.

Johnson, D. E., *Introduction to Filter Theory*, Prentice-Hall, Englewood Cliffs, N.J., 1976.

Kendall, L. Su., *Active Network Synthesis*, McGraw-Hill, New York, 1965.

Kudsia, C. M. and O'Donovan, M. V., *Microwave Filters for Communications Systems*, Artech House, Norwood, MA, 1974.

Kuo, F. F., *Network Analysis and Synthesis*, John Wiley, New York, 1962.

Rhodes, J. D., *Theory of Electrical Filters*, John Wiley, London, 1976.

Saal, R. and Ulbrich, E., 'On the design of filters by synthesis', *IRE Trans. on Circuit Theory*, December 1958.

Scanlan, J. O. and Levy, R., *Circuit Theory*, Vol. 1, Oliver and Boyd, Edinburgh, 1970.

Storch, L., 'Synthesis of constant-time-delay ladder networks using Bessel polynomials', *Proc. IRE*, Vol. 42, pp. 1666–1675, November 1954.

Temes, G. C. and La Patra, J. W., *Introduction to Circuit Synthesis and Design*, McGraw-Hill, New York, 1977.

Temes, G. C. and Mitra, S. K., *Modern Filter Theory and Design*, John Wiley, New York, 1973.

Thomson, W. E., 'Networks with maximally flat frequency characteristics', *Proc. IEE*, Part 3, Vol. 96, pp. 487–490, November 1949.

Tuttle, D. F. Jr, *Network Synthesis*, Vol. 1, John Wiley, New York, 1958.

Van Valkenburg, M. E., *Introduction to Modern Network Synthesis*, John Wiley, New York, 1960.

Wai-Kai Chen, *Theory and Design of Broadband Matching Networks*, Pergamon Press, Oxford, 1976.

Weinberg, L., *Network Analysis and Synthesis*, McGraw-Hill, New York, 1962.

Zverev, A. I., *Handbook of Filter Synthesis*, John Wiley, New York, 1967.

INDEX